SpringerWienNewYork

Sponsored by the
European Association of Neurosurgical Societies

Advances
and Technical Standards
in Neurosurgery

Vol. 31

Edited by

J. D. Pickard, Cambridge (Editor-in-Chief),
N. Akalan, Ankara, C. Di Rocco, Roma,
V. V. Dolenc, Ljubljana, R. Fahlbusch, Erlangen,
J. Lobo Antunes, Lisbon, J. J. Mooij, Groningen,
M. Sindou, Lyon, C. A. F. Tulleken, Utrecht

SpringerWienNewYork

With 84 partly coloured Figures

© 2006 Springer-Verlag/Wien
Printed in Austria

SpringerWienNewYork is a part of Springer Science Business Media
springeronline.com

Library of Congress Catalogue Card Number 74-10499

Typesetting: Asco Typesetters, Hong Kong
Printing: Druckerei Theiss GmbH, 9431 St. Stefan, Austria, www.theiss.at

Printed on acid-free and chlorine-free bleached paper

SPIN: 11538387

ISSN 0095-4829
ISBN-10 3-211-28253-X SpringerWienNewYork
ISBN-13 978-3-211-28253-3 SpringerWienNewYork

Preface

As an addition to the European postgraduate training system for young neurosurgeons, we began to publish in 1974 this series of *Advances and Technical Standards in Neurosurgery* which was later sponsored by the European Association of Neurosurgical Societies.

This series was first discussed in 1972 at a combined meeting of the Italian and German Neurosurgical Societies in Taormina, the founding fathers of the series being Jean Brihaye, Bernard Pertuiset, Fritz Loew and Hugo Krayenbuhl. Thus were established the principles of European co-operation which have been born from the European spirit, flourished in the European Association, and have been associated throughout with this series.

The fact that the English language is now the international medium for communication at European scientific conferences is a great asset in terms of mutual understanding. Therefore we have decided to publish all contributions in English, regardless of the native language of the authors.

All contributions are submitted to the entire editorial board before publication of any volume for scrutiny and suggestions for revision.

Our series is not intended to compete with the publications of original scientific papers in other neurosurgical journals. Our intention is, rather, to present fields of neurosurgery and related areas in which important recent advances have been made. The contributions are written by specialists in the given fields and constitute the first part of each volume.

In the second part of each volume, we publish detailed descriptions of standard operative procedures and in depth reviews of established knowledge in all aspects of neurosurgery, furnished by experienced clinicians. This part is intended primarily to assist young neurosurgeons in their postgraduate training. However, we are convinced that it will also be useful to experienced, fully trained neurosurgeons.

We hope therefore that surgeons not only in Europe, but also throughout the world, will profit by this series of *Advances and Technical Standards in Neurosurgery*.

The Editors

Contents

Advances

Gene Technology Based Therapies in the Brain. T. Wirth[1,4] and S. Ylä-Herttuala[1,2,3], [1] A. I. Virtanen Institute, University of Kuopio, Kuopio, Finland, [2] Department of Medicine, University of Kuopio and Gene Therapy Unit, Kuopio, Finland, [3] Gene Therapy Unit, Kuopio University Hospital, Kuopio, Finland, [4] Ark Therapeutics Oy, Kuopio, Finland

Contents

Technical Standards

Anatomy of the Orbit and its Surgical Approach. G. HAYEK, PH. MERCIER, and H. D. FOURNIER, Laboratory of Anatomy, Faculty of Medicine, University of Angers, Angers, France

Neurosurgical Concepts and Approaches for Orbital Tumours. J. C. MARCHAL and T. CIVIT, Department of Neurosurgery, Hôpital Central, Nancy University Hospital, Nancy Cedex, France

Endoscopic Third Ventriculostomy in the Treatment of Hydrocephalus in Pediatric Patients. C. Di Rocco[1], G. Cinalli[2], L. Massimi[1], P. Spennato[2], E. Cianciulli[3], and G. Tamburrini[1], [1]Pediatric Neurosurgical Unit, Catholic University Medical School, Rome, Italy, [2]Neuroendoscopy Unit, Department of Pediatric Neurosurgery, Santobono-Pausilipon Children's Hospital, Naples, Italy, [3]Department of Pediatric Neuroradiology, Santobono-Pausilipon Children's Hospital, Naples, Italy

**Minimally Invasive Procedures for the Treatment of Failed Back Surgery
Syndrome.** P. MAVROCORDATOS and A. CAHANA, Department of Anesthesiology,
Pharmacology and Intensive Care, Geneva, Switzerland

Surgical Anatomy of Calvarial Skin and Bones—With Particular Reference to Neurosurgical Approaches. H. D. FOURNIER, V. DELLIÈRE, J. B. GOURRAUD, and PH. MERCIER, Laboratory of Anatomy, Faculty of Medicine, University of Angers, Angers, France

List of Contributors

Cahana, A., Department of Anesthesiology, Pharmacology and Intensive Care, Geneva, Switzerland

Cianciulli, E., Department of Pediatric Neuroradiology, Santobono-Pausilipon Children's Hospital, Naples, Italy

Cinalli, G., Neuroendoscopy Unit, Department of Pediatric Neurosurgery, Santobono-Pausilipon Children's Hospital, Naples, Italy

Civit, T., Department of Neurosurgery, Hôpital Central, Nancy University Hospital, Nancy Cedex, France

Dellière, V., Laboratory of Anatomy, Faculty of Medicine, University of Angers, Angers, France

Di Rocco, C., Pediatric Neurosurgical Unit, Catholic University Medical School, Rome, Italy

Fournier, H. D., Laboratory of Anatomy, Faculty of Medicine, University of Angers, Angers, France

Gourraud, J. B., Laboratory of Anatomy, Faculty of Medicine, University of Angers, Angers, France

Hayek, G., Laboratory of Anatomy, Faculty of Medicine, University of Angers, Angers, France

Marchal, J. C., Department of Neurosurgery, Hôpital Central, Nancy University Hospital, Nancy Cedex, France

Massimi, L., Pediatric Neurosurgical Unit, Catholic University Medical School, Rome, Italy

Mavrocordatos, P., Department of Anesthesiology, Pharmacology and Intensive Care, Geneva, Switzerland

Mercier, Ph., Laboratory of Anatomy, Faculty of Medicine, University of Angers, Angers, France

Spennato, P., Neuroendoscopy Unit, Department of Pediatric Neurosurgery, Santobono-Pausilipon Children's Hospital, Naples, Italy

Tamburrini, G., Pediatric Neurosurgical Unit, Catholic University Medical School, Rome, Italy

Wirth, T., Ark Therapeutics Oy, Kuopio, Finland

Ylä-Herttuala, S., A. I. Virtanen Institute, University of Kuopio, Kuopio, Finland

Advances

Gene Technology Based Therapies in the Brain

T. Wirth[1,4] and S. Ylä-Herttuala[1,2,3]

[1] A. I. Virtanen Institute, University of Kuopio, Kuopio, Finland
[2] Department of Medicine, University of Kuopio, Kuopio, Finland
[3] Gene Therapy Unit, Kuopio University Hospital, Kuopio, Finland
[4] Ark Therapeutics Oy, Kuopio, Finland

With 4 Figures

Contents

Abstract

Gene therapy potentially represents one of the most important developments in modern medicine. Gene therapy, especially of cancer, has created exciting and elusive areas of therapeutic research in the past decade. In fact, the first gene therapy performed in a human was not against cancer but was performed to a 14 year old child suffering from adenosine deaminase (ADA) deficiency. In addition to cancer gene therapy there are many other diseases and disorders where gene therapy holds exciting and promising opportunities. These include amongst others gene therapy within the central nervous system and the cardiovascular system. Improvements of the efficiency and safety of gene therapy is the major goal of gene therapy development. After the death of Jesse Gelsinger, the first patient in whom death could be directly linked to the viral vector used for the treatment, ethical doubts were raised about the feasibility of gene therapy in humans. Therefore, the ability to direct gene transfer vectors to specific target cells is also a crucial task to be solved and will be important not only to achieve a therapeutic effect but also to limit potential adverse effects.

Keywords: Gene therapy; viral vectors; Parkinson's disease; Alzheimer's disease; brain tumours; cerebral vasospasm.

Introduction to Gene Therapy: The Past, Present and Future

Scientific understanding of the molecular basis of life increased dramatically after Oswald T. Avery's discovery in 1944 that deoxyribonucleic acid (DNA) was the "transforming principle" – the secret code of life. Then Francis Crick and James Watson described the "double helix" structure of DNA in 1953. The process, however, by which DNA replicates itself during cellular reproduction, or how DNA expresses its genetic information, was still a mystery in the late 1950s. A little less than 20 years after Oswald T. Avery's discovery, Marshall Nirenberg and his colleagues in 1962 deciphered UUU (one three-unit batch of uracil, which was a "code word" for identifying phenylalanine) as the first word in the chemical dictionary of life. Nearly 30 years after Nirenberg's breakthrough, in 1990 the first clinical study involving gene transfer was commenced (Mountain, 2000) and it contributed to the start of a whole new industrial area – biotechnology.

A four-year old girl called Ashanti de Silva became the first gene therapy patient on September 14, 1990 at the NIH Clinical Center. She had adenosine deaminase (ADA) deficiency, a genetic disease which left her defenceless against infections. White blood cells were taken from her blood, and the normal genes for making adenosine deaminase were inserted into them. Afterwards the corrected cells were re-injected back into her circulation. Unfortunately, the effects of Ashanti's gene therapy were not clearly demonstrated due to simultaneous enzyme replacement therapy with polyethylene glycol adenine deaminase (PEG-ADA), which she had to take as a back up.

Since the commencement of the first clinical trial, the field has grown rapidly. Today there are close to 1000 ongoing gene therapy clinical trials worldwide (Edelstein, 2004), most of which are targeted against cancer.

Table 1. *Most common clinical targets for gene therapy. Edelstein, 2004. Gene therapy clinical trials worldwide 1989–2004 – an overview. Copyright John Wiley & Sons Limited. Reproduced with permission*

Indications	Gene therapy clinical trials	
	Number	%
Cancer diseases	656	66,5
Monogenic diseases	93	9,4
Vascular diseases	80	8,1
Infectious diseases	65	6,6
Other diseases	29	2,9
Gene marking	52	5,3
Healthy volunteers	12	1,2
Total	987	

Table 1 lists the most common clinical targets for gene therapy (http://www.wiley.co.uk/genmed/clinical).

Unfortunately, gene therapy clinical trials experienced one drawback after another as several clinical trials failed to show efficacy (Scollay, 2001).

In September 1999, the worst case scenario for gene therapy became reality, when 18 year old Jesse Gelsinger took part in a gene therapy clinical trial at the University of Pennsylvania in Philadelphia. He suffered from a partial deficiency of ornithine transcarbamylase (OTC), a liver enzyme that is required for the removal of excessive nitrogen from amino acids and proteins. Four days after treatment, Jesse Gelsinger died because of multiorgan failure. He was the first patient in whom death could be directly linked to the viral vector used for the treatment. A little later, in April of the following year the journal Science published an article from Maria Cavazzana-Calvo *et al.* (Cavazzana-Calvo, 2000) where they reported the first definitive cure of disease by gene therapy. Three young children suffering from the fatal X-linked SCID-XI syndrome had developed a functional immune system after gene therapy treatment. After that success several more patients have been treated using the same gene therapy strategy. Some years later 2 out of 11 treated patients had developed a leukaemia-like disease obviously as a result of the use of the murine leukaemia virus (MLV) vector (Hacein-Bey-Abina, 2003). After the tragedy of Jessie Gelsinger's death the number of approved clinical trials have decreased worldwide (Fig. 1).

Nevertheless, despite these drawbacks gene therapy research and development itself has never stopped, or slowed down. As a result of that, on October the 16th 2003, China became the first country to approve the commercial production of a gene therapy. Shenzhen SiBiono GenTech (Shen-

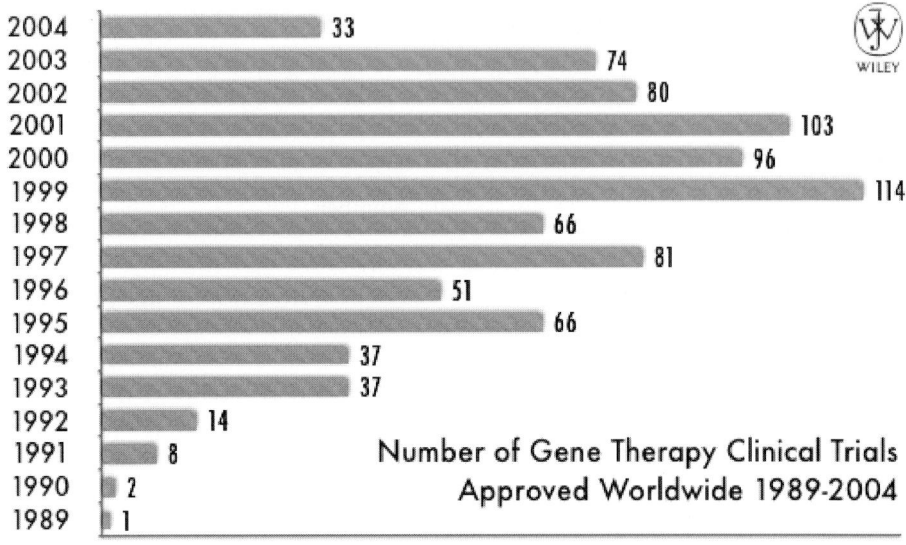

Year	Value
2004	33
2003	74
2002	80
2001	103
2000	96
1999	114
1998	66
1997	81
1996	51
1995	66
1994	37
1993	37
1992	14
1991	8
1990	2
1989	1

Number of Gene Therapy Clinical Trials
Approved Worldwide 1989-2004

The Journal of Gene Medicine, © 2004 John Wiley and Sons Ltd   www.wiley.co.uk/genmed/clinical

Fig. 1. Decrease in the number of approved gene therapy clinical trials after the year 1999. Edelstein, 2004. Gene therapy clinical trials worldwide 1989–2004 – an overview. Copyright John Wiley & Sons Limited. Reproduced with permission

zhen, China), obtained a drug license from the State Food and Drug Administration of China (SFDA; Beijing, China) for its recombinant Ad-p53 gene therapy (Gendicine) for head and neck squamous cell carcinoma (HNSCC). At the same time there are some very promising ongoing gene therapy clinical trials worldwide for the treatment of diseases such as tissue ischemia (Morishita, 2004), cancer (Trask, 2000, Prados, 2003, Lamont, 2000, Immonen, 2004), haemophilia A or B (Monahan, & White, 2002) and Parkinson's disease (Howard, 2003) with potential for the launching into the market. But so far the American Food and Drug Administration (FDA) has not yet approved any human gene therapy product for sale.

Regarding the variety of areas where gene therapy could be applicable, one notes that only a few are in fact directed to diseases of the central nervous system (CNS). These areas include treatment of brain tumours, e.g. glioblastoma, and degenerative conditions, e.g. Alzheimer's disease and Parkinson's disease, and ischemic brain diseases.

Potential Areas for Gene Therapy in the Brain

The "tenacious start" of gene therapy for neurological diseases is not really surprising, since gene therapy to the brain faces unique obstacles in addition to those one faces with gene therapy in general. Evaluation of the appropriate vector, route of administration, efficiency of the transgene expression, and immune response against gene transfer vectors being used are some of the problems one faces with gene therapy. In addition to those, gene therapy to the brain has to overcome obstacles such as the blood brain barrier and the limited space within the brain, which restricts the volume of gene transfer vectors that can be injected if administered locally.

Also, the lack of appropriate animal models for neurodegenerative disease such as Parkinson's disease, or Alzheimer's disease has been a problem. That again posed an obstacle that hindered gene therapy to move into the clinic. However, in recent years tremendous strides have been made in developing appropriate animal models of human neurodegenerative diseases, and along the development of these animal models the movement of gene therapy from benchside to the clinic has been justified.

Feasible areas of gene therapy in the brain include Alzheimer's disease (Mattson, 2004), ischemic brain diseases (Zadeh, & Guha, 2003), Parkinson's disease (Samii, 2004), and brain tumours (Abeloff, 2000), of which gliomas have been the subject of the largest number of gene therapy strategies (Chiocca, 2003). Also, epilepsy (Gutierrez-Delicado & Serratosa, 2004) amyotrophic lateral sclerosis (ALS), a motor neuron disease, (Weiss, 2004), lysosomal storage disease (LSD) (Futerman, & van Meer, 2004), and Huntington's disease (Hogarth, 2003) have been subject of numerous promising gene therapy strategies. For example, approaches such as the fibrinogen-galanin encoding adeno-associated viral vector gene therapy (Haberman, 2003), the gene transfer of the Neuropeptide Y (Richichi, 2004), as well as the gene transfer of the aspartoacylase (ASPA) gene (Seki, 2004, McPhee, 2005) has been used in studies against epilepsy. However, because of limited space, this review will be focusing only onto the first four diseases mentioned above. The reader is referred to the following references for more information about gene therapy in epilepsy (McCown, 2004), ALS (Boillée & Cleveland, 2004, Bruijn, 2004, Alisky & Davidson, 2000, Azzouz, 2004, Azzous, 2000, Pompl, 2003, Ascadi, 2002, Kaspar, 2003, Wang, 2002), LSD (Kaye & Sena-Esteves, 2002, Cabrera-Salazar, 2002, Eto & Ohashi, 2002), and Huntington's disease (McBride, 2003, Bemelmans, 1999, Bachoud-Levi, 2000, Bachoud-Levi, 1998, MacMillan, 1994).

Gene Therapy for Parkinson's Disease

Diseases which are commonly related to aging have received major interest worldwide. From an epidemiological perspective Parkinson's disease and Alzheimer's disease share an increasing prevalence with aging, whereas clinically they are characterized by different clinical symptoms and molecular etiology with very limited potential for cure at present (Winkler, 1998).

The pathology of Parkinson's disease reveals prominent loss of dopaminergic neurons, especially in the substantia nigra, usually in connection with the formation of extracellular inclusions, termed Lewy bodies. Clinical symptoms usually do not appear in adults until when about 80% of striatal dopamine and 50% of nigral neurons are lost (Samii, 2004).

Attaining focal, sustained physiologic delivery of L-Dopa or dopamine, and preventing further death of dopaminergic neurons has been the main focus of gene therapy of Parkinson's disease (Finkelstein, 2001). This has been achieved by mainly two different strategies. One strategy is to provide localized growth factors to sustain dopaminergic neurons, preventing them from undergoing apoptosis. The neuroprotective effect of growth factors has been demonstrated in several studies (Eberhardt & Schulz 2004). Among these growth factors, glial cell line-derived neurotrophic factor (GDNF) is one of the most promising candidates for gene therapy of Parkinson's disease. Studies using intracerebral injections of the recombinant GDNF protein have shown that GDNF can provide almost complete protection of nigral dopamine neurons against 6-hydroxydopamine (6-OHDA) – or MPTP-induced damage in rodents and non-human primates, promote axonal sprouting and regrowth of lesioned dopamine neurons, and stimulate dopamine turnover and function in neurons spared by the lesion (Björklund, 1997; Gash, 1998; Kordower, 2000).

Another approach to sustain physiological delivery of L-Dopa, or dopamin is replacing/supplementing critical enzymes in the dopaminergic pathway. The three most relevant enzymes for dopamin production are 1) tyrosin hydroxylase, the rate limiting enzyme in the synthesis of dopamin, 2) GTP cyclohydrase (GCH), to generate more tetrahydrobiopterin (a essential cofactor for TH) and 3) aromatic amino acid decarboxylase (AADC), an enzyme that converts L-Dopa to dopamine. It became clear, that these enzymes represent potential targets for gene therapy and therefore they have also been subjected to a lot of research (During, 1994, Lampela, 2002, Sun, 2003; Eberling, 2003; Sanchez Pernaut, 2001; Shen, 2000). Also the delivery of the gene encoding for glutamate decarboxylase (GAD) has been subjected to research (Luo, 2002). Currently, there are two ongoing gene therapy clinical trials regarding Parkinson's disease. One uses the approach of subthalamic GAD gene transfer; the other uses the approach of intrastriatal gene transfer of ADDC (www.gemcris.od. nih.gov).

Gene Therapy for Alzheimer's Disease

Alzheimer's disease (AD) is the most common cause of dementia (Palmer, 2002). Dementia is a collective name for progressive degenerative brain syndromes which affect memory, thinking, behaviour and emotion. The precise mechanisms that lead to this disease are not fully understood and many genetic, cellular and molecular irregularities are implicated. Central to the disease, however, is the altered proteolytic processing of the amyloid precursor protein (APP) resulting in the production and aggregation of neurotoxic forms of amyloid β-peptide (Aβ) (Mattson, M. P. 2004). In addition, neuropathological examination of AD brains reveals neuronal and synaptic loss and neurofibrillary tangles. The progression of AD is slow, starting with mild memory problems and ending with severe intellectual impairment. It is the cognitive areas of the brain that are the first to be affected from this disease leading, amongst other things, to memory loss and behavioural abnormalities. It then spreads to the parts of the brain that control movement. Eventually, the loss of brain function becomes so severe that it can be the primary cause of death (Brown, 2003).

Several specific neurotransmitter systems are regularly and substantially altered in AD brains. One of the most prominent systems affected in the course of AD are the cholinergic neurons of the nucleus basalis magnocellularis (NBM) (Winkler, 1998). Several gene therapy approaches have been documented to be promising in experimental animal models. In this regard, greatest interest as a potential gene therapy approach, and also subject of two ongoing clinical trials (www.gemcris.od.nih.gov), is the use of nerve growth factor (NGF) as a neuroprotective molecule (Tuszynski, 2002, Wu, 2004, Winkler, 1998, Tuszynski, 1998). NGF has been shown to be able to prevent the death of cholinergic neurons after axotomy, and that it was also able to reverse spontaneous age-related morphological and behavioural decline in rat (Kromer, 1987, Fisher, 1987). In addition to the use of NGF for the treatment of AD, there are studies about the feasibility of neprilysin (NEP) gene transfer for the treatment of Alzheimer's disease. Marr and colleagues (Marr, 2004) demonstrated that injection of NEP expressing lentiviruses into the hippocampus of transgenic mice led to an approximate 50% reduction in the number of amyloid plaques.

Gene Therapy for Vascular Brain Diseases

There are several potentially feasible applications of gene therapy for the treatment of vascular brain diseases. One application is the prevention of vasospasm after subarachnoid hemorrhage (SAH). Another application is the stimulation of growth of collateral blood vessels in the area of ischemia, and third, stabilization of atherosclerotic plaques, inhibition of thrombosis, and prevention of restenosis after angioplasty of the carotid and posterior circulation arteries (Toyoda, 2003).

Vasospasm after SAH typically occurs slowly several days after subarachnoid hemorrhage (Dietrich, 2000), and therefore the timing for gene therapy seems feasible, since maximal expression of the transgene occurs usually a few days after gene transduction with viral vectors. In addition, the risk of vasospasm after SAH is transient (Lüders, 2000). Thus, even with current available vectors, which provide transient transduction, gene therapy for the prevention of vasospasm after subarachnoid hemorrhage may be achievable. Gene therapy to prevent vasospasm after SAH has been done, for instance, using endothelial NOS (eNOS) gene transfer. Endothelial NOS improved NO-mediated relaxation *in vitro* after experimental SAH (Onoue, 1998), but did not demonstrate a therapeutical effect *in vivo* after intracisternal injection of adenovirus containing the gene for eNOS in dogs (Stoodley, 2000), even though an increase of cerebral blood flow could be demonstrated in rats after intracisternal injections of replication-defective adenovirus containing the gene for eNOS (Lüders, 2000). Compared to eNOS, injection of adenovirus containing the gene for human extracellular superoxide dismutase (ECSOD) into the cisterna magna 30 minutes after induction of experimental SAH, reduced cerebral vasospasm after subarachnoid hemorrhage in rabbits (Watanabe, 2003).

Preservation of cerebral circulation and prevention of cerebral infarction could be achieved by stimulation of growth of collateral blood vessels. A variety of growth factors have been reported to induce angiogenesis in different experimental animal models (Ylä-Herttuala & Martin, 2000) and have shown to be therapeutically effective, many of them being used in clinical trials of gene therapy. Growth factors that induce angiogenesis include vascular endothelial growth factors (VEGFs), basic fibroblast growth factor (bFGF) (Ylä-Herttuala & Alitalo, 2003), and hepatocyte growth factor (HGF) (Morishita, 2004). For angiogenesis in the brain, Yukawa (Yukawa, 2000) demonstrated that adenoviral gene delivery of bFGF into the cerebrospinal fluid (CSF) induced angiogenesis in the bilateral paraventricular region in rat brains.

Regarding ischemic stroke, the initial damage after stroke is not a feasible target for gene therapy. That's because the therapeutic window is (at most) only a few hours after onset of ischemic stroke. At the same time the expression of the transgene requires hours to days, depending on the vector used. For that reason gene therapy strategies can be used only for prevention of succeeding damages in the ischemic penumbra. Shimamura *et al.* (Shimamura, 2004) demonstrated that gene transfer of HGF into the brain resulted in attenuation of brain ischemic injury even if HGF was transduced 24 hours after the ischemic event. Also, the expression of genes such as Bcl-2 (Yenari, 2003), the 72-kD inducible heat shock protein (HSP72) (Hoehn, 2001), or the cyclooxygenase-1 (COX-1) (Lin, 2002) have also been demonstrated to reduce ischemic injury.

Some strategies focus on inhibition of genes that are expressed and believed to be harmful after ischemic stroke, such as the interleukin-1 receptor (Yang, 1997). In addition, genes such as interleukin-10, transforming growth factor-β1 (TGF-β1), glial cell line-derived neurotropic factor (GDNF) (Shirakura, 2004), and nerve growth factor (NGF) (Shirakura,

2004), have been targets for gene therapy (Shimamura, 2004). Fibroblastic growth factor-2 (FGF-2) has been shown to decrease brain injury after cerebral ischemia when administered systemically (Bethel, 1997). As a result to that Shigeru *et al.* developed a systemic gene therapy using macrophages infiltrating the infarct to deliver and express FGF-2 (Shigeru, *et al.* 2004).

So far none of the above mentioned strategies have reached clinical trials.

Gene Therapy for Brain Tumours

Regarding brain tumours, malignant gliomas have been the primary target for gene therapy. The average life expectancy of a patient diagnosed with glioblastoma multiforme is 10 months after diagnoses (Ammirati, 1987). Several therapeutic approaches to treat cancer are limited in their success because of the lack of specificity of the drugs used for therapy. The therapeutic index of several cytotoxic drugs is very narrow, which limits the possibility to reach effective tissue concentrations. In addition, some of the difficulties encountered include inaccessibility to resective surgery because of the anatomical location of the tumour and because of infiltration of tumour cells into surrounding tissues. For that reason gene therapy of brain tumour has been one of the most exciting and elusive areas of therapeutic research in the past decade. It potentially represents one of the most important developments in the treatment of brain tumours. However, gene delivery to brain tumours is a formidable obstacle. Transduction rates $> 5\%$ of the tumour mass are difficult to achieve (Puumalainen, 1998a), even in experimental tumours. For that reason the therapeutic effect must not be limited only to transduced cells, but it must be able to exert a therapeutic effect on neighbouring, non-transduced, cells as well (bystander effect).

One of the most studied gene therapy strategies in the treatment of malignant gliomas is the combination of thymidine kinase and Ganciclovir. This approach has also been called 'suicide' gene therapy as the non-toxic pro-drug is converted in transduced cells into a toxic molecule, which can kill tumour cells (Fecci, 2002, Puumalainen, 1998b, Smitt, 2003, Sandmair, 2000a, Sandmair, 2000b). Currently, there are 7 ongoing clinical trials regarding thymidine kinase ganciclovir therapy (www.gemcris.od. nih.gov) either alone or in combination with other gene therapy strategies. Even though promising results regarding HSV-tk/ganciclovir therapy have been obtained in human trials using adenoviral vectors (Immonen, 2004) there have been also failures regarding therapeutic efficacy of HSV-tk/ ganciclovir therapy when a retroviral vector was used for gene transfer (Rainov, 2000, Shand, 1999).

In addition to thymidine kinase ganciclovir treatment there are three other well characterized pro-drug activating systems that have been used in experimental animal models in the treatment of gliomas. These are the Escherichia coli cytosine deaminase/5-fluorocytosin (CD/5-FC) (Miller, 2002), the rat cytochrome P450 2B1/cyclophosphamide (CPA) (Manome, 1996, Ichikawa, 2001) and the Escherichia coli reductase/CB1954 system (Friedlos, 1998, Weedon, 2000, Palmer, 2004). (Connors, 1995)

Other gene therapy strategies are triggering apoptosis in tumour cells via tumour suppressor genes (such as p53, Fas, ras, TNF-α and caspases) (Shimoura & Hamada, 2003), inhibition of angiogenesis (Puduvalli, 2004, Kirsch, 2000, Tanaka, 1998), augmentation of extracellular matrix protein expression (Lakka, 2003, Mohanam, 2002), modulation of the immune system (Friese, 2003, Yamanaka, 2003, Witham, 2003, and Yang, 2004), eradication of the tumour via oncolytic viruses (Rainov & Ren, 2003, Gromeier & Wimmer, 2001, Lou, 2004), and the use of small interfering RNA (siRNA) (Uchida, 2004), ribozymes (Ge, 1995), and antisense oligonucleotides (Gondi, 2004, Datta, 2004). Also, the generation of fusion proteins that are expressed on the surface of cell membranes and capable of binding a specific ligand could be used for the treatment of brain tumours. We recently constructed and demonstrated the functionality of two different avidin-fusion proteins *in vitro* and *in vivo* using viral vector expression systems in rat brain (Lehtolainen, *et al.* 2002, and Lehtolainen, *et al.* 2003).

In vivo studies have demonstrated that avidin-fusion protein expressed in rat malignant glioma cells were capable of binding biotinylated molecules administered either locally to the brain or systemically into the right carotid artery. Systemically administered biotinylated ligands targeted with high specificity to the intracerebral tumours of rats that were expressing the fusion protein. This again could be achieved by local gene transfer of the target tissue with the fusion protein, followed by *i.v.* administration of a biotinylated drug (Lehtolainen, 2003). These results suggest, that local gene transfer of the fusion protein to target tumour may offer a novel tool for the delivery of biotinylated molecules *in vitro* and *in vivo* for therapeutic and imaging purposes, offering a possibility for an enhanced local effect and a decreased systemic exposure to toxic therapeutic compounds or the imaging agents.

Challenges of Gene Therapy in the Brain

As mentioned earlier, gene therapy to the brain faces unique difficulties in addition to the general issues one faces with gene therapy. A major limiting factor is the delivery of the gene to the brain. The brain is surrounded by the blood brain barrier (BBB), which most gene expression vectors do not naturally cross. The BBB is a capillary barrier that results from a continuous layer of endothelial cells bound together with tight junctions. The endothelial barrier excludes molecules from the brain based on electric

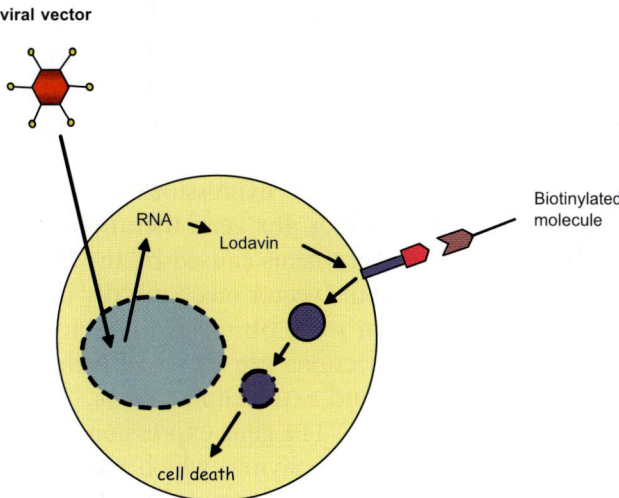

Fig. 2. The principle of biotin binding fusion proteins expressed on the cell surface of the target cell. The target cell/tissue is transduced with an appropriate gene transfer vector containing the gene for Avidin-fusion protein. The cell synthesizes the protein and transports it to the cell membrane. After binding of a biotinylated molecule the Avidin-fusion protein is endocytosed into the cytoplasm of the cell with the biotinylated molecule. The biotinylated molecule is released inside the cytoplasm and the Av-fusion protein is transported back to the cell surface, ready to bind another biotinylated molecule

charge, lipid solubility and molecular weight. Special transport systems are available at brain capillaries for glucose, amino acids, amines, purines, nucleosides, and organic acids. All other materials must cross between endothelial cells (paracellular route) or across cytoplasm (transcellular route) to move from the capillary blood into the tissue (Haluska & Anthony, 2004). Molecules of greater than >500 kDa do not pass in general the BBB. In result to that, under normal conditions large biological molecules such as antibodies and complexes such as viruses have no or very little access across the BBB (Neuwelt, *et al.* 1995, Pardridge, 2002). However, there are viruses that naturally do cross the BBB. One example of these is the Semliki Forest Virus (Fazakerley, 2004).

There are three main strategies of gene expression vector delivery into the brain that have been studied in animal models: 1) Stereotactic inoculation of the gene expression vector into the brain (Qureshi, 2000), 2) intrathecal or intraventricular administration of the gene expression vector (Shimamura, 2003), and 3) intravascular application of the gene expression vector (Rainov, 1999). Of those three methods the stereotactic inoculation of gene expression vector by burrhole is the most commonly used strategy. So far, only the stereotactic inoculation or the craniotomy based inoculation of gene expression vector, including injection into the wall of the tumour cavity, and the intrathecal injection methods, have reached clinical trials.

The intracerebral injection of gene expression vectors is the simplest approach for local gene therapy (e.g. for gene therapy of brain tumours) and an easy way of solving the problems caused by the BBB. In addition, it has the advantage of targeting the vector mechanically into the treatment area. However, there still remain some obstacles to overcome. For example, direct intraparenchymal injections are limited by the small volumes that can be injected into focal and extracellular areas. Also, diffusion of the gene transfer vector is very low. The gene expression vectors do not significantly penetrate into brain parenchyma, which means that the transduced area may be restricted to only a few micrometers. (Puumalainen, 1998b, Rainov & Kramm, 2001, Hsich, 2002) However, a recently developed method to improve the tissue distribution of macromolecules, such as viruses, or liposomes to the brain is the bulk flow convection-enhanced infusion that maintains a pressure gradient during interstitial infusion (Bobo, 1994, Saito, 2004, Nguyen, *et al.* 2003).

The delivery of genes into the brain via the transvascular route has been attempted through BBB disruption using a intracarotid infusion of hyperosmolar solutions and vasoactive compounds. One of the earliest techniques and the first to be used in humans was the injection of a sugar solution into arteries of the neck (Neuwelt, 1980, Greg, 2002). The idea of using hyperosmolar solution is that the resulting high sugar concentration in the capillaries sucks water out of the endothelial cells, shrinking them and opening gaps between cells. The disadvantage of this approach, however, is that it requires arterial access, and the disruption of the BBB may lead to chronic neuropathological changes in the brain. Blood proteins such as albumin are toxic to the brain cells and BBB disruption allows blood components to enter the brain (Schlachetzki, *et al.* 2004). Another strategy to deliver genes through the BBB is the use of certain endogenous transport systems within the BBB. The capillary endothelium, which forms the BBB, expresses receptor-mediated transcytosis systems for certain endogenous peptides, such as insulin and transferrin, (Pardridge, 2002a+b). This strategy has been mainly used in the context of non-viral gene transfer vectors, e.g. with liposomes. Liposomes covered with peptides or antibodies that bind to a specific transcytotic receptor on the endothelium of the BBB are also often referred to as "Trojan horses". These "Trojan horses" have been mainly developed and used for cancer treatment (Pardridge, 2002a+b).

In addition to the problems related to the BBB there is a second major impediment that remains to be overcome when using viral gene transfer vectors: the immune system. Immune-mediated vector toxicity has been reported with a broad range of viral vectors, including herpes simplex viruses (Wood, 1994, Bowers, 2003, Wakimoto, 2003), adenoviruses and adeno-associated viruses (Lowenstein & Castro, 2003, Joos & Chirmule, 2003, Sun, 2003, Byrnes, 1995, Kajiwara, 2000), and retroviruses (Rainov, 2000b). It has been shown that in animals intravascular injection of viral vectors induces the release of cytokines, interleukins, activates macrophages, induces T-cell and B-cell responses, induces viral neutralizing antibodies, and induces the activation of the endothelium (Lowenstein, 2004). The majority of immunological studies regarding viral vectors have been done with adenoviruses and HSV-1. For example, Wood *et al.* (Wood, 1994) described a strong inflammatory response, characterized by diffuse up-regulation of major histocompatibility complex class I antigens and the activation of microglia after stereotactic injection of a defective HSV-1 vector into rat brain. In general, an immune response can be generated against both, the virion and the proteins expressed by the viral genome.

The extent of inflammatory and immune response to other viral vectors such as with alphaviruses, and adeno-associated viruses injected into the brain remains to be elucidated in more detail.

Targeting of the gene transfer vectors to target cells and avoidance of the transduction of unwanted non-target cells is a general problem in gene transfer based therapies. Regarding brain gene therapy several approaches have been tackled in order to target gene transfer vectors to neuronal/cancer cells. These include amongst other things the use of a) tissue specific promoters such as the human PDGF-beta, the neuron specific enolase or the glial fibrillary acidic protein promoter (Liu, 2004, Jakobsson, 2003), b) antibody based targeting (e.g. liposomes) or re-targeting (e.g. adenovirus) of the gene transfer vectors (Zhang, 2004, Miller, 1998), or c) conditionally replicating viruses (Gomez-Manzano, 2004, Markert, 2000). Figure 3 gives an example of how adenoviruses for example can be modified in order to make them cell or tissue type specific.

Gene Transfer Vectors

There are two main types of gene delivery vectors: viral and non-viral vectors. Retroviruses/lentiviruses, recombinant herpes simplex virus, adenoviruses, and adeno-associated viruses are the most common viral vectors that have been used for the delivery of genes into the CNS. More recently, there have been studies also about the possible use of Baculoviruses (Lehtolainen, 2002, Tani, 2003) Semliki Forest viruses (SFV) (Lundstrom, 2001), Sindbis virus (Ehrengruber, 2002) and recombinant Simian virus-40 (SV40) (Cordelier, 2003) for gene therapy in the brain.

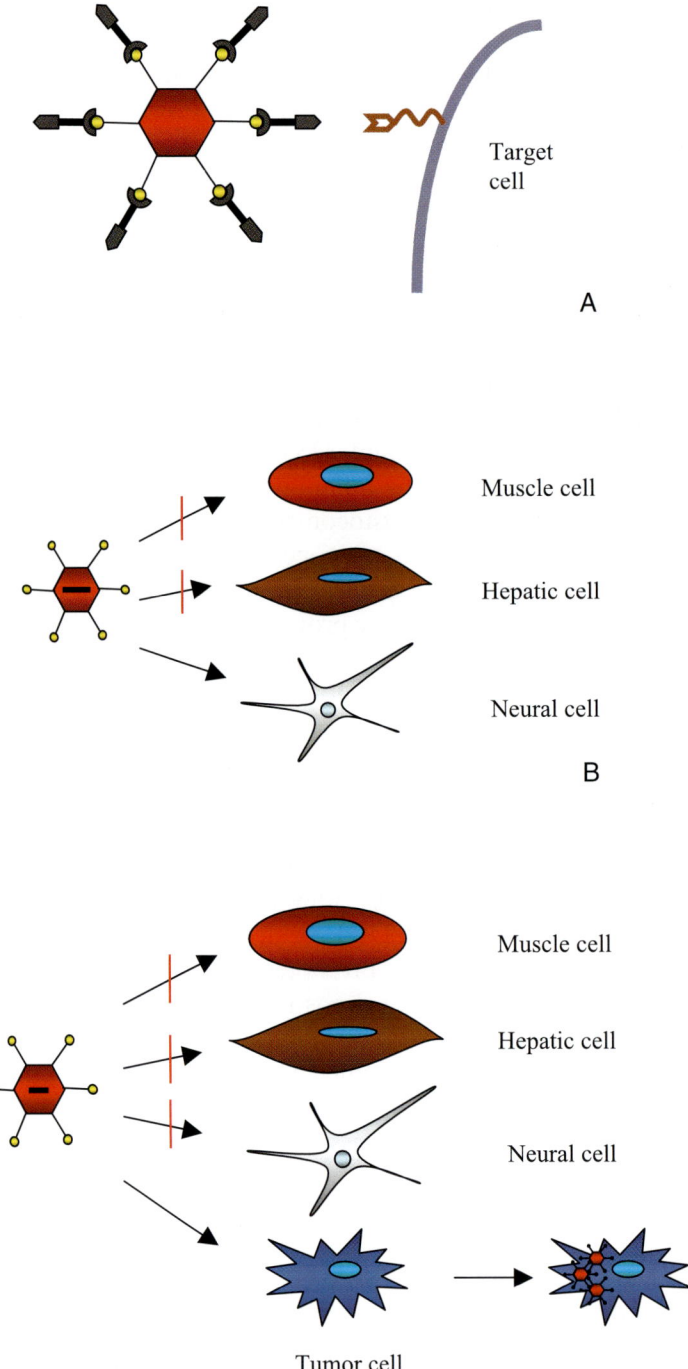

Target cell

A

Muscle cell

Hepatic cell

Neural cell

B

Muscle cell

Hepatic cell

Neural cell

Tumor cell

C

Fig. 3. Targeting of adenoviruses can be achieved by (A) re-directing the vector capsid to new cellular receptors using molecular adaptors, such bi-specific antibodies, through (B) placing the transgene under the control of a cell type-specific promoter, or through (C) genetically modifying them into conditionally replicating viruses

Viral Gene Transfer Vectors

Retroviruses

DNA can be introduced into cells using retrovirus vectors. Retroviruses allow stable integration of expressed genes. The retroviridae are a large group of viruses associated with many diseases ranging from completely benign infections to fatal conditions such as HIV and tumours caused by oncogenic viruses (Coffin, 1990). Brain tumours are theoretically suitable for retrovirus-mediated gene transfer, since retroviruses only infect proliferating cells, while normal, generally non-dividing brain tissue remains intact (Miller, 1990). However, the failure of a phase III trial where 248 patients with newly diagnosed, previously untreated glioblastoma multiforme were treated by retrovirus-mediated transduction of glioblastoma cells with the HSV-tk gene and subsequent systemic treatment with ganciclovir, was a drawback for the retroviral vector. Especially, since the failure of that trial was mainly attributed to poor rate of delivery of the HSV-tk gene (Rainov, 2000a).

Recently, retroviral vectors based on lentiviruses (such as the human immunodeficiency virus) have been developed that are capable of transducing also non-dividing cells in a long lasting manner (Naldini, 1996, Kirik & Bjorklund, 2003, Jakobsson, 2003). Since retroviruses are integrating vectors one concern has been the possibility of random integration of foreign DNA into target cells, carrying the potential risk of insertional mutagenesis, the perturbation of other genes involved in growth control, or inactivation of tumour suppressor genes (Temin, 1990). However, especially in cancer gene therapy the risk of insertional mutagenesis is only of minimal concern. Despite some safety and ethical concerns about the use of lentiviruses, they seem to be feasible gene transfer vectors to the brain (Van den Haute, 2003, Marr, 2003, Watson & Wolfe, 2003, Koponen, 2003, Regulier, 2002).

Herpes Simplex Virus-1 (HSV-1)

HSV's have some characteristics that render them particularly suitable for use as neuronal vectors. They are neurotropic, and hence, infect neurons efficiently. In addition, they can accommodate large inserts, since approximately half of its genome is composed of non-essential genes that can be replaced with heterologous genes. Another interesting feature of HSV is that it can be transported retrogradely in neurons and transferred across synapses. (Simonato, 2000) Replication defective HSV-1 vectors are produced by deleting all, or a combination, of the five immediate-early genes (ICP0, ICO4, ICO22, ICP27 and ICP47) (Thomas, 2003), which are required for lytic infection and expression of all other viral proteins. Vectors derived from HSV-1 still remain capable to infect a wide range of cell types – including neurons in the CNS.

A number of attenuated strains have been developed. Thymidine kinase/Ganciclovir therapy for brain tumour has been successfully applied with HSV-tk viruses transfecting the tumour to express the thymidine kinase gene (Todo, 2000, Moriuchi, 1998, Kramm, 1996). Disadvantages with the use of HSV-1 vectors are lytic infections and potential neurotoxicity (Zlokovic & Apuzzo, 1997). Because HSV-1 do not have a long lasting gene expression, their use is mainly restricted to brain tumours.

Adenoviruses

Adenoviruses are large, double stranded DNA viruses which can carry large fragments of foreign DNA. Adenoviruses exist extrachromosomally within the cell although the DNA migrate into the cell nuclei. Adenoviruses have a known tropism for pulmonary and intestinal epithelial cells; they are not neurotoxic and are linked to only minor diseases in human. Adenoviruses have a broad host and cell range. They are capable of high-efficiency gene delivery into a variety of organs, including lung, skeletal muscle, heart, liver, blood vessels and the central nervous system (Sandmair, 2000b). For these reasons recombinant adenoviral vectors have been extensively used in experimental models, as well as in clinical protocols. Because of their broad host and cell range adenoviruses have been modified by different means to make them specific to certain cell types (Fig. 3). One approach involves genetically modifying the fibre knob, through which attachment of the virus to the cell receptor and entry into the cell occurs. A second approach is immunological modification of the adenovirus tropism using bi-specific molecules that on one side bind to the fibre knob or the penton base of the adenovirus, and on the other side bind to the cell surface receptor, different from the viral receptor (Wickham, 2003).

Also, specificity of adenoviruses has been achieved using tissue specific promoters, such as the human synapsin 1 gene promoter for neuron specific transgene expression (Kugler, 2003), or more recently, using conditionally replicating adenoviruses (Steinwaerder, 2001). The use of genes such as the HSV-thymidine kinase gene that specifically targets dividing/tumour cells has also been studied extensively (Moolten, 1994).

Adeno-Associated Virus (AAV's)

So far, eight distinct AAV serotypes have been identified which infect different cell types with different efficiency (Thomas, 2003). However, most recombinant AAV vectors have been derived from AAV2 and most of the in vivo studies have been performed using AAV2 vectors containing the strong cytomegalovirus immediate-early (CMV) promoter (Tenenbaum, 2004). AAV vectors have been shown to transduce a broad range of neural cells. Transduction efficiency, however, varies markedly from one region to another.

AAV vectors are integrating vectors. Wild-type AAV integrates exclusively into a single site on human chromosome 19, whereas it appears that AAV recombinants integrate much less efficiently and more randomly (Balague, 1997). AAV vectors have been shown to give sustained transgene expression upon in vivo administration. Expression of homologous genes has been detected two years after injection in mice and several months after injection in dogs, primates, and man. (Mountain 2000) AAV vectors can transfer genes efficiently to both quiescent and proliferating cells. The main disadvantages of AAV vectors are the small insert size they can accommodate and the use of helper viruses in the manufacturing process, which caused problems, such as low titre, contamination and costly purification procedures. However, progress has been made in manufacturing process, which allows high-titre production without helper viruses. (Ferrari, 1997, Snyder & Flotte 2002, During, 2003)

Currently, there are two ongoing brain gene therapy clinical trials where AAV vectors are used for the treatment of Parkinson's disease (www.gemcris.od.nih.gov).

Non-Viral Vectors

Viral vectors have been shown to be efficient gene transfer tools. Nevertheless, drawbacks, such as the bloodstream's rapid clearance of viral vectors (when injected systemically), their immunogenic and inflammatory potential, together with certain safety concerns, urged the development of new synthetic gene delivery vectors. (Poly)cationic carriers and cationic lipids have been studied extensively as alternatives for viral vectors (da Cruz, 2004, Anderson, 2003, Lesage, 2002, Goldman, 1997). The (poly)cationic carriers possess groups which are protonated at physiological pH. The electrostatic attraction between the cationic charged polymer and the negatively charged DNA results in a particular complex – the polyplex, which is the transduction reagent. As with (poly)cationic carriers, cationic lipids posses additionally a hydrophobic group, which ensures that the cationic lipids assemble into bi-layer vesicles on dispersion in aqueous media (Brown, 2001). A great advantage of non-viral vectors is that they can be produced more easily than viral vectors. However, compared to viral gene transfer vectors these non-viral vectors are facing different types of problems, such as binding to plasma proteins or blood cells, which can lead to aggregates and clogging of capillaries (Ogris & Wagner, 2002).

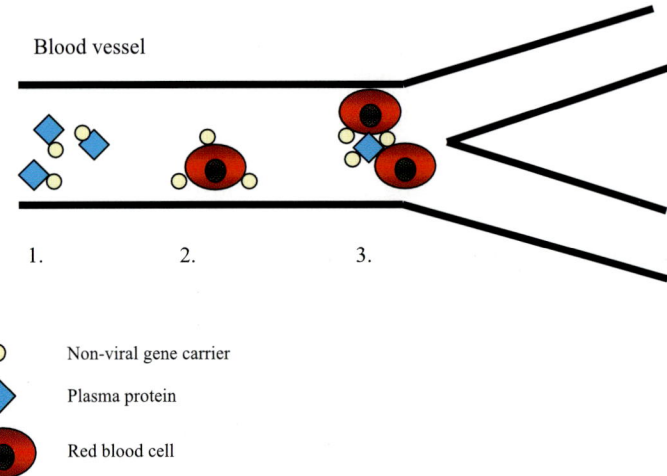

Fig. 4. Obstacles for positively charged non-viral gene transfer vectors within the blood circulation. *(1)* They can bind to plasma proteins or *(2)* blood cells, and *(3)* aggregates can clog capillaries. In addition, one of the major drawbacks of non-viral vectors is their low transducing efficiency in vivo (Ogris, 2002)

Ethics

Gene therapy raises many questions among the society. They raise concern about their safety in humans and their offspring, their environment safety, and their impact on the status within the society; Is it going to be a treatment modality only accessible for a certain group of people (people with a high social status) or is it going to be accessible for everyone? Opinions and points of views about gene therapy vary from one extreme to another. Cultural as well as religious points of views have strong impact on these standpoints.

Several questions have to be asked in order to justify gene therapy in humans; Questions such as which are the diseases where gene therapy is ethically acceptable? It appears that gene therapy is more tolerated for life-threatening diseases (e.g. diseases like cancer or AIDS) than e.g. in the correction of a learning disorders (Rabino, 2003). Also, somatic gene therapy appears to be more tolerated than germline gene therapy. Several questions have to be asked in order to justify gene therapy in humans, like which are the diseases where gene therapy is ethically acceptable? Where the use of gene therapy in the treatment of a genetic disease (e.g. cancer) might be ethically justified, how about when dealing with genetic 'disorders'? Would it be ethically acceptable to practise gene therapy on people with Down's syndrome? What is the justification of using gene therapy in those people?

In addition to issues raised above there are also technical issues concerning the justification of gene therapy in humans. For example, what are the technical details of the DNA and vector to be used? The technical aspects involved, risks endeavoured by the patient, and the fear of human genetic engineering are some of the major reasons why human gene therapy experiments have long been delayed.

The use of viral gene transfer vectors, such as lentiviruses raises scepticism about the safety of these vectors. Non-viral vectors are not yet efficient enough, but have gained better acceptance in the society. It looks like gene therapy of brain tumours will be ethically acceptable whereas the use of genetically modified stem cells may be a much more difficult topic. However, the normal principles of good clinical research apply in the conduct of the ethical evaluation of gene therapy protocols as well. The integrity and free will of a patient should be respected, all available information for the informed consent should be given, and the safety of an individual must be the first concern of the treatment protocol.

Concluding Remarks

There is extensive research going on in the field of gene therapy and especially malignant glioma, which has been subject of an increasing interest as a possible target for it. AdHSV-tk/ganciclovir gene therapy is one of the research lines with some promising results in early clinical trials that need to be confirmed in larger patient series (Immonen, 2004). However, there still remain obstacles that have to be overcome, especially when talking about gene therapy into the brain. Gene transfer vectors have to be able to cross the BBB. The induction of immune response for some vectors must be avoided, and the production of viral vectors in large scale has to be optimized. Nevertheless, there is no doubt that none of these problems mentioned above are problems that can not be resolved.

Acknowledgement

We thank Prof. Matti Vapalahti, M.D., Ph.D. and Kar Airenne, Ph.D. for critical reading of the manuscript.

References

1. Abeloff MD *et al* (2000) Clinical oncology, 2nd edn. Churchill, Livinstone
2. Alisky JA, Davidson BL (2000) Gene therapy for amyotrophic lateral sclerosis and other motor neuron diseases. Hum Gene Ther 11(17): 2315–2330
3. Ammirati M *et al* (1987) Effect of the extent of surgical resection on survival and quality of life in patients with supratentorial glioblastomas and anaplastic astrocytomas. Neurosurgery 21(2): 201–206

4. Anderson DM *et al* (2003) Stability of mRNA/cationic lipid lipoplex in human and rat cerebrospinal fluid: methods and evidence for non-viral mRNA gene delivery into the central nervous system. Hum Gene Ther 14(3): 191–202

5. Ascadi G *et al* (2002) Increased survival and function of SOD 1 mice after glial-cell derived neurotrophic factor gene therapy. Hum Gene Ther 13(9): 1047–1059

6. Azzous M *et al* (2000) Increased motoneuron survival and improved neuromuscular function in transgenic ALS mice after intraspinal injection of adeno-associated virus encoding Bcl-2. Hum Mol Genet 9(5): 803–811

7. Azzous M *et al* (2004) VEGF delivery with retrogradely transported lentivector prolongs survival in a mouse ALS model. Nature 429: 413–417

8. Bachoud-Levi AC *et al* (2000) Neuroprotective gene therapy for Huntington's disease using a polymer encapsulated BHK cell line engineered to secrete human CNTF. Hum Gene Ther 11(12): 1723–1729

9. Bachoud-Levi AC *et al* (1998) Prospective for cell and gene therapy in Huntington's disease. Prog Brain Res 117: 511–524

10. Bemelmans AP *et al* (1999) Brain-derived neurotrophic factors-mediated protection of striatal neurons in an excitotoxic rat model of Huntington's disease, demonstrated by adenoviral gene transfer. Hum Gene Ther 10(18): 2987–2997

11. Bethel A *et al* (1997) Intravenous basic fibroblast growth factor-2 decreases brain injury resulting from focal ischemia in cats. Stroke 28: 609–615

12. Bobo RH *et al* (1994) Convection-enhanced delivery of macromolecules in the brain. Proc Natl Acad Sci 91: 2076–2080

13. Boillée S and Cleveland DW (2004) Gene therapy for ALS delivers. Trends Neurosci 27(5): 235–238

14. Bowers WJ *et al* (2003) Immune responses to replication-defective HSV-1 type vectors within the CNS: implication for gene therapy. Gene Ther 10(11): 941–945

15. Brown M (2003) Gene therapy success for Alzheimer's. Drug Discov Today 8(11): 474–5

16. Brown MD *et al* (2001) Gene delivery with synthetic (non-viral) carriers. Int J Pharm 229: 1–21

17. Bruijn LI *et al* (2004) Unraveling the mechanisms involved in motor neuron degeneration in ALS. Annu Rev Neurosci 27: 723–749

18. Byrnes AP *et al* (1995) Adenovirus gene transfer causes inflammation in the brain. Neuroscience 66(4): 1015–1024

19. Cabrera-Salazar MA *et al* (2002) Gene therapy for the lysosomal storage disorders. Curr Opin Mol Ther 4(4): 349–358

20. Cavazzana-Calvo M *et al* (2000) Gene therapy of a human severe combined immunodeficiency (SCID)-XI disease. Science 288: 669–672

21. Chiocca EA (2003) Gene therapy: a primer for neurosurgens. Neurosurgery 53(2): 364–373

22. Coffin M (1990) Retroviridae and their replication. In: Fields BN, Knipe DM (eds) Virology. Raven Press Ltd, New York, p 1437–1500

23. Connors TA (1995) The choice of prodrugs for gene directed enzyme pro-
    drug therapy of cancer. Gene Ther 2: 702–709
24. Cordelier P et al (2003) Inhibition of HIV-1 in the central nervous system by
    IFN-alpha2 delivered by an SV40 vector. J Interferon Cytokin Res 14(16):
    1525–1533
25. Da Cruz MT et al (2004) Improving lipoplex-mediated gene transfer into C6
    glioma cells and primary neurons. Exp Neurol 187(1): 65–75
26. Datta K et al (2004) Senzitazing glioma cells to cisplatin by abrogating
    the p53 response with antisense oligonucleotides. Cancer Gene Ther 11(8):
    525–531
27. Dietrich HH and Dacey RG (2000) Molecular keys to the problems of cere-
    bral vasospasm. Neurosurgery 46(3): 517–29
28. Durin MJ et al (2003) Development and optimization of adeno-associated
    virus vector transfer into the central nervous system. Methods Mol Med 76:
    221–236
29. During MJ et al (1994) Long-term behavioral recovery in parkinsonian rats
    by an HSV vector expressing tyrosine hydroxylase. Science 266: 1399–403
30. Eberhardt O, Schulz JB (2004) Gene therapy in Parkinson's disease. Cell
    Tissue Res 63(4): 758–769
31. Eberling JL et al (2003) In vivo PET imaging of gene expression in parkinso-
    nian monkeys. Mol Ther 8(6): 873–5
32. Edelstein ML et al (2004) Gene therapy clinical trials worldwide 1989–2004
    – an overview. J Gene Med 6(6): 597–602
33. Ehrengruber MU (2002) Alphaviral vectors for gene transfer into neurons.
    Mol Neurobiol 26: 183–201
34. Eto Y, Ohashi T (2002) Novel treatment for neuronopathic lysosomal stor-
    age diseases – cell therapy/gene therapy. Curr Mol Med 2(1): 83–89
35. Fazakerley JK (2004) Semliki forest virus infection of laboratory mice: a
    model to study the pathogenesis of viral encephalitis. Arch Virol Suppl 18:
    179–190
36. Fecci PE, Gromeier M, Sampson JH (2000) Viruses in the treatment of brain
    tumors. Neuroimaging Clin N Am 12(4): 553–570
37. Ferrari FK et al (1997) New developments in the generation of Ad-free, high
    titer rAAV gene theraby vectors. Nat Med 3(11): 1295–1297
38. Finkelstein R, Baughman RW, and Steele FR (2001) Harvesting the neural
    gene therapy fruit. Mol Ther 3(1): 3–7
39. Fisher W (1987) Amelioration of cholinergic neuron atrophy and spatial
    memory impairment in aged rats by nerve growth factor. Nature 329: 65–68
40. Friedlos F et al (1998) Gene directed enzyme prodrug therapy: quantitative
    bystander cytotoxicity and DNA damage induced by CB 1954 in cells
    expressing bacterial nitroreductase. Gene Ther 5(1): 105–112
41. Friese MA et al (2003) MICA/NKG2D-mediated immunogene therapy of
    experimental gliomas. Cancer Res 63(24): 8996–9006
42. Futerman AH, van Meer G (2004) The cell biology of lysosomal storage dis-
    orders. Nat Rev Mol Cell Biol 5(7): 554–565
43. Gash DM, Gerhardt GA, Hoffer BJ (1998) Effects of glial cell line-derived

neurotrophic factor on the nigrostriatal dopamine system in rodents and nonhuman primates. Adv Pharmacol 42: 911–915

44. Ge L *et al* (1995) Gene therapeutic approaches to primary and metastatic brain tumors: II. Ribozyme-mediated suppression of CD44 expression. J Neurooncol 26(3): 251–257

45. Goldman CK *et al* (1997) In vitro and in vivo gene delivery mediated by a synthetic polycationic amino polymer. Nat Biotech 15(5): 462–466

46. Gomez-Manzano C *et al* (2004) A novel E1A–E1B mutant adenovirus induces glioma regression in vivo. Oncogene 23(10): 1821–1828

47. Gondi CS *et al* (2004) Adenovirus-mediated expression of antisense urokinase plasminogen activator receptor and antisense cathepsin B inhibits tumor growth, invasion, and angiogenesis in gliomas. Cancer Res 64: 4069–4077

48. Greg M (2002) Breaking down barriers. Science 297(5587): 1116–1118

49. Gromeier M, Wimmer E (2001) Viruses for the treatment of malignant glioma. Curr Opin Mol Ther 3(5): 503–508

50. Gutierrez-Delicado E, Serratosa JM (2004) Genetics of epilepsies. Curr Opin Neurol 17(2): 147–153

51. Haberman RP *et al* (2003) Attenuation of seizures and neuronal death by adeno-associated virus vector galanin expression and secretion. Nat Med 9(8): 1076–1080

52. Hacein-Bey-Abina S *et al* (2003) A serious adverse event after successful gene therapy for X-linked severe combined immunodeficiency. N Engl J Med 348(3): 255–6

53. Hagell P *et al* (2002) Dyskinesias following neural transplantation in Parkinsons disease. Nat Neurosci 5: 627–628

54. Haluska M, Anthony ML (2004) Osmotic blood-brain barrier modification for the treatment of malignant brain tumors. CJON 8(3): 263–267

55. Hoehn B *et al* (2001) Overexpression of HSP72 after induction of experimental stroke protects neurons from ischemic damage. J Cereb Blood Flow Metab 21(11): 1303–1309

56. Hogarth P (2003) Huntington's disease: a decade beyond gene discovery. Curr Neurol Neurosci Rep 3(4): 279–284

57. Howard K (2003) First Parkinson gene therapy trial launches. Nat.Biotech 21(10): 1117–8

58. Hsich G, Sena-Esteves M, Breakefield XO (2002) Critical issues in gene therapy for neurologic disease. Hum Gene Ther 13(5): 579–604

59. Ichikawa T *et al* (2001) Intraneoplastic polymer-based delivery of cyclophosphamide for intratumoral bioconversion by a replicating oncolytic viral vector. Cancer Res 61(3): 864–868

60. Immonen A *et al* (2004) AdvHSV-tk gene therapy with intravenous ganciclovir improves survival in human malignant glioma: a randomized, controlled study. Mol Ther 10(5): 967–72

61. Jakobsson J *et al* (2003) Targeted transgene expression in rat brain using lentiviral vectors. J Neurosci Res 73(6): 876–885

62. Jakobsson J *et al* (2003) Targeted transgene expression in rat using lentiviral vectors. J Neurosci Res 73(6): 876–885

63. Joos K, Chirmule N (2003) Immunity to adenovirus and adeno-associated viral vectors: implications for gene therapy. Gene Ther 10(11): 955–963
64. Kajiwara K *et al* (2000) Humoral immune response to adenovirus vectors in the brain. J Neuroimmunol 103(3): 253–265
65. Kaspar BK *et al* (2003) Retrograde viral delivery of IGF-1 prolongs survival in a mouse model. Science 301: 839–842
66. Kaye EM, Sena-Esteves M (2002) Gene therapy for the central nervous system in the lysosomal storage disorders. Neurol Clin 20(3): 879–901
67. Kirik D, Bjorklund A (2003) Modeling CNS neurodegeneration by over-expression of disease causing proteins using viral vectors. Trends Neurosci 26(7): 386–392
68. Kirsch M *et al* (2000) Anti-angiogenic treatment strategies for malignant brain tumors. J Neurooncol 50: 149–163
69. Koponen JK *et al* (2003) Doxycycline-regulated lentiviral vector system with a novel reverse transactivator rtTA2S-M2 shows a tight control of gene expression in vitro and in vivo. Gene Ther 10(6): 459–466
70. Kordower J *et al* (2000) Neurodegeneration prevented by lentiviral vector delivery of GDNF in primate models of Parkinson's disease. Science 290: 767–773
71. Kramm CM *et al* (1996) Long-term survival in a rodent model of disseminated brain tumors by combined intrathecal delivery of herpes vectors and ganciclovir treatment. Hum Gene Ther 7(16): 1989–1994
72. Kromer LF (1987) Nerve growth factor treatment after brain injury prevents neuronal death. Science 235(4785): 214–216
73. Kugler S *et al* (2003) Human synapsin 1 gene promoter confers highly neuron-specific long-term transgene expression from an adenoviral vector in the adult rat brain depending on the transduced area. Gene Ther 10(4): 337–347
74. Lakka SS *et al* (2003) Synergistic down-regulation of urokinase plasminogen activator receptor and matrix metalloproteinase-9 in SNB19 glioblastoma cells efficiently inhibits glioma cell invasion, angiogenesis, and tumor growth. Cancer Res 63(10): 2454–2461
75. Lamont JP *et al* (2000) A prospective phase II trial of ONYX-015 adenovirus and chemotherapy in recurrent squamous cell carcinoma of the head and neck (the Baylor experience). Ann Surg Oncol 7(8): 588–592
76. Lampela P *et al* (2002) The use of low-molecular-weight PEIs as gene carriers in the monkey fibroblastoma and rabbit smooth muscle cell culture. J Gene Med 4(2): 205–214
77. Lehtolainen P *et al* (2002) Baculovirus exhibits restricted cell type specificity in rat brain: a comparison of baculovirus- and adenovirus mediated intracerebral gene transfer in vivo. Gene Ther 9(24): 1693–1699
78. Lehtolainen P *et al* (2002) Clining and characterization of Scavidin, a fusion protein for the targeted delivery of biotinylated molecules. J Biol Chem 277(10): 8545–8550
79. Lehtolainen P *et al* (2003) Targeting of biotinylated compounds to its target

tissue using a low-density lipoprotein receptor-avidin fusion protein. Gene Ther 10(25): 2090–2097

80. Lesage D *et al* (2002) Evaluation and optimization on DNA delivery into gliosarcoma 9L cells by a cholesterol-based cationic liposome. Biochim Biophys Acta 1564(2): 393–402

81. Lin H *et al* (2002) Cyclooxygenase-1 and bicistronic cyclooxygenase-1/ prostacyclin synthase gene transfer protects against ischemic cerebral infarction. Circulation 105(16): 1962–1969

82. Liu BH *et al* (2004) CMV enhancer/human PDGF-beta promoter for neuron specific transgene expression. Gene Ther 11(1): 52–60

83. Lou E (2004) Oncolytic viral therapy and immunotherapy of malignant brain tumors: two potential new approaches of translational research. Ann Med 36(1): 2–8

84. Lowenstein PR, Castro MG (2003) inflammation and adaptive immune responses to adenoviral vectors injected into the brain: peculiarities, mechanisms, and consequences. Gene Ther 10(11): 946–954

85. Lowenstein PR (2004) Immunological needles in the gene therapy haystack: applying a genetic paradigm to gene therapy. Gene Ther 11(1): 1–3

86. Lüders JC *et al* (2000) Adenoviral gene transfer of nitric oxide synthase increases cerebral blood flow in rats. Neurosurgery 47(5): 1206–1217

87. Lundstrom K *et al* (2001) Semliki Forest virus vectors: efficient vehicles for in vitro and in vivo gene delivery. FEBS Lett 504(3): 99–103

88. Luo J *et al* (2002) Subthalamic GAD gene therapy in a Parkinson's disease rat model. Science 298: 425–429

89. MacMillan JC *et al* (1994) Clinical considerations in gene therapy of Huntington's disease. Gene Ther 1[Suppl] 1: 88

90. McCown TJ (2004) The clinical potential of antiepileptic gene therapy. Expert Opin Biol Ther 4(11): 1771–1776

91. McPhee SW *et al* (2005) Effects of AAV-2-mediated aspartoacylase gene transfer in the tremor rat model of Canavan disease. Brain Res Mol Brain Res 135(1–2): 112–121

92. Manome Y *et al* (1996) Gene therapy for malignant gliomas using replication incompetent retroviral and adenoviral vectors encoding the cytochrome P450 2B1 gene together with cyclophosphamide. Gene Ther 3(6): 513–520

93. Markert JM *et al* (2000) Conditionally replicating herpes simplex virus mutant, GS07 for the treatment of malignant glioma: results of a phase I trial. Gene Ther 7(10): 867–874

94. Marr RA *et al* (2003) Neprilysin gene transfer reduces human amyloid pathology in transgenic mice. J Neurosci 23(6): 1992–1996

95. Marr RA *et al* (2004) Neprilysin regulates amyloid Beta peptide levels. J Mol Neurosci 22(1–2): 5–11

96. Mattson MP (2004) Pathways towards and away from Alzheimer's disease. Nature 430: 631–639

97. McBride JL *et al* (2003) Structural and functional neuroprotection in a rat model of Huntington's disease by viral gene transfer of GDNF. Exp Neurol 181(2): 213–223

98. Miller CR *et al* (1998) Differential susceptibility of primary and established human glioma cells to adenovirus infection: targeting via the epidermal growth factor receptor achieves fibre receptor-independent gene transfer. Cancer Res 58(24): 5738–5748

99. Miller CR *et al* (2002) Intratumoral 5-fluorouracil produced by cytosine deaminase/5-fluorocytosin gene therapy is effective for experimental human glioblastoma. Cancer Res 62(3): 773–780

100. Miller DG *et al* (1990) Gene transfer by retrovirus vectors occurs only in cells that are actively replicating at the time of infection. Mol Cell Biol 10: 4239–4242

101. Mohanam S *et al* (2002) Modulation of invasive properties of human glioblastoma cells stably expressing amino-terminal fragment of urokinase-type plasminogen activator. Oncogene 21(51): 7824–7830

102. Monahan PE, White GC 2nd (2002) Hemophilia gene therapy: update. Curr Opin Hematol 9(5): 430–436

103. Moolten FL (1994) Drug sensitivity ("suicide") genes for selective cancer chemotherapy. Cancer Gene Ther 1: 279–287

104. Morishita R *et al* (2004) Safety Evaluation of Clinical Gene Therapy Using Hepatocyte Growth Factor to Treat Peripheral Arterial Disease. Hypertension Jul 6 [Epub ahead of print]

105. Morishita R *et al* (2004) Therapeutic angiogenesis using hepatocyte growth factor (HGF). Curr Gene Ther 4(2): 199–206

106. Moriuchi S *et al* (1998) Enhanced tumor cell killing in the presence of ganciclovir by herpes simplex virus type 1 vector-directed coexpression of human tumor necrosis factor-alpha and herpes simplex virus thymidine kinase. Cancer Res 58(24): 5731–5737

107. Mountain A (2000) Gene therapy: the first decade. TIBTECH 18: 119–128

108. Naldini L *et al* (1996) In vivo gene delivery and stable transduction of non-dividing cells by a lentiviral vector. Science 272: 263–267

109. Neuwelt EA *et al* (1995) Gene replacement therapy in the central nervous system – viral vector-mediated therapy of global neurodegenerative disease. Behav Brain Sci 18(1): 1–9

110. Neuwelt G *et al* (1980) Reversible osmotic blood-brain barrier disruption in humans: implications for the chemotherapy of malignant brain tumors. Neurosurgery 7(1): 44–52

111. Nguyen TT *et al* (2003) Convective distribution of macromolecules in the primate brain demonstrated using computerized tomography and magnetic resonance imaging. J Neurosurg 98(3): 584–590

112. Ogris M, Wagner E (2002) Targeting tumors with non-viral gene delivery systems. Drug Discov Today 7(8): 479–485

113. Onoue H *et al* (1998) Expression and function of recombinant endothelial nitric oxide synthase gene in canine basilar artery after experimental subarachnoid hemorrhage. Stroke 29: 1959–1966

114. Palmer AM *et al* (2002) Pharmacotherapy for Alzheimer's disease: progress and prospects. Trends Pharmacol Sci 23: 426–433

115. Palmer DH *et al* (2004) Virus-directed enzyme prodrug therapy: intratumoral administration of a replication-deficient adenovirus encoding nitroreductase to patients with resectable liver cancer. J Clin Oncol 22(9): 1546–1552

116. Pardridge WM (2002) Drug and gene delivery to the brain: the vascular route. Neuron 36: 555–558

117. Pardridge WM (2002a) drug and gene delivery to the brain: the vascular route. Neuron 36(4): 555–558

118. Pardridge WM (2002b) Drug and gene targeting to the brain with molecular Trojan horses. Nat Rev Drug Discov 1(2): 131–139

119. Pompl PN *et al* (2003) A therapeutic role for cyclooxygenase-2 inhibitors in a transgenic mouse model of amyotrophic lateral sclerosis. FASEB 17(6): 725–727

120. Prados MD *et al* (2003) Treatment of progressive or recurrent glioblastoma multiforme in adults with herpes simplex virus thymidine kinase gene vector-producer cells followed by intravenous ganciclovir administration: a phase I/II multi-institutional trial. J Neurooncol 65(3): 269–278

121. Puduvalli VK (2004) Inhibition of angiogenesis as a therapeutic strategy against brain tumors. Cancer Treat Res 117: 307–336

122. Puumalainen AM *et al* (1998b) Beta-galactosidase gene transfer to human malignant glioma in vivo using replication-deficient retroviruses and adenoviruses. Hum Gene Ther 9(12): 1769–1774

123. Puumalainen AM, Vapalahti M, Ylä-Herttuala S (1998a) Gene therapy for malignant glioma patients. Adv Exp Med Biol 451: 505–509

124. Qureshi NH *et al* (2000) Multicolumn infusion of gene therapy cells into human brain tumors: technical report. Neurosurgery 46(3): 663–668; discussion 668–669

125. Rabino I (2003) Gene therapy: ethical issues. Theor Med 24: 31–58

126. Rainov NG (2000a) A phase III clinical evaluation of herpes simplex virus type 1 thymidine kinase ganciclovir gene therapy as an adjuvant to surgical resection and radiation in adults with previously untreated glioblastoma multiforme. Hum Gene Ther 11(17): 2389–2401

127. Rainov NG *et al* (2000b) Immune response induced by retrovirus-mediated HSV-tk/GCV pharmacogene therapy in patient with glioblastoma multiforme. Gene Ther 7(21): 1853–1858

128. Rainov NG, Kramm CM (2001) Vector delivery methods and targeting strategies for gene therapy of brain tumors. Curr Gene Ther 1: 367–383

129. Rainov NG, Ren H (2003) Oncolytic viruses for treatment of malignant brain tumours. Acta Neurochir [Suppl] 88: 113–123

130. Rainov NG *et al* (1999) Intraarterial delivery of adenovirus vectors and liposome-DNA complexes to experimental brain neoplasms. Hum Gene Ther 10(2): 311–318

131. Regulier E *et al* (2002) Dose-dependent neuroprotective effect of ciliary neurotrophic factor delivered via tetracycline-regulated lentiviral vectors in the quinolinic acid rat model of Huntington's disease. Hum Gene Ther 13(16): 1981–1990

132. Richichi C *et al* (2004) Anticonvulsant and antiepileptogenic effects mediated by adeno-associated virus vector neuropeptide Y expression in the rat hippocampus. J Neurosci 24(12): 3051–3059

133. Saito R *et al* (2003) Distribution of liposomes into brain and rat brain tumor models by convection-enhanced delivery monitored with magnetic resonance imaging. Cancer Res 64: 2572–2579

134. Samii A, Nutt JG, Ransom BR (2004) Parkinson's disease. Lancet 363: 1783–93

135. Sanchez-Pernaut R *et al* (2001) Functional effect of adeno-associated virus mediated gene transfer of aromatic L-amino acid decarboxylase into the striatum of 6-OHDA-lesioned rats. Mol Ther 4(4): 324–330

136. Sandmair A-M *et al* (2000a) Thymidine kinase gene therapy for human malignant glioma using replication-deficient retroviruses or adenoviruses. Human Gene Ther 11(16): 2197–2205

137. Sandmair A-M (2000b) Gene therapy for malignant glioma. Thesis work. University of Kuopio

138. Schlachetzki F *et al* (2004) Gene therapy of the brain. Neurology 62: 1275–1281

139. Scollay R (2001) Gene therapy: a brief overview of the past, present and future. Ann NY Acad Sci 953: 26–30

140. Seki T *et al* (2004) Adenoviral gene transfer of aspartoacylase ameliorates tonic convulsions of spontaneously epileptic rats. Neurochem Int 45(1): 171–178

141. Shand N *et al* (1999) A phase 1–2 clinical trial of gene therapy for recurrent glioblastoma multiforme by tumor transduction with the herpes simplex thymidine kinase gene followed by ganciclovir. GLI328 European-Canadian Study Group. Hum Gene Therapy 10(14): 2325–2335

142. Shen Y *et al* (2000) Triple transduction with adeno-associated virus vectors expressing tyrosine hydroxylase, aromatic-L-amino-acid decarboxylase, and GTP cyclohydrolase I for gene therapy of Parkinson's disease. Hum Gene Ther 11(11): 1509–1519

143. Shigeru T *et al* (2004) Infiltrating macrophages as in vivo targets for intravenous gene delivery in cerebral infarction. Stroke Published online before print

144. Shimamura M *et al* (2003) HJV-envelope vector for gene transfer into central nervous system. Biochem Biophys Res Commun 300(2): 464–471

145. Shimamura M *et al* (2004) Novel therapeutic strategy to treat brain ischemia: overexpression of hepatocyte growth factor gene reduced ischemic injury without cerebral edema in rat model. Circulation 109(3): 424–431

146. Shimoura N, Hamada H (2003) Gene therapy using an adenovirus vector for apoptosis-related genes is a highly effective therapeutic modality for killing glioma cells. Curr Gene Ther 3(2): 147–153

147. Shirakura M *et al* (2004) Postischemic administration of Sendai virus vector carrying neurotrophic factor genes prevents delayed neuronal death in gerbils. Gene Ther 11(9): 784–790

148. Simonato M *et al* (2000) Gene transfer into neurons for the molecular anal-

ysis of behaviour: focus on herpes simplex vectors. Trends Neurosci 23(5): 183–190

149. Smitt PS *et al* (2003) Treatment of relapsed malignant glioma with an adenoviral vector containing the herpes simplex thymidine kinase gene followed by ganciclovir. Mol Ther 7(6): 851–858

150. Snyder RO, Flotte TR (2002) Production of clinical grade recombinant adeno-associated virus vectors. Curr Opin Biotechnol 13(5): 418–423

151. Steinbach JP and Weller M (2002) Mechanisms of apoptosis in central nervous system tumor: application to theory. Curr Neurol Neurosci Rep 2(3): 246–253

152. Steinwaerder DS *et al* (2001) Tumor-specific gene expression in hepatic metastases by a replication-activated adenovirus vector. Nat Med 7(2): 240–243

153. Stoodley M *et al* (2000) Effect of adenovirus-mediated nitric oxide synthase gene transfer on vasospasm after experimental subarachnoid hemorrhage. Neurosurgery 46(5): 1193–1205

154. Sun JY *et al* (2003) Immune responses to adeno-associated virus and its recombinant vectors. Gene Ther 10(11): 964–976

155. Sun M *et al* (2003) Correction of a rat model of Parkinson's disease by coexpression of tyrosine hydroxylase and aromatic amino acid decarboxylase from a helper virus-free herpes simplex virus type 1 vector. Hum Gene Ther 14(5): 415–424

156. Tanaka T *et al* (1998) Viral vector-targeted antiangiogenic gene therapy utilizing an agiostatin complementary DNA. Cancer Res 58(15): 3362–3369

157. Tani H *et al* (2003) In vitro and in vivo gene delivery by recombinant Baculoviruses. J Virol 77(18): 9799–9808

158. Temin HM (1990) Safety considerations in somatic gene therapy of human disease with retroviral vectors. Human Gene Ther 1: 111–123

159. Tenenbaum L *et al* (2004) Recombinant AAV-mediated gene delivery to the central nervous system. J Gene Med 6: 212–222

160. Thomas CE *et al* (2003) Progress and problems with the use of viral vectors for gene therapy. Nature 4: 346–358

161. Todo T *et al* (2000) Evaluation of ganciclovir-mediated enhancement of the antitumoral effect in oncolytic, multimediated herpes simplex virus type 1 (G207) therapy of brain tumors. Cancer Gene Ther 7(6): 939–946

162. Toyoda K *et al* (2003) Gene therapy for cerebral vascular disease: update 2003. Br J Pharmacol 139: 139–147

163. Trask TW *et al* (2000) Phase I study of adenoviral delivery of the HSV-tk gene and ganciclovir administration in patients with current malignant brain tumors. Mol Ther 1(2): 195–203

164. Tuszynski MH *et al* (1998) Targeted intraparenchymal delivery of human NGF by gene transfer to the primate forebrain for 3 months does not accelerate β-amyloid plaque deposition. Exp Neurol 154: 573–582

165. Tuszynski MH (2002) Growth-factor gene therapy for neurodegenerative disorders. Lancet Neurol 1: 51–57

166. Uchida H *et al* (2004) Adenovirus-mediated transfer of siRNA against sur-

viving induced apoptosis and attenuated tumor cell growth in vitro and in vivo. Mol Ther 10(1): 162–171

167. Van der Haute C *et al* (2003) Lentiviral vector-mediated delivery of short hairpin RNA results in persistent knockdown of gene expression in mouse brain. Hum Gene Ther 14(18): 1799–1807

168. Wakimoto H *et al* (2003) Effects of innate immunity on herpes simplex virus and its ability to kill tumor cells. Gene Ther 10(11): 983–990

169. Wang LJ *et al* (2002) Neuroprotective effects of glial cell-line derived neurotrophic factor mediated by adeno-associated virus in a transgenic animal model of amyotrophic lateral sclerosis. J Neurosci 22(16): 6920–6928

170. Watanabe Y *et al* (2003) Gene transfer of extracellular superoxide dismutase reduces cerebral vasospasm after subarachnoid hemorrhage. Stroke 34(2): 434–440

171. Watson DJ, Wolfe JH (2003) Lentiviral vectors for gene transfer to the central nervous system. Applications in lysosomal storage disease animal models. Methods Mol Med 76: 383–403

172. Weedon SJ *et al* (2000) Sensitisation of human carcinoma cells to the prodrug CB1954 by adenovirus vector-mediated expression of E coli nitroreductase. Int J Cancer 86(6): 848–854

173. Weiss MD *et al* (2004) Current pharmacological management of amyotrophic [corrected] lateral sclerosis and a role rational polypharmacy. Expert Opin Pharmacother 5(4): 735–746.

174. Wickham TJ (2003) Ligand-directed targeting of genes to the site of disease. Nat Med 9(1): 135–139

175. Wildner O (1999) In situ use of suicide genes for therapy of brain tumours. Ann Med 31(6): 421–429

176. Winkler J *et al* (1998) Cholinergic strategies for Alzheimer's disease. J Mol Med 76: 555–567

177. Witham TF *et al* (2003) Expression of a soluble transforming growth factor-beta (TGFbeta) receptor reduces tumorigenicity by regulation natural killer (NK) cell activity against 9L gliosarcoma in vivo. J Neurooncol 64(1–2): 55–61

178. Wood MJ *et al* (1994) Immunological consequences of HSV-1-mediated gene transfer into the CNS. Gene Ther [Suppl] 1: 82

179. Wu K *et al* (2004) The effects of rAAV2-mediated NGF gene delivery in adult and aged rats. Mol Ther 9(2): 262–269

180. Yamanaka R *et al* (2003) Induction of antitumor immunological response by an intratumoral injection of dendritic cells pulsed with genetically engineered Semliki Forest virus to produce interleukin-18 combined with the systemic administration of interleukin-12. J Neurosurg 99(4): 746–753

181. Yang GY *et al* (1997) Overexpression of interleukin-1 receptor antagonist in the mouse brain reduces ischemic brain injury. Brain Res 751(2): 181–188

182. Yang SY *et al* (2004) Gene therapy of rat malignant gliomas using neural stem stells expressing IL-12. DNA Cell Biol 23(6): 381–389

183. Yenari MA *et al* (2003) Gene therapy and hypothermia for stroke treatment. Ann N Y Acad Sci 992: 54–68; discussion 79–81

184. Ylä-Herttuala S, Alitalo K (2003) Gene transfer as a tool to induce therapeutic vascular growth. Nat Med 9(6): 694–701
185. Ylä-Herttuala S and Martin JF (2000) Cardiovascular gene therapy. Lancet 355: 213–222
186. Yukawa H *et al* (2000) Adenoviral gene transfer of basic fibroblast growth factor promotes angiogenesis in rat brain. Gene Ther 7: 942–949
187. Zadeh G and Guha A (2003) Angiogenesis in nervous system disorders. Neurosurgery 53(6): 1362–1374
188. Zhang Y *et al* (2004) Intravenous RNA interference gene therapy targeting the human epidermal growth factor receptor prolongs survival in intracranial brain tumour. Clin Cancer Res 10(11): 3667–3677
189. Zlokovic BV and Apuzzo M (1997) Cellular and molecular neurosurgery: pathways from concept to reality – part II: vector systems and delivery methodologies for gene therapy of the central nervous system. Neurosurgery 40(4): 1992–1998

Technical Standards

Anatomy of the Orbit and its Surgical Approach

G. Hayek, Ph. Mercier, and H. D. Fournier

Laboratory of Anatomy, Faculty of Medicine, University of Angers,
Angers, France

With 16 Figures

Contents

Abstract

A micro anatomical and surgical study of the orbit was conducted on cadaver specimens. First of all we reviewed the anatomy of the orbit with special emphasis on microanatomical structures. Three neurosurgical approches are then described with all structures encountered along these routes. The superior approach which provides a good access to the superior part of the orbit is the only route which can explore all parts of the optic nerve even in the optic canal. The lateral compartment of the orbit could be exposed by the lateral approach above or below the lateral rectus muscle. It is the only route that could give access to the inferior part of the orbit. The supero lateral approach is the largest route and has advantages of the two preceding routes. It gives access to the superior part of the orbit but not the optic canal and gives also a good exposition to the lateral part of the orbit but less than the lateral route in the inferior part. These approaches could be used to remove all intra orbital lesions apart from those located in the infero medial part of the orbit.

Keywords: Anatomy; orbit; orbital anatomy; orbital approach; orbital tumor; surgical approach.

Introduction

Even though it represents a confined space, the orbit is amenable to a variety of exploratory surgical techniques. The diversity of possible techniques is due to its location at the juncture of various different anatomical regions. Containing the eyeball and located between the face and the cranium, the orbit is the meeting place of tissues and organs that are the focus of a range of different specialties, including ophthalmology, otorhinolaryngology and neurosurgery. The difficulties associated with approaching the orbit are related to its relatively small volume, its irregular, four-sided pyramid shape and to its situation embedded in the craniofacial structures.

It is essential to be intimately familiar with the microanatomy of the orbit before undertaking surgery in this region. In this chapter we have undertaken detailed study of the microsurgical anatomy of the orbit in cadaver specimens. After a literature review, we have described three neurosurgical approaches to the orbit on the basis of this anatomical study with emphasis on microanatomical structures that could be encountered along these routes.

Anatomy of the Orbit

The Orbital Cavity (Fig. 1)

The orbits are two cavities located symmetrically on either side of the sagittal plane at the root of the nose. The shape of each is that of a four-sided pyramid with its axis set off from the sagittal plane by an angle of 20°.

The roof is thin and concave in a downward direction. It can be separated into two laminae by the frontal sinus: there is the lacrimal fossa in the anterolateral part, and the anteromedial part houses the fovea into which the trochlea of the superior oblique muscle is inserted.

The floor separates the orbital cavity from the maxillary sinus. It is

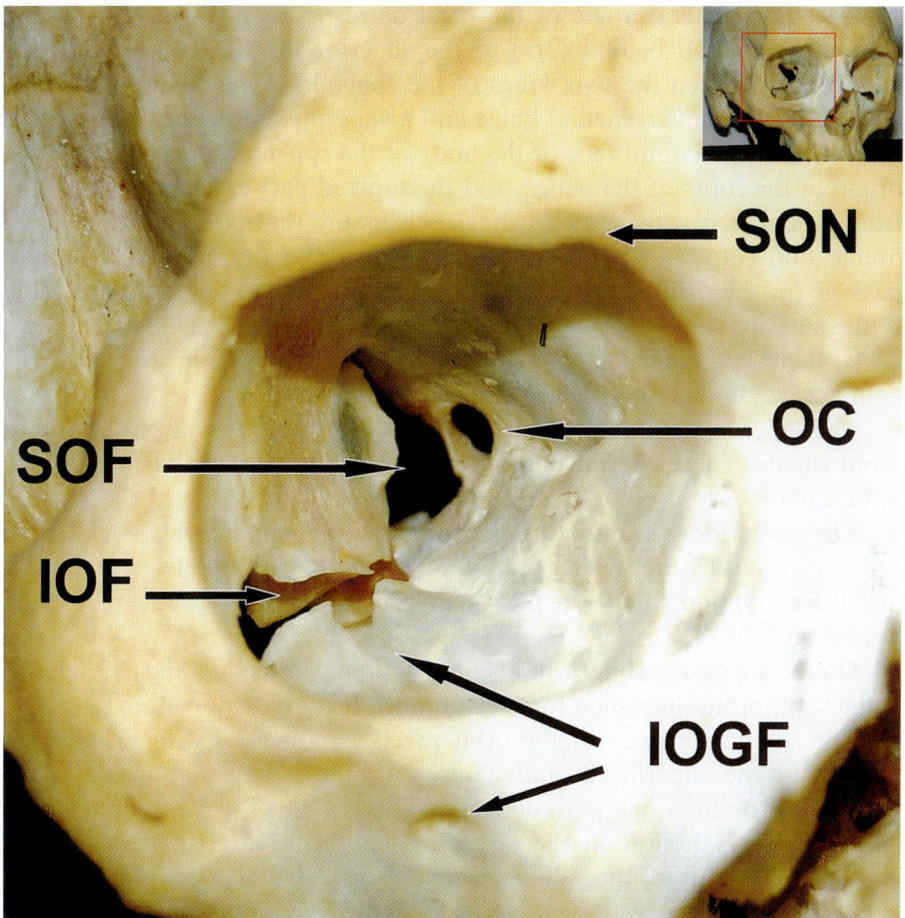

Fig. 1. Photograph of anterior aspect of the right orbital cavity showing optic canal (*OC*), superior orbital fissure (*SOF*), inferior orbital fissure (*IOF*), infra orbital groove and foramen (*IOGF*) and supra orbital notch (*SON*)

traversed by the infraorbital groove that runs from the back towards the front until it changes into a canal for the maxillary nerve.

The lateral wall is oblique outside and in front. The posterior two-thirds is formed by the greater wing of sphenoid with the superior orbital fissure at the top and the inferior fissure at the bottom. This wall is very thick, especially at the front (the lateral pillar), and it separates the orbit from the cerebral temporal fossa behind and from the temporal fossa (which houses the temporal muscle) in front.

The thinnest of the walls of the orbit is the slightly sagittal medial wall that has, in its forward portion, the lacrimal groove that subsequently turns into the nasal canal.

The superolateral angle of the orbit corresponds in front (1/3) to the lacrimal fossa and behind (2/3) to the superior orbital fissure. This is a dehiscence between the two wings of the sphenoid bone in the shape of a comma, with an inferomedial bulge and a superolateral taper. The bulging end matches the lateral face of the body of the sphenoid bone between the origin of the roots of the wings, and the tapered part extends as far as the frontal bone between the two wings. This fissure represents a line of communication between the middle cerebral fossa and the orbit, providing a passage for the orbital nerves (but not the optic nerve) and corresponding to the anterior wall of the cavernous sinus. At the junction between the two parts, there is a small bony protruberance on the lower lip to which the common tendinous ring (Zinn's tendon) is attached.

The superomedial angle is perforated by the anterior and posterior ethmoid canals. The posterior end is continuous with the medial wall of the optic canal.

The posterior two-thirds of the inferolateral angle correspond to the inferior orbital fissure, which provides a communication between the orbit and the pterygomaxillary (or pterygopalatine) fossa; this is covered by the periosteum.

On the superior orbital rim at the junction (1/3 medial and 2/3 lateral) is the supraorbital foramen for the supraorbital nerve and vessels.

The summit or apex of the orbit precisely coincides with the bulging portion of the superior orbital fissure. A little above and inside is the exocranial foramen of the optic canal. This canal, of 6–12 millimeters in length, forms a hollow at the origin of the small wing (between its two roots) on the body of the sphenoid bone. This canal is a site of communication between orbit and the anterior fossa of the cranium. It gives passage to the optic nerve with its meningeal sheath and for the ophthalmic artery traversing below this nerve from the inside to the outside.

The Orbital Fascia or Periorbita (Fig. 2)

This corresponds to the orbital periosteum. Its bone attachment is very loose apart from at points around the optic canal and the superior orbital fissure where it is continuous with the dura mater. In front, it continues into the cranial periosteum on the orbital rim to which it is very strongly attached. Here it sends out extensions towards the peripheral tarsal rim to form the orbital septum, which delineates the orbit in front and separates the intraorbital fatty tissue from the orbicular muscle of the eye. Inside, it is attached to the posterior lacrimal crest and on top, it is traversed by the levator palpebrae superior muscle. The periorbita thus surrounds the contents of the orbit, forms a bridge over the top, and closes the inferior orbital fissure. It is perforated by the various vessels and nerves of the orbit.

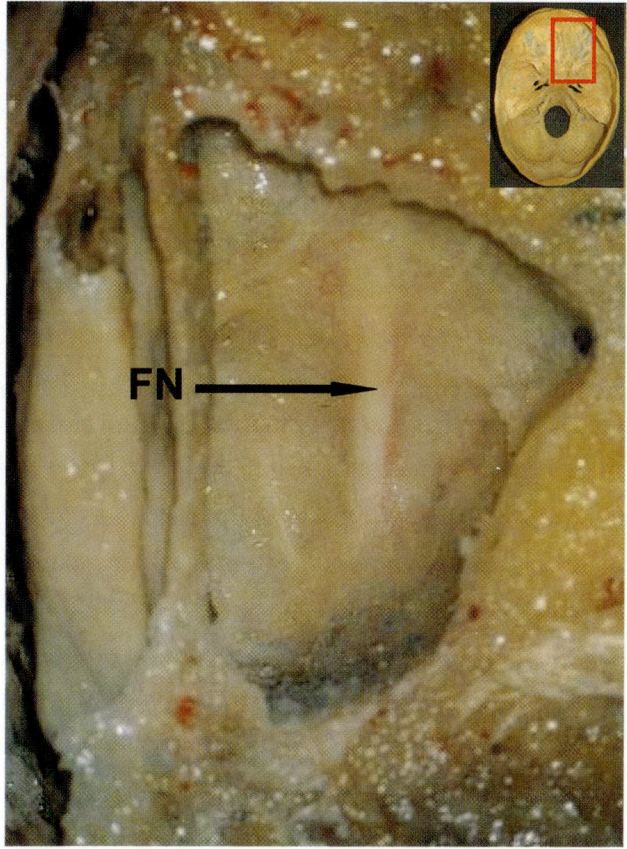

Fig. 2. Photograph of superior aspect of the right orbit after roof removal showing the transparent orbital fascia (periorbita) and underneath the frontal nerve

Orbital Contents

The orbit can be split into two parts, an anterior part containing the eyeball and a posterior compartment containing the muscles, the vessels and the nerves supplying the eyeball, all supported in a cellular, fatty matrix, the so-called adipose body of the orbit.

The eyeball does not touch any of the walls but is suspended at a distance of 6 mm outside and 11 mm inside. Its anterior pole is at a tangent to a straight line joining the upper and lower rims of the orbit, and it projects out beyond a line joining the medial and lateral edges, especially towards the outside. Finally, the anteroposterior axis of the eyeball (which is precisely sagittal) forms an angle of 20° with the axis of the orbit, oblique in front and outside.

From the optic nerve as far as the sclero-corneal junction, the eyeball is covered by a two-layer fascia (Tenon's capsule) with parietal and visceral sheets separating it from the orbital fatty tissue. There is a virtual space between the two sheets (the episcleral space) which forms a sort of lubricated joint system to facilitate the movements of the eye. The fascia is fused behind with the capsule of the optic nerve and in front with the sclera where it joins the cornea. In its anterior part, it is perforated by the muscles of the eye. The fascia turns back over these muscles to create their aponeurotic sheath.

Orbital Muscles (Fig. 3)

The orbit contains seven muscles, the first being the levator palpebrae superior muscle and the other six controlling the eye movements: four rectus muscles (superior, inferior, lateral and medial) and two oblique muscles (superior and inferior).

The levator palpebrae superior is a fine, triangular muscle, which originates above and in front of the optic canal at which point it is fine and tendinous although it sharply broadens out and assumes a more muscular character. It runs along the upper wall of the orbit just above the superior rectus muscle (covering its medial edge). It terminates in an anterior tendon that spreads out in the form of a large fascia, which extends out to the eyelid. The edges of this fascia form extensions, including a lateral one which traverses the lacrimal gland between its palpebral and orbital parts and goes on to attach to the fronto-zygomatic suture.

Rectus muscles: these four muscles form a conical space that is closed in front by the eyeball. They arise in the common annular tendon (Zinn's tendon); this tendon is located on the body of sphenoid near the infraoptic tubercle, and it surrounds the superior, medial and inferior edges of the optic canal, and then continues across the inferomedial part of the superior orbital fissure before inserting on a tubercle of the greater wing. It subsequently splits into four lamellae arranged at right angles to one another, from which the four rectus muscles arise respectively. The superolateral and inferomedial ligaments are solid but the other two are perforated: the one in the superomedial band lets the optic nerve and the ophthalmic artery through, and the other, which is larger, stretches between the inferomedial and superolateral bands passing through the inferolateral band. This opening called the common tendinous ring (Zinn's ring) or the oculomotor foramen corresponds to the bulging end of the superior orbital fissure and provides a passage for the nasociliary nerve, both branches of the oculomotor nerve, the abductor nerve and the sympathetic root of the ciliary ganglion. The superior ophthalmic vein can also pass through or above this opening, and the inferior ophthalmic vein may pass inside or below it.

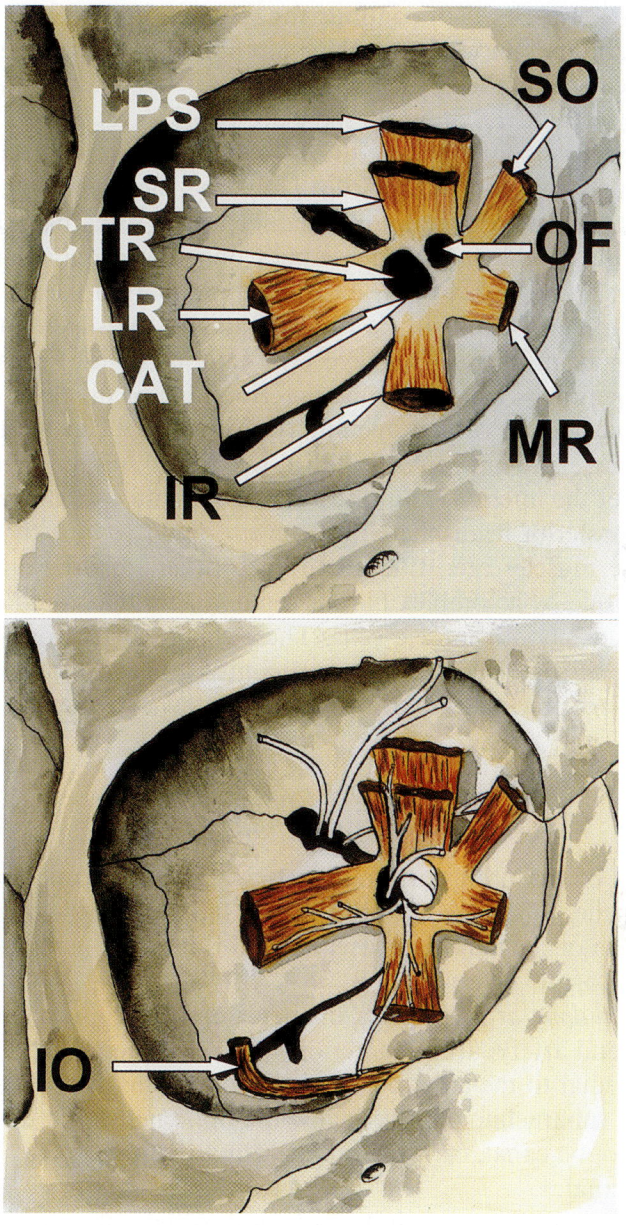

Fig. 3. Artist's drawing of right orbital cavity with extra ocular muscles showing common annular tendon (*CAT*), common tendinous ring (*CTR*), optic foramen (*OF*), levator palpebrae superior muscle (*LPS*), the four recti muscles: superior (*SR*), medial (*MR*), inferior (*IR*), lateral (*LR*), superior oblique muscle (*SO*) and inferior oblique muscle (*IO*)

The rectus muscles then continue for four centimeters in a forward direction to terminate in tendons, which are attached to the anterior part of the sclera near the limbus.

The oblique muscles, of which there are two.

The superior oblique muscle arises as a short tendon attached inside and above the optic foramen. It runs along the superomedial angle of the orbit and then becomes tendinous again when it turns back at an acute angle over the trochlea. It then becomes once more muscular and turns backwards in a lateral direction, skirts the upper part of the eyeball passing under the superior rectus muscle to terminate on the superolateral side of the posterior hemisphere of the eye. The inferior oblique muscle, shorter than the superior, is located on the anterior edge of the floor of the orbit and arises outside the orbital opening of the lacrimal canal before passing outside, behind and upwards. It skirts the lower surface of the eyeball, passing under the inferior rectus muscle to terminate on the inferior, lateral side of the posterior hemisphere of the eye.

All these muscles are attached to each other, to orbital fascia or to Tenon's capsule by a complex fibrous septa system that could contain vessels, nerves or smooth muscle cells. These septa can be considered as an important accessory locomotor system contributing to the motility of eye and explaining some motility disturbances in blow out fractures of the orbit (24).

The Arteries of the Orbit (Figs. 4–6)

The arteries of the orbit correspond essentially to the ophthalmic artery and its branches, although the orbit is also supplied by the infraorbital artery, a branch of the maxillary artery which is itself the terminal branch of the external carotid artery.

The ophthalmic artery, with a diameter of 1.5 mm, is a branch of the internal carotid artery, arising anteriorly where it emerges from the cavernous sinus medial to the anterior clinoid process. This artery makes its way through the subarachnoid space below the optic nerve and then continues on into the optic canal, carrying on laterally to perforate the sheath at the exit of the canal.

In the orbital cavity, it is initially lateral to the optic nerve and medial to the ciliary ganglion. Next, obliquely and accompanied by the nasociliary nerve, it crosses the top side of the optic nerve below the superior rectus muscle, ultimately reaching the medial orbital wall. From there it makes its way forward between the superior oblique and the medial rectus muscles, passes under the trochlea and then climbs back up again to pass between the orbital rim and the medial palpebral ligament. It terminates by splitting into two different arteries, the supratrochlear artery and the angular artery

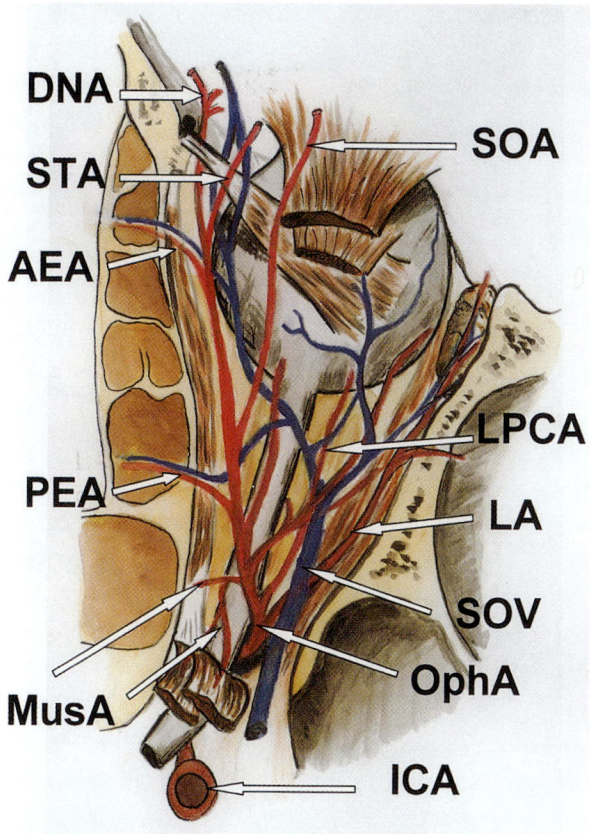

Fig. 4. Artist's drawing of superior view of right orbit showing ophthalmic artery and its branches. *ICA* Internal carotid artery, *OphA* ophthalmic artery, *SOV* superior ophthalmic vein, *LA* lacrymal artery, *PEA* posterior ethmoidal artery, *AEA* anterior ethmoidal artery, *LPCA* long posterior ciliary artery, *SOA* supra orbital artery, *STA* supra trochlear artery, *DNA* dorsal nasal artery, *MusA* muscle artery

(which forms an anastomosis with the dorsal artery of the nose). It should be noted that in about 15% of subjects, the ophthalmic artery passes underneath optic nerve.

Collateral branches of the ophthalmic artery
Variable in number from 10 to 19, most of these arise in the intraorbital segment of the artery. Anatomical variations are very common.

The central artery of the retina (Fig. 7)—one of the smallest—is present in all cases. It arises from the ophthalmic artery in 50% of subjects, and from one of its branches (the posterior long ciliary artery) in the other 50%. It often arises below and outside the optic nerve and then skirts over its lower side before penetrating at a distance of 10–15 mm from the poste-

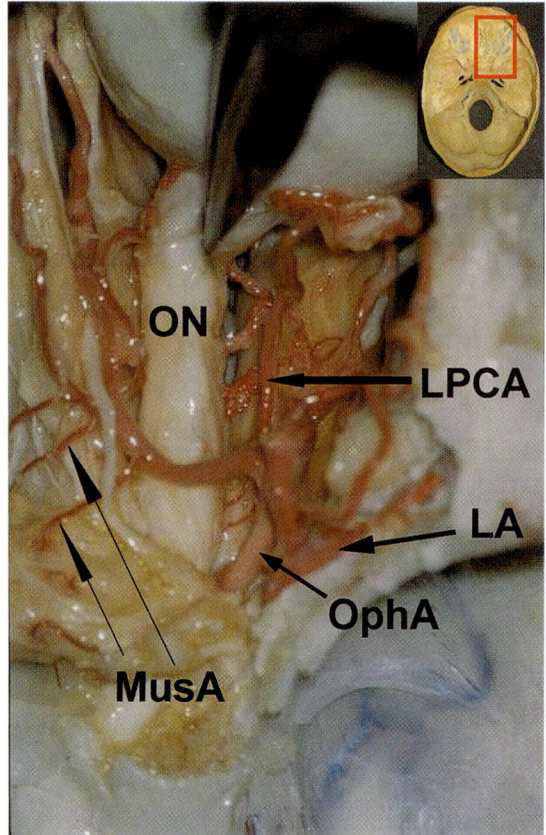

Fig. 5. Photograph of superior view of the right orbit, the roof and orbital fat have been removed and superior muscles were sectioned to show the optic nerve (*ON*) surrounded by ophthalmic artery (*OphA*) and some of its branches: lacrimal artery (*LA*), long posterior ciliary artery (*LPCA*) and muscles arteries (*MusA*)

rior pole of the eyeball. It then continues as far as the papilla where it splits to form its terminal branches.

The lacrimal artery, one of the largest branches, arises near the exit of the optic canal above and outside the optic nerve. It passes forwards, upwards and in a lateral direction, coming out of the cone to continue to the lateral wall of the orbit where it carries on with the lacrimal nerve above the lateral rectus muscle as far as the lacrimal gland. This artery gives rise to one or two zygomatic branches. One of these passes across the zygomatico-temporal foramen and forms an anastomosis with the deep temporal arteries. The lacrimal artery gives a recurrent anastomotic branch that passes through the lateral part of the superior orbital fissure to rejoin a branch of the middle meningeal artery.

Muscular branches are numerous. These often arise from one or two

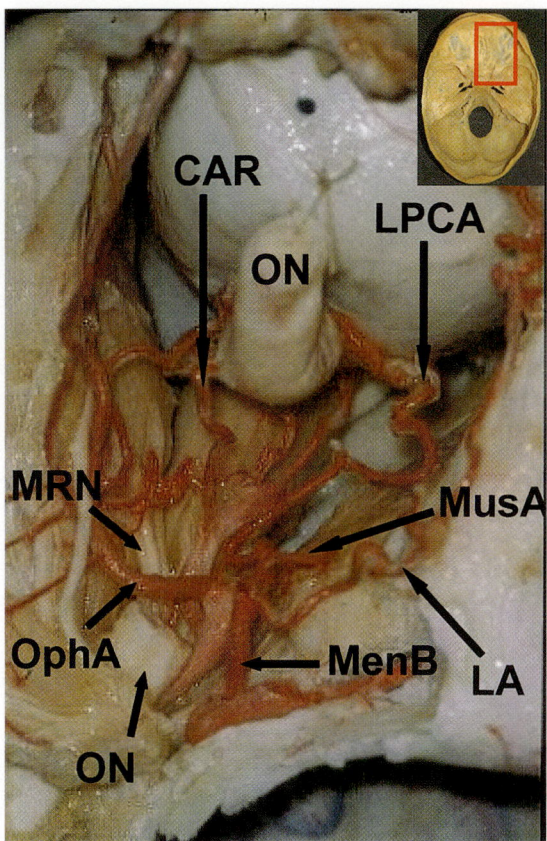

Fig. 6. Superior view of the right orbit, the roof and orbital fat have been removed and superior muscles with optic nerve were sectioned to show branches of the ophthalmic artery (*OphA*) situated under the optic nerve: *CAR* central artery of retina, *LPCA* long posterior ciliary artery, *MusA* muscle artery, *LA* lacrimal artery, *MenB* meningial branch. In this view we can also see the nerve of the medial rectus muscle (*MRN*)

arterial trunks. The inferior muscular artery—one of the largest of the branches of the ophthalmic artery—is the most commonly found. Other muscular branches exist, arising in the ophthalmic artery or one of its branches.

The ciliary arteries can be divided into three different groups:

The posterior long ciliary arteries, commonly two in number, arise in the ophthalmic artery at the point at which it crosses over the optic nerve. These enter the sclera not far from where the optic nerve enters it.

The posterior short ciliary arteries, seven in number, pass in a forward direction around the optic nerve.

The anterior ciliary arteries arise from muscular branches and pass in front, over the tendons of the rectus muscles.

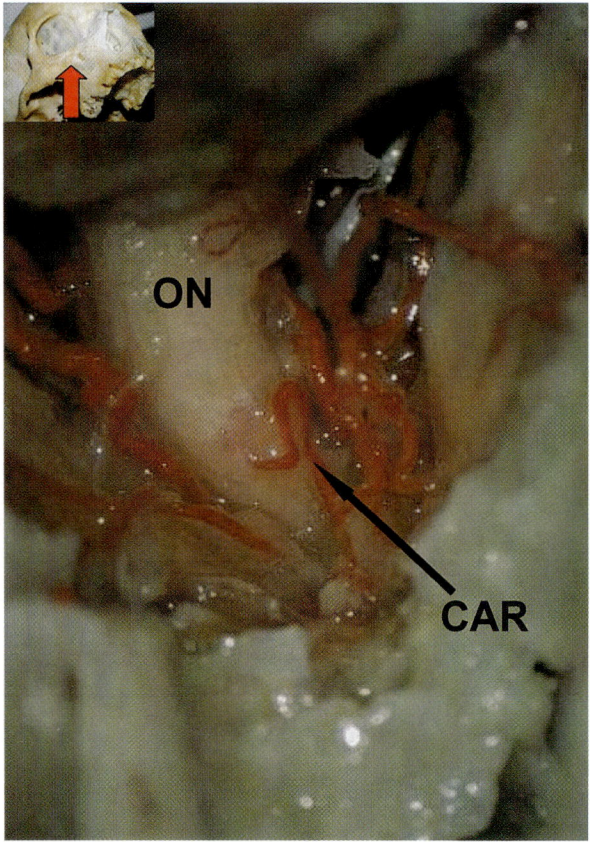

Fig. 7. Photograph of inferior aspect of the right orbit that was opened by removing its floor to show the central artery of retina (*CAR*) penetrating optic nerve (*ON*)

The supraorbital artery (which is absent in 12% of subjects) arises from the ophthalmic artery, often in its medio-optic part. It passes up and forwards, comes out of the cone between the levator palpebrae superior muscle and the superior oblique muscle, and then meets the supraorbital nerve, which it accompanies between the levator muscle and the periorbita until the supraorbital incisure or foramen. Inside the orbit, it gives rise to muscular branches and, in some cases, the supratrochlear artery.

The posterior ethmoidal artery arises within the muscular cone medial and above the optic nerve then leaves the cone between the superior oblique muscle underneath and the levator muscle above to carry on over the trochlear nerve towards the posterior ethmoid canal.

The anterior ethmoidal artery, which is present more often than the preceding vessel, arises near the anterior ethmoid canal: when it enters this canal, it is accompanied by the nerve of the same name.

The meningeal branch is a small branch that passes behind, across the superior orbital fissure, into the middle cerebral fossa to form an anastomosis with the middle and accessory meningeal arteries.

The medial palpebral arteries, two in number (the superior and the inferior), arise from the ophthalmic artery below the trochlea.

The supratrochlear artery, a terminal branch of the ophthalmic artery, leaves the orbit in the superomedial part together with the nerve of the same name.

The dorsal artery of the nose, another terminal branch of the ophthalmic artery, emerges from the orbit between the trochlea and the medial palpebral ligament.

Veins of the Orbit

There is a very dense venous network in the orbit, organized around the two ophthalmic veins that drain into the cavernous sinus. These veins are valve-less. Periorbital drainage also occurs towards the facial system via the angular vein. Thus, for the venous system, as with the arterial system, the orbit is a site of anastomosis between the endocranial and exocranial systems.

The superior ophthalmic vein (Fig. 8), a large-caliber vein present in all subjects, constitutes the orbit's main venous axis. It is formed by the union behind the trochlea of two rami, the first from the frontal veins and the other from the angular vein. This vessel then crosses the orbit from the front towards the back accompanying the artery and passing under the superior rectus muscle. Throughout its path, it receives a great number of collateral tributaries, including ethmoidal, muscular, ciliary, vorticose (from the choroid), lacrimal, palpebral, conjunctival and episcleral rami, and the central vein of the retina. This last, the caliber of which is very small, first traverses the optic nerve behind the lamina cribrosa sclerae then carries on into the subarachnoid space before piercing the dura mater and either rejoining the superior ophthalmic vein or passing directly into the cavernous sinus.

When it reaches the apex of the orbital pyramid, the superior ophthalmic vein insinuates itself between insertions of the lateral and superior rectus muscles and leaves the orbit via the enlarged portion of the superior orbital fissure outside the common tendinous ring. It terminates at the face of the cavernous sinus.

The inferior ophthalmic vein (Fig. 9) that is not present in all subjects, is the result of a venous anastomosis in the anterior inferomedial part of the orbit. It receives rami from muscles, the lacrimal sac and the eyelids. It carries on behind, above the inferior rectus muscle, whence it often rejoins the superior ophthalmic vein, although in some subjects, it carries on to the

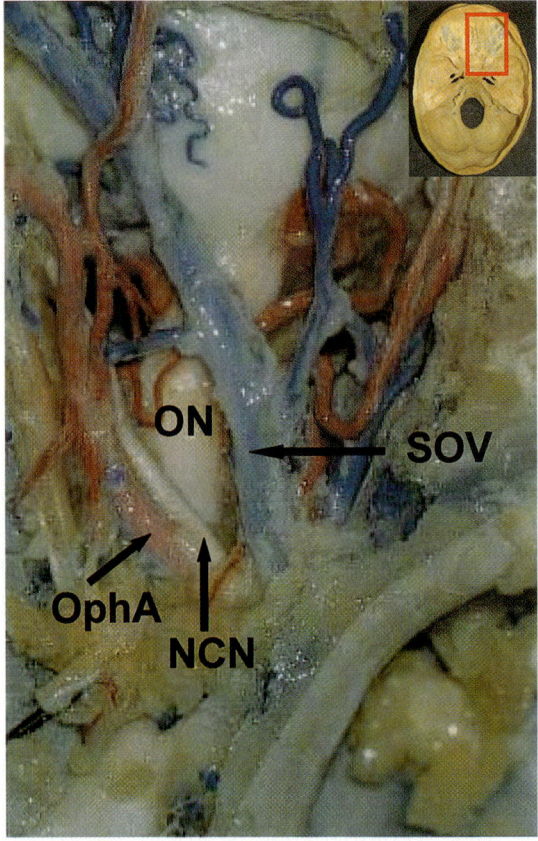

Fig. 8. Superior view of the right orbit after section of superior muscles to see structures passing between these muscles and optic nerve (*ON*): superior optic veine (*SOV*), naso ciliary nerve (*NCN*) and ophthalmic artery (*OphA*)

cavernous sinus as a distinct vessel. It communicates with the pterygoid plexus by small veins crossing the walls of the orbit.

Nerves of the Orbit (Figs. 10, 11)

The orbit contains a huge number of nervous structures of various types (as defined by physiological function).

There include:
- a component of the central nervous system: the optic nerve;
- three motor nerves: the third, fourth and sixth cranial nerves;
- a sensitive nerve: the ophthalmic nerve, a branch of the fifth cranial nerve
- an autonomic center: the ciliary ganglion.

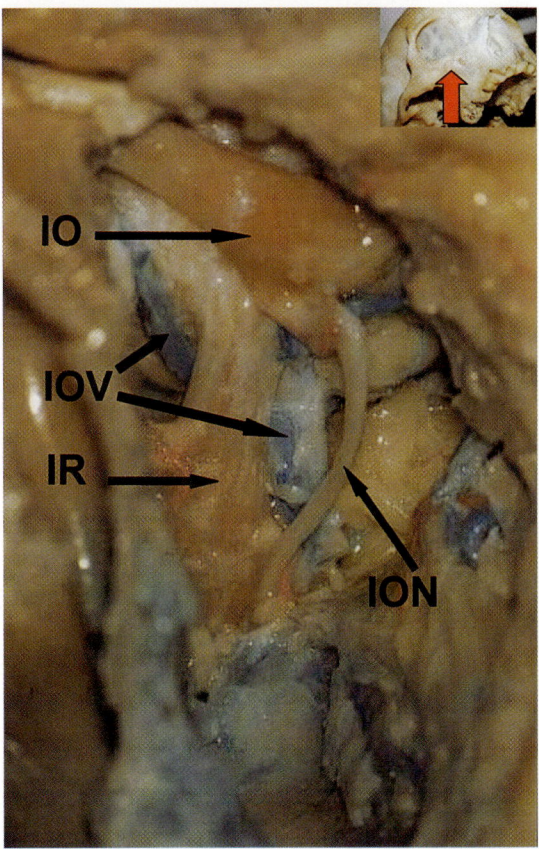

Fig. 9. Inferior view of the right orbit after opening of its floor and fascia showing inferior oblique muscle (*IO*), inferior rectus muscle (*IR*), nerve of the inferior oblique muscle (*ION*) and inferior ophthalmic vein (*IOV*)

The optic nerve (Fig. 12), about 4.5 cm in length and with a diameter of 3 mm, leaves the eyeball inside and below its posterior pole and carries on behind and inside. It is conventionally divided into three different parts, namely intraorbital, intracanicular and intracranial segments.

The intraorbital segment is about 30 mm long and follows a sinuous trajectory, which provides reserve length so that the eyeball can be moved without damaging the nerve. In this segment, relationships are made with:
– the muscles of the orbit, firstly at a distance from the nerve and sepa-
 rated from it by a mass of fatty tissue (through which the ciliary nerves
 and vessels), then coming closer nearer the point of entry of the optic
 canal where the nervous sheath is attached to the tendinous fibers of the
 superior oblique muscle, the medial rectus muscle and the superior rectus
 muscle;

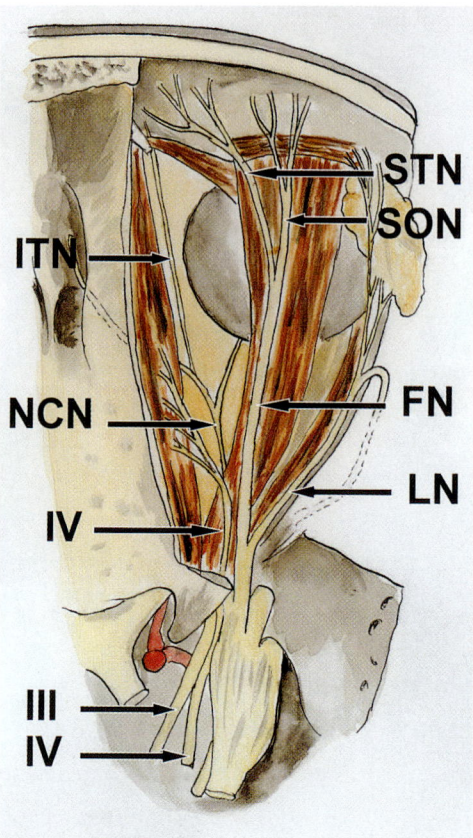

Fig. 10. Artist's drawing of superior view of the right orbit showing orbital nerves. *ITN* Infra trochlear nerve, *NCN* naso ciliary nerve, *IV* trochlear nerve, *III* oculomotor nerve, *LN* lacrymal nerve, *FN* frontal nerve, *STN* supra trochlear nerve, *SON* supra orbital nerve

– the ophthalmic artery which crosses over the nerve;
– the ciliary ganglion, juxtaposed with its lateral surface at the union of the anterior 2/3 and the posterior 1/3, and located between it and the lateral rectus muscle.

The intracanicular segment is 5 mm long and the nerve is here accompanied underneath and outside by the ophthalmic artery. Just in front of the canal, the ophthalmic artery and the nasociliary nerve carry on medially and in a forward direction above the optic nerve; in contrast, the nerve supplying the medial rectus muscle (which comes from the inferior branch of the third cranial nerve) passes below the optic nerve in a medial direction (Fig. 6). In this canal, the optic nerve is separated on the inside from

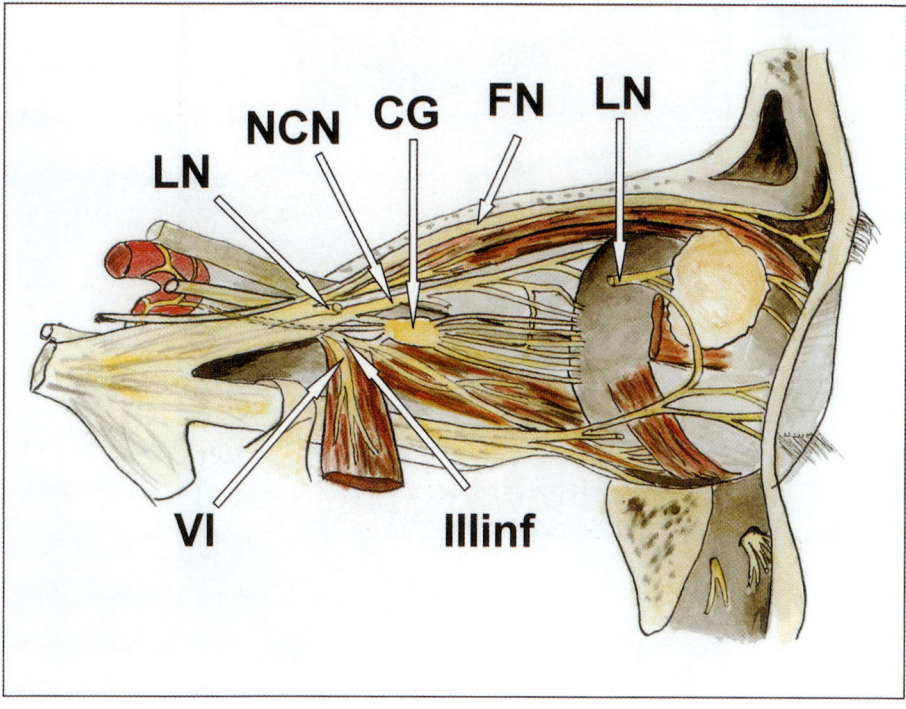

Fig. 11. Artist's drawing of lateral view of the right orbit showing orbital nerves. *LN* Lacrymal nerve, *NCN* naso ciliary nerve, *CG* ciliary ganglion, *FN* frontal nerve, *VI* abductor nerve, *IIIinf* inferior division of oculomotor nerve

the sphenoidal sinus and the posterior ethmoidal cells by a very thin lamella of bone.

The intracranial segment is 10 mm long, and passes behind and inside as far as the optic chiasm.

Blood is supplied to the intracanalicular segment of the optic nerve by recurrent branches of the ophthalmic artery. The intraorbital segment is supplied by the posterior ciliary arteries and the central artery of the retina, and drainage is via the central vein of the retina.

The oculomotor nerve supplies all the extraocular muscles apart from the superior oblique muscle and the lateral rectus muscle; its parasympathetic motor contingent also innervates—via the ciliary ganglion—the intraocular muscles (sphincter of the iris and the ciliary muscle).

Before entering the orbit, this nerve splits to form two terminal rami (superior and inferior) which penetrate the orbit across the bulging medial part of the superior orbital fissure inside the common tendinous ring. At this point, the nasociliary nerve is located between them on the inside with

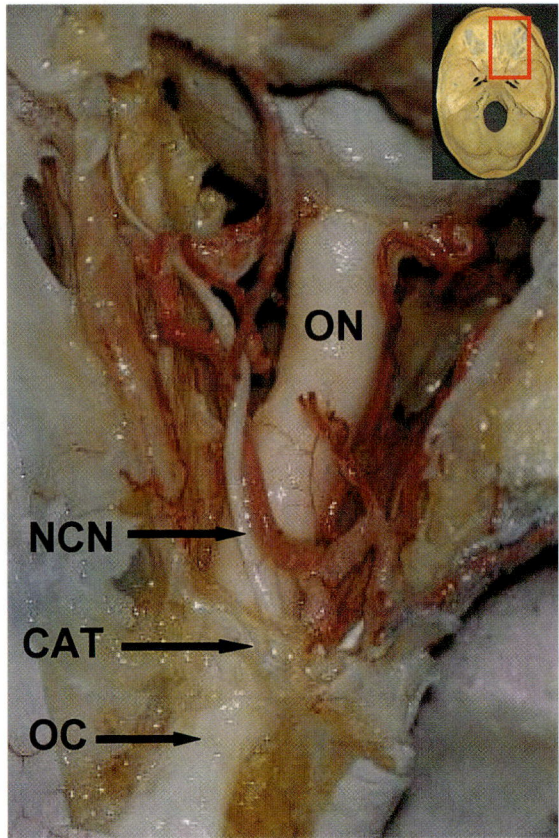

Fig. 12. Photograph of superior view of the right orbit, the roof and orbital fat have been removed, optic canal (*OC*) was opened, and superior muscles were sectioned to show the optic nerve (*ON*) in all its pathway: *CAT* common annular tendon, *NCN* naso ciliary nerve

the abductor nerve on the outside. The two branches then enter the muscular cone and diverge away from one another.

The superior, smaller-caliber branch climbs up the lateral side of the optic nerve and splits to form four or five rami that innervate the superior rectus muscle and, via a perforating ramus, the levator palpebrae superior.

The inferior branch is initially located below and outside the optic nerve and then spreads out over the upper surface of the inferior rectus muscle, splitting to form three branches. The first of these passes below the optic nerve on its way to the medial rectus muscle; the second travels outside towards the inferior rectus muscle; and the third (the longest), carries on in front between the inferior rectus muscle and the lateral rectus muscle on its way to the inferior oblique muscle (Fig. 9). From this branch

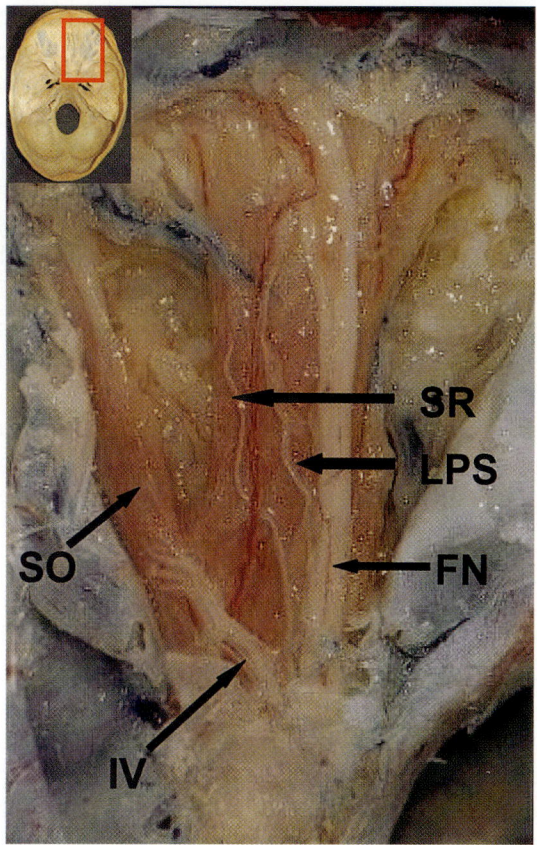

Fig. 13. Photograph of superior view of the right orbit after removal of the roof and opening of periorbita to see superior muscles and superficial nerves: *IV* trochlear nerve, *FN* frontal nerve, *SO* superior oblique muscle, *SR* superior rectus muscle, *LPS* levator palpebrae superior muscle

arises a short branch (sometimes actually composed of two or three distinct branches) to the ciliary ganglion to form its parasympathetic or motor root. After a synapse at this ganglion, fibers of the third cranial nerve intertwine with sympathetic and sensitive fibers to constitute the short ciliary nerves, which travel to the ciliary muscle and the sphincter of the iris.

The trochlear nerve (Fig. 13) enters the orbit across superior orbital fissure where its tapered and bulging portions come together. It passes outside the common tendinous ring above the orbital muscles and inside the frontal nerve (which is far more bulky). Inside the orbit, it carries on medially above the origin of the levator palpebrae superior, to reach the superior oblique muscle on its orbital side.

The abductor nerve starts in the medial part of the superior orbital fissure inside the common tendinous ring outside the branches of the oculomotor nerve. It then spreads out over the lateral rectus muscle and splits to form four or five branches, which carry on into the muscle.

The ophthalmic nerve, a superior branch of the trigeminal nerve, is exclusively sensitive. It innervates the eyeball, the lacrimal gland, the conjunctiva, part of the mucosa of the nasal cavity, and the skin of the nose, forehead and scalp. This nerve is the smallest of the three branches of the trigeminal nerve. After its passage into the lateral wall of the cavernous sinus but before it enters the orbit, it splits to form three branches, namely (going from the outside to the inside) the lacrimal nerve, the frontal nerve and the nasociliary nerve.

– *The lacrimal nerve*, the branch with the smallest caliber, enters the orbit via the lateral part of the superior orbital fissure and remains outside the cone. Then, together with the lacrimal artery, it travels along the superolateral edge of the orbit above the lateral rectus muscle and, from the zygomaticotemporal nerve (a branch of the maxillary nerve), it receives a branch which contains parasympathetic secretomotor fibers coming from the pterygopalatine ganglion on their way to the lacrimal gland. As the nerve crosses the gland, it sends out numerous branches to ensure its nervous supply, and then it perforates the orbital septum and terminates in the skin of the upper eyelid. In some subjects, the lacrimal nerve is missing, in which case its role is fulfilled by the zygomaticotemporal nerve.

– *The frontal nerve*, the largest of the branches of the ophthalmic nerve, enters the orbit through the tapered part of the superior orbital fissure, above the muscles, between the lacrimal nerve outside and the trochlear nerve inside. Outside the cone, it insinuates between the levator palpebrae superior muscle and the periorbita. Halfway along, it splits to form a small medial branch called the supratrochlear nerve, and a large lateral branch called the supraorbital nerve. The first passes above the trochlea of the superior oblique muscle and distributes to the medial 1/3 of the upper eyelid and the conjunctiva. The second passes into the supraorbital incisure and innervates the middle 1/3 of the upper eyelid and the conjunctiva.

– *The nasociliary nerve*, the most medial of the branches of the ophthalmic nerve, is intermediate in size to the frontal and lacrimal nerves, and is the only one to reach the eyeball. It is also the only one to pass through the common tendinous ring, which it does inside the two branches of the oculomotor nerve, just above the sympathetic root of the ciliary ganglion. Afterwards, together with the ophthalmic artery, it crosses the optic nerve before travelling obliquely between the medial rectus muscle below and the superior rectus and superior oblique muscles above, as far as the medial wall of the orbit. At the level of the anterior ethmoidal foramen, it splits to form two branches:

– the anterior ethmoidal nerve, medial, which crosses the canal of the same name with the corresponding artery, and then passes over the cribriform plate of the ethmoid bone.

– the infratrochlear nerve, lateral, continues in the direction of the common trunk. Under the trochlea of the superior oblique muscle, it splits to form rami going to the mucosae (the medial part of the conjunctiva and the lacrimal ducts) and the skin (the medial part of the eyelid and the root of the nose).

Along its route, the nasociliary nerve sends out three major collateral branches, from behind forwards:

– the communicating branch of the ciliary ganglion which leaves the nasociliary nerve at the point at which it enters the orbit; it contains the fibers for corneal sensitivity as well as the sympathetic fibers responsible for dilatation of the iris which are supplied to the trigeminal nerve from the cervico-trigeminal anastomosis;

– the long ciliary nerves, of which there may be either two or three, leave the nasociliary nerve where it crosses the optic nerve; they join up with the short ciliary nerves (from the ciliary ganglion), perforate the sclera and terminate in the ciliary body, the iris and the cornea; they usually contain the sympathetic fibers responsible for dilatation of the pupil;

– the posterior ethmoidal nerve crosses the canal of the same name and distributes to the sphenoidal sinus and posterior ethmoidal cells.

The ciliary ganglion, is a reddish-gray, pinhead-sized (2 mm × 1 mm) structure which is somewhat flattened out along its major horizontal axis. It is located near the apex of the orbit (at the junction of the anterior 2/3 and the posterior 1/3) in loose fatty tissue between the optic nerve and the lateral rectus muscle, usually lateral to the ophthalmic artery.

There are three roots:

– the motor or parasympathetic root which comes from the inferior branch of the third cranial nerve (by the inferior oblique branch). Its fibers mainly innervate the ciliary muscle and, to a lesser extent, the sphincter of the iris;

– the sympathetic root, a branch of the carotid plexus which enters the orbit via the common tendinous ring;

– the sensitive root, a long and fine fiber which rejoins the nasociliary nerve where it enters the orbit. This supplies the eye and the cornea.

The 8–10 filamentous branches of the ganglion are called the short ciliary nerves and these make their way in a forward direction around the optic nerve, together with ciliary arteries and the long ciliary nerves.

Lacrimal Gland

This comprises two different, laterally continuous parts: the upper, broader orbital part, and the lower, smaller palpebral part. The orbital part is accommodated in the lacrimal fossa of the zygomatic process of the frontal bone. It is located above the levator muscle (and laterally, above the lateral rectus muscle). Its lower surface is attached to the sheath of the levator muscle and its upper surface to the orbital periosteum; its anterior edge is in contact with the orbital septum and its posterior edge is attached to the orbital fatty tissue.

The palpebral part, which is in the form of two or three lobules, extends out below the fascia of the levator muscle in the lateral part of the upper eyelid. It is attached to the superior conjunctival fornix.

Approach Routes to the Orbit

Orbit's pyramidal shape affords five possible anatomical approach routes, namely the anterior, superior, inferior, lateral and medial faces [22, 23]. The neurosurgeon does not need to be familiar with either the anterior approaches (which tend to be restricted to ophthalmologic applications, essentially to access small, anterior neoplasms) nor, as a rule, the medial and inferior approaches (which are the domain of the Ear, Nose and Throat specialist working in collaboration with the ophthalmologist). On the other hand, the neurosurgeon is often called upon to collaborate on techniques based on an approach via either the lateral or the superior routes [9, 19].

The huge number of monographs about the various techniques tends to lead to confusion, even though they generally differ in only relatively minor details. The goal must always be to obtain optimum access while inflicting as little damage as possible, with damage considered in both functional and esthetic terms. It is beyond debate now that no single method is ideal for exploration of the orbit in all cases: the technique to be used has to be chosen in the light of the nature, the location and the size of the lesion.

In neurosurgery, three possible approach routes should be considered namely the lateral approach, the superior approach and a hybrid superolateral approach.

Incision

The type of incision to be made will depend on the approach route chosen although a temporofrontal incision is suitable for all three alternatives. For a lateral approach, certain experts prefer an incision in the shape of an elongated S starting at the eyebrow and descending along the lateral edge of the orbit before curving behind along the upper edge of the zygomatic arch. This option is shorter but tends to be more disfiguring. It is important to go no further than 4 centimeters beyond the canthus to preclude damage to the frontotemporal branch of the facial nerve [26]. Both the fascia and the temporal muscle are eventually detached to expose the lateral wall of the orbit.

The temporofrontal incision begins just in front of the tragus at the level of the upper edge of the zygomatic arch and is then carried on in an upward direction. At the level of the temporal line, the incision curves around towards the hair line past the median line. Some experts even prefer bilateral incision [1, 12, 20]. Subsequent detachment of the scalp should preserve the temporofrontal branch of the facial nerve (located between the galea aponeurotica and the temporal fascia) and should avoid in front the bulky temporal muscle mass in order to obtain full access. The detachment procedure starts under the galea and then, 4 cm from the edge of the orbit, an incision is made in the superficial layer of the temporal fascia (which is at this point divided into two layers) in order to pass dissection between the two layers [33]. This leaves the superficial layer in contact with the scalp, and the nerve branch between the two intact. In order to make absolutely sure that this branch is not damaged, it is even better to leave the two layers in contact with the galea, and make dissection between the fascia and the muscle [2]. The fascia and the periosteum are then released from the edge of the orbit and the zygomatic arch. After the fascia and the muscle have been incised at a point 1 cm from the superior temporal line, this muscle can be detached (using a raspatory in order to protect its vasculature [32]) and laid under and behind in order to expose the whole pterional region and the entire lateral wall of the orbit.

The Lateral Approach Route

Exploration of the orbit via this approach is safe in terms of both functional and esthetic parameters. This special anatomical situation led the earliest surgeons to choose this route. From an anatomical point of view, it should be remembered that the anterior quarter of the lateral wall is thick, and that it only thins out over a length of about one centimeter up to the bottom of the temporal fossa. Then, it becomes thicker again forming the angle at the meeting point between the orbit, the middle cerebral fossa and the temporal fossa. This bony junction has to be disrupted before the apex of the orbit can be accessed.

The essential differences between the many variants of this technique that have been described correspond to whether or not the anterior edge of the orbit has to be sectioned and whether the corner junction has to be disrupted or not.

In general, distinction can be made between two different types of lateral approach [8, 9]: the first (Krönlein's operation and its derivatives) involves deliberate section of the edge of the orbit so that, at the end of the operation, the original architecture has to be meticulously reconstructed (osteoplastic techniques); in the second type, the edge of the orbit is not encroached upon but the posterior corner is disrupted (non-osteoplastic techniques).

Osteoplastic Techniques

In 1888, Krönlein [25] was the first to describe this approach route which was subsequently modified to a greater or lesser extent by a number of surgeons [5, 6, 26, 29]. Once the periorbita has been detached from the internal surface of the orbit, the bone flap is trimmed in the lateral wall by making two straight cuts at the edge of the orbit, one above the frontozygomatic suture and the other at the level of the upper edge of the zygomatic arch. These cuts can be made using an oscillating saw, shears, a craniotome or a very fine drill. They are taken as far back as possible and then the flap is fractured at the rearmost point. This flap exposes the anterior part of the orbit. In order to expose the posterior part, it is necessary to disrupt the junction corner using a drill to expose the anterior aspect of the temporal and frontal dura mater, and open up the superior orbital fissure. This approach can also be extended downwards as far as the inferior orbital fissure.

Non-Osteoplastic Techniques

These techniques are first and foremost designed for the relief of orbital hyperpressure but they can also be used for tumors (located laterally in the middle or posterior part of the orbit) to which access could be hindered by the presence of the rim of the orbit. The best way to proceed is to make the key hole behind the zygomatic process of the frontal bone. Starting from this hole and using a drill, the thick orbito-cranial corner of the greater wing of sphenoid is resectioned to expose the frontotemporal dura mater and open the lateral end of the superior orbital fissure. In addition, the lateral wall of the orbit is resectioned as far in the forward direction as possible. Here again, resection can be continued into the greater wing of sphenoid as far as the inferior orbital fissure.

After bony opening in the two techniques, the periorbita is opened with a T-shaped incision (Fig. 14) with one branch parallel to the orbital axis and the other parallel to the rim of the orbit. This permits upward and forward exposure of the lacrimal gland and the orbital fatty tissue. The first feature to be distinguished is the lateral rectus muscle that is most obvious in the posterior part of the orbit where there is no fatty tissue. Once this has been identified, two microsurgical routes are possible [3], a superior route above the muscle, and an inferior one below it. The orbital fatty tissue is dissected as far as the lesion and then kept apart using self retaining retractor. The lateral rectus muscle can be kept out of the way with a suture but traumatic retraction can result in paralysis of this muscle.

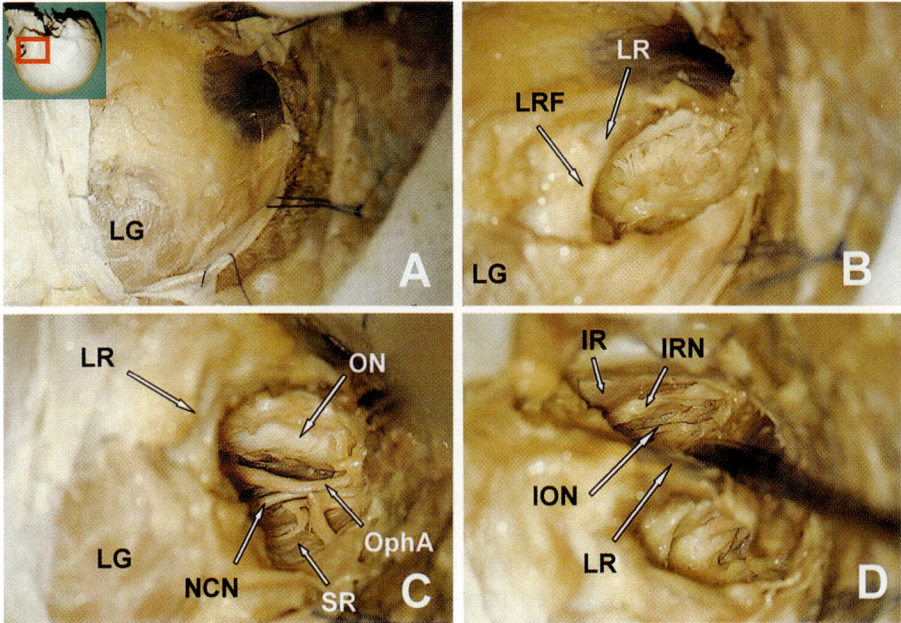

Fig. 14. Cadaver dissection showing anatomic structures encountered through a lateral approach to the right orbit in the surgical position. (A) View after the lateral wall has been removed and the periorbita opened showing lacrymal gland (*LG*) in front and above with orbital fat. (B) Microscopic view after fat has been lacerated showing lateral rectus muscle (*LR*) and its fascia reflecting on the eyeball (*LRF*). (C) Fat has been removed in the posterior part of the cone to show the optic nerve (*ON*) with ophthalmic artery (*OphA*) and nasociliary nerve (*NCN*) crossing over optic nerve from outside to inside below superior rectus muscle (*SR*). (D) the inferior route, the lateral rectus muscle has been superioly retracted showing under the optic nerve the origine of inferior rectus muscle (*IR*) with its nerve (*IRN*) and nerve of the inferior oblique muscle (*ION*)

Passing above the lateral rectus muscle, the first structures encountered are the lacrimal artery and nerve, which follow the upper edge of the muscle before they reach the lacrimal gland. In the anterior part of the muscular cone, no other features are encountered before arriving near the optic nerve to find the posterolateral ciliary artery and the short ciliary nerves coming from the ciliary ganglion located on the lateral side of the optic nerve. Behind as one approaches the apex, there is the superior branch of the third cranial nerve which gives off a branch to the superior rectus muscle that it encroaches upon at the meeting point of the middle third and the posterior third. It also sends out a branch to the levator palpebrae superior muscle that traverses the superior rectus muscle. Any attempt to approach the apex of the muscular cone via this route will be hindered by the pres-

ence of the superior ophthalmic vein which traverses the upper surface
of the optic nerve and carries on behind to leave the muscular cone between the superior rectus muscle and the lateral rectus muscle in contact
with the common annular tendon outside of Zinn's ring, to reach the cavernous sinus.

Passing below the lateral rectus muscle in the anterior part of the cone,
there are no large structures before the posterolateral ciliary artery and the
ciliary ganglion with the short ciliary nerves close to the optic nerve. These
nervous elements are usually submerged in the orbital fatty tissue and may
be damaged in the course of surgery. Behind, towards the apex, there is the
inferior branch of the third cranial nerve which sends out three branches:
one of these travels down towards the medial rectus muscle after passing
under the optic nerve; the second is for the inferior rectus muscle which it
joins at the meeting point of the middle third and the posterior third; and
the third branch carries on in front along the inferior rectus muscle, ultimately reaching the inferior oblique muscle. This last branch being so
long can easily be damaged when lesions, which extend downwards are
being excised, although such damage is usually without any obvious clinical symptoms. Finally, the inferior ophthalmic vein leaves the muscular
cone between the lateral and inferior rectus muscles outside Zinn's ring.

A central route in the lateral approach has also been described [16] but
lateral rectus muscle must to be disinserted.

The Superior Approach Route

Use of the frontal route to access the orbit was systematically employed by
Dandy [14] and Naffziger [30]. This approach was then used by other surgeons as reported in a number of articles.

More recently, use of this route has become more widespread as a result
of the work of Bachs [4] (a single flap technique) and Karagjozow [21] (a
double flap technique). Both of these techniques afford excellent exposure
but they are difficult to perform and take a long time. In practice, in the
Bach's operation [4], frontal orbital rim and the roof are sectioned without
being able to see the dura mater, which is blindly declined, across the frontal hole. In the Karagjozow technique [21], the passage of the Gigli saws
across the orbital fissures is blind and there is a significant risk of trauma
[12]. Finally, it should be noted that there is the possibility of a limited superior approach across a frontal sinus of large size [11].

Our technique involves creating an initial frontal flap then detaching
the dura mater from the roof of the orbit. The orbital periosteum is then
detached passing under orbital rim and as far as possible under the roof.
The supraorbital neuro vascular bundle should be entirely released and
when it is embedded in a canal, the canal has to be converted into a groove

by means of two sagittal cuts made with a gouge. A flexible sheet slid into the orbit will protect orbital structures and it is sometimes necessary to displace the frontal lobe using a self-retaining retractor. The second flap is then created from two cuts, one medial and the other lateral, in the upper orbital rim. These can be made using an oscillating saw, shears, a craniotome or a minidrill. These cuts are carried on as far back as possible into the roof before it is broken and folded down. The rest of the roof can then be resected using a gouge forceps and, if necessary, the optic canal can be opened using a diamond drill. In this case, attention must be paid to keeping the region cool enough to prevent thermal damage to the optic nerve. If the optic canal is opened, the optic nerve can be seen at the summit of the orbit penetrating into the orbit through the common annular tendon (Zinn's tendon). The periorbita is found in the shape of a T with one branch parallel to the orbit's axis and the other to the edge of the orbit. The two flaps are detached and then suspended laterally. The orbital structures that can be seen at this stage (Fig. 15A) are located at the level of the apex (the rest being covered by the fatty tissue). In front of the optic nerve can be seen the origins of the levator palpebrae superior muscle and the superior rectus muscle. Still in front of this nerve but a little further in is the superior oblique muscle. Outside the optic nerve, there is the broad,

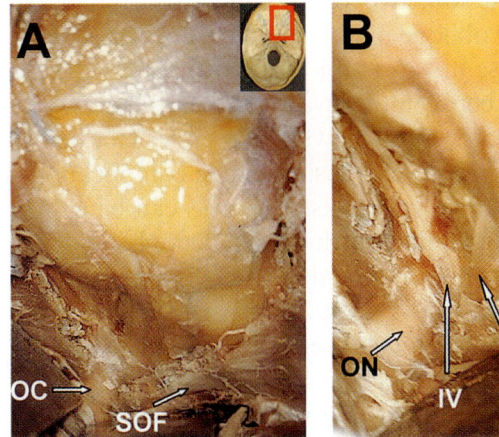
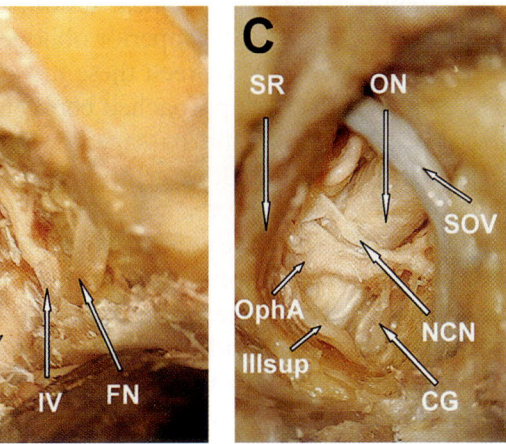

Fig. 15. Cadaver dissection showing anatomic structures encountered through a superior approach to the right orbit in the surgical position. (A) view after roof removal and opening of the optic canal (*OC*) and the superior orbital fissure (*SOF*). (B) Periorbita has been opened thereby it could be seen optic nerve (*ON*), frontal nerve (*FN*) and trochlear nerve (*IV*). (C) The lateral route with superior muscles retracted medially and fat has been removed to see the three structures crossing above the optic nerve (*ON*): superior ophthalmic vein (*SOV*), nasociliary nerve (*NCN*) and ophthalmic artery (*OphA*). We can also see at the apex the superior division of the oculomotor nerve (*IIIsup*) and on the lateral side of optic nerve the ciliary ganglion (*CG*)

flat frontal nerve making its way forwards (Fig. 15B); this is often readily perceived through the periorbita and it represents a useful anatomical landmark which is easy to find. It enters the orbit via the tapered part of the superior orbital fissure. Between this nerve and the optic nerve, as observed from this perspective, is the trochlear nerve. This extremely fine nerve carries on in front and medially making a angle of 45° with the frontal nerve and crossing over the optic nerve at the point at which it traverses the common annular tendon. It soon rejoins the upper edge of the superior oblique muscle. It is for this reason that opening the common annular tendon is associated with the risk of sectioning this nerve unless it has been identified and dissected out at the time of the opening procedure.

Throughout its path, the frontal nerve is above the lateral edge of the superior muscles (the levator palpebrae superior and the superior rectus muscles). In the anterior part of the orbit, it is less visible since it descends into the fatty tissue. This tissue can be dissected out to follow its path and discover the point at which it divides to form the supraorbital nerve (which follows the same path as the frontal nerve) and the supratrochlear nerve (which follows a medially concave curve and carries on towards the anteromedial part, passing above the superior oblique). At the same time, the whole path of the superior muscles is exposed; the levator which covers the medial edge of the rectus.

To approach an intraconical lesion, two possibilities exist: either the medial route, between the superior oblique muscle and the superior muscles or a lateral approach between these muscles and the lateral rectus muscle. The cone may also be approached between the levator and the superior rectus muscle [7, 31] but this passage is very narrow. The muscles can be put aside but this must be done gently to prevent postoperative complications.

Medially, retracting the superior muscles outside, access can be gained to the medial part of the cone although the fourth cranial nerve behind represents a major obstacle. It is important to recognize three elements which traverse this part of the orbit close to the optic nerve passing above the nerve from its lateral side as far as the medial region. These elements are (moving from the back towards the front) (Fig. 15C): the ophthalmic artery, the nasociliary nerve and the ophthalmic vein. In this medial part, it is possible to go right down as far as the level of medial rectus muscle. Beyond this point, the obstacle is this muscle's nerve which comes from the third cranial nerve passing under the optic nerve. If necessary, this exploration can be continued by opening the annular tendon along with the dura mater to completely expose the optic nerve. As has already been pointed out, this extension ought to be undertaken with care to avoid section of the trochlear nerve that passes just above in contact with the tendon.

Laterally, it is possible via this route to reach the lateral rectus muscle without any problem. Nevertheless, it is important to take care with the abductor nerve which is located on the internal side of the posterior half of this muscle. In this lateral part of the muscular cone, the lacrimal artery can be seen, climbing up from the ophthalmic artery to rejoin, a little way above the upper edge of the lateral rectus muscle, the lacrimal nerve which comes from the completely lateral part of the superior orbital fissure. These two elements (artery and nerve) carry on in front towards the lacrimal gland. In the posterior part of the cone and towards its apex, the surgeon may be hindered by the superior ophthalmic vein that comes from the medial region, traverses above the optic nerve and carries on towards the lateral part of the superior orbital fissure. It can be dissected medially and pushed towards the outside but this is associated with a risk of damage to the ciliary nerves and arteries.

An approach between the two superior muscles is achieved by pushing the muscles aside, the levator palpebrae to the medial side, and the superior rectus to the lateral side. The frontal nerve can be pushed out medially together with the levator but this will hinder access to the apex. Only the middle third of the upper part of the cone can be exposed [7, 31]. To approach the apex, the frontal nerve has to be pushed aside laterally which means that it has to be dissected, which inevitably entails a risk of damaging it. In addition, pushing these muscles aside may lead to resection of the branch supplying the levator muscle (coming from the superior branch of the third cranial nerve) which traverses the superior rectus muscle. The structures exposed by this approach are the superior ophthalmic vein, the ciliary nerves and arteries, the nasociliary nerve and the ophthalmic artery with its branches supplying the superior muscles. All these structures mean that this route is particularly difficult and fraught with risk, even if it is the most direct to the middle third of the upper part of the cone [31].

The Hybrid Lateral/Superior Approach

This hybrid approach involves the roof and the lateral wall of the orbit. It was, in effect, first envisaged by Karagjozov [21] when he extended the superior approach to encroach into the lateral wall and then other surgeons operated using the same route with some minor modifications [1]. As with the superior route it could be based on either a double flap or a single flap encroaching on the edge of the orbit. Single-flap methods are difficult to perform and the dura mater cannot be under control. In consequence, we prefer a double-flap technique. The first flap is fronto-pteriono-temporal and after detachment of the dura mater from the smaller wing of the sphenoid bone, the roof of the orbit and the middle cerebral fossa, we resect the thick orbito-cranial part of the greater wing of the sphenoid bone as far as the opening of the lateral end of the superior orbital fissure. From there, we resect a small part of the roof and the lateral wall of the orbit. The frontal

and temporal lobes are reclined using an autostatic retractor and the orbital structures are protected with a flexible sheet. It is then possible, using
an oscillating saw, a craniotome or a mini drill to create a second flap
encroaching on the superolateral edge of the orbit including as much as
possible of the roof and lateral wall. The rest of these walls is resected using
a gouge forceps.

In practice, as with the lateral approach, it is not necessary to create
a genuine craniotomy—a simple key hole behind the zygomatic process
of the frontal bone suffices. From this hole, using a drill, the thick orbito-
cranial part of the greater wing of the sphenoid bone is resected to expose
the frontotemporal dura mater and open up the lateral end of the superior
orbital fissure. Resection is then extended forwards into the lateral wall of
the orbit, and upwards into the roof as far as possible in the medial direction. The edge of the orbit is then laterally cut at the level of the upper edge
of the zygomatic arch then as far inside as possible. The saw cuts should
join the preceding openings in the lateral and upper walls.

For certain lesions, and as with the lateral approaches which preserve
the edge of the orbit, it is not even necessary to section orbital rim (pterional route [18] or postero lateral approach [10, 13]). The value of this route
is that it preserves the orbital arch without the surgeon's access to the anterior part of the muscular cone being hindered. On the other hand, the over-
hanging edge of the orbit may block access to the apex.

Using a microscope, two anatomical markers can be identified (Fig.
16A): in front, in the superolateral part, the lacrimal gland (located from
the perspective of this route, in the middle of the anterior edge of orbital
structures) and behind, the distal tapered part of the superior orbital fissure. The periorbita is cut into between these two landmarks passing inside
the lacrimal gland. A second incision is then made perpendicular to the first
and parallel to orbital rim. The two flaps of the periorbita are then detached and suspended.

Along the first incision can be seen the lacrimal neurovascular bundle
(Fig. 16B) going from the superior orbital fissure towards the lacrimal
gland. Near and inside the superior orbital fissure, the large frontal nerve
is easily seen. Outside and underneath the lacrimal bundle, the lateral rectus muscle can be seen. Finally, under the frontal nerve are found the levator and superior rectus muscles (Fig. 16C).

An incision is made into the fatty tissue and laceration is performed
above and inside the lacrimal bundle. In front (inside the lacrimal gland),
the eyeball is soon reached (Fig. 16D). Passing downward on the surface
leads to the optic nerve, which is surrounded by the ciliary nerves and vessels. As with the superior route, in the posterior part of the cone there are
three structures which cross over the optic nerve, namely (moving from the
front towards the back) the superior ophthalmic vein, the nasociliary nerve

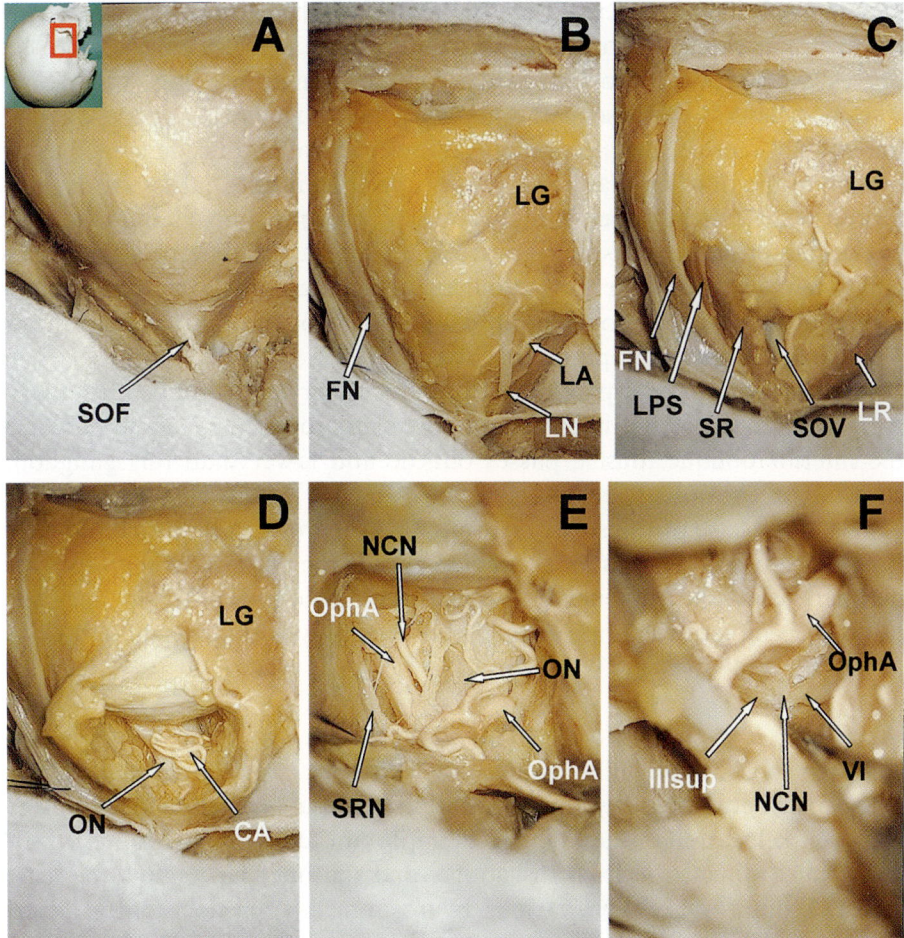

Fig. 16. Cadaver dissection showing anatomic structures encountered through a supero lateral approach to the right orbit in the surgical position. (A) Photograph after bony opening showing back the lateral extremity of the superior orbital fissure (*SOF*). (B) Periorbita has been opened and we can see the landmarks: medially frontal nerve (*FN*), laterally the lacrimal nerve (*LN*) and artery (*LA*) and in the front the lacrimal gland (*LG*). (C) Fat has been dissected to show under the frontal nerve (*FN*) the levator palpebrae superior muscle (*LPS*) and underneath the superior rectus muscle (*SR*); lateraly we can see the lateral rectus muscle (*LR*) and between these two rectus muscles the superior ophthalmic vein (*SOV*). (D) Medially and just below the lacrimal gland (*LG*) the eyeball is discovered with the optic nerve (*ON*) surrounded by ciliary arteries (*CA*). (E) Toward the apex we can see the ophthalmic artery (*OphA*) with its branches crossing over the optic nerve (*ON*) with the naso ciliary nerve (*NCN*). We can also see the nerve of superior rectus muscle (*SRN*). (F) Behind the ophthalmic artery (*OphA*) at the apex we can see the nerves which enter the orbit via the common tendinous ring: abductor nerve (*VI*), naso ciliary nerve (*NCN*) and the superior division of the oculomotor nerve (*IIIsup*), the inferior division being hidden by the ophthalmic artery

and the ophthalmic artery (Fig. 16E). Behind this, the apex is reached. Between this artery in front and the origin of the muscles can be seen the nerves which enter the orbit via the common tendinous ring (Fig. 16F). The most lateral of these is the abductor nerve which is applied against the internal face of the lateral rectus muscle; above and inside the superior branch of the third cranial nerve passes upwards, inside and forwards, juxtaposing against the superior rectus muscle where it splits giving a branch to the levator palpebrae superior muscle; medially, the extremely fine nasociliary nerve crosses over the optic nerve to accompany the ophthalmic artery and ultimately (below) the inferior branch of the third cranial nerve which is hidden by the ophthalmic artery.

Below the proximal part of the ophthalmic artery, is found the ciliary ganglion applied to the inferolateral side of the optic nerve. The nerve of the inferior oblique muscle passes outside and lower than this ganglion, carrying on forwards to reach its muscle.

As in the case of the superior route, this superolateral approach makes it possible to explore most of the regions inside the orbit other than the medial part below the optic nerve. It makes it possible to explore the apex but not the opening of the optic canal (or at least only with great difficulty).

Discussion

The choice between these alternative approach routes must be made on the basis of the location of the lesion. It is important to choose the route, which is the most direct, the least disruptive, and the one that entails the least esthetic compromise. The various neurosurgical approach routes afford access to all areas of the orbit except the inferomedial part. It could be approached by an infero lateral approach [17] but it is more accessible via a medial ENT-type approach.

If it can be avoided, the orbital edges should not be systematically resectioned for esthetic reasons. In contrast, there should be no hesitation when it comes to opening this edge if it is a question of better exposure or to prevent inopportune displacements.

The intraorbital fatty tissue can hinder access to orbital lesions but even if it is in the way, it must not be damaged. It can be lacerated and retracted by self-retained retractor [22, 23, 27] but it should not be excised (at the risk of postoperative enophthalmos). The surgeon should be aware that this fatty tissue is more abundant in the front where it is both extra- and intra-conical. It then steadily thins out as it approaches the apex where it is exclusively intraconical. It is not always easy to pinpoint the various anatomical structures because of the presence of this fatty tissue, and even the eyeball itself can sometimes be confused with a pathological mass. During installation, it is therefore useful to leave access to the eye; pressure thereon can help define its location.

We will mention a few anatomical markers, which might be useful. The lacrimal gland is easy to locate because of its relatively firm texture and pinkish color. It is not covered by the fatty tissue and lies directly on top of the eyeball so it can be followed as far as the eyeball to return back into the cone. This gland receives the neurovascular lacrimal bundle that can thus be located and which runs along the upper edge of the lateral rectus muscle. The muscles that form the orbital cone are easier to detect behind than in front since the fatty tissue is entirely intraconical at this point. Finally, the frontal nerve is very easily identified in the posterior half since it is located immediately below the periorbita, and it can be followed out in front by dissecting it out of the fatty matrix. It passes immediately over the levator palpebrae muscle.

Familiarity with the microanatomy of the orbit is essential if the various complications that this type of surgery can entail are to be avoided. The most serious of these is blindness which can only result from major and direct damage to either the optic nerve or the vessels supplying the retina. The nerve can be damaged as a result of either rough displacement or during its dissection when operating on lesions located nearby. The central artery of the retina enters into the dura mater on the nerve's medial side at a distance of 8–15 mm behind the eyeball. Therefore, when operating on posterior and medial lesions, special attention must be paid to this artery [28]. Impaired oculomotor function is the second type of complication that can be temporary or definitive, depending on the cause. The etiology may be muscular, due to damage inflicted when dissecting a lesion very close or actually involving the muscle tissue, or following traumatic displacement of the muscle. Impaired oculomotor function may also be due to nervous damage. All the branches of the oculomotor nerve (apart from that going to the inferior oblique muscle) have fairly short paths between the tendinous ring and their respective muscles (junction of the posterior third and the middle third). They can be damaged individually during surgery to remove lesions at the apex or in the posterior third of the cone. Similarly for the abductor nerve which rejoins the lateral rectus muscle in the middle of the internal side. In contrast, the branch of the inferior oblique muscle can be damaged in surgery on the anterior two-thirds if it involves the inferolateral part of the muscular cone. Such involvement rarely has any clinical repercussions. Finally, the path of the trochlear nerve is fairly short. It enters the orbit via the medial end of the tapered part of the superior orbital fissure and then passes above and in contact with the annular tendon before immediately rejoining the upper edge of the superior oblique muscle near the apex. It can be sectioned if the tendinous ring is opened which is why it should be located and dissected out beforehand. However, even when this is done as carefully as possible, damage to this nerve is almost inevitable when this type of dissection is attempted [28]. Otherwise, dissection close to the optic nerve (especially via the inferior lateral route) entails a risk of damage to the ciliary ganglion or nerves, resulting in Bernard-Horner syndrome.

The advantage of the lateral approach is that it is less invasive because opening of the cranial cavity is avoided although in this case, access is limited to the anterior part of the orbit. For posterior lesions, the orbitotemporal junction will have to be resected. Even after enlargement, it is difficult to explore the apical region via this route, and quite impossible to access the optic canal. On the other hand, it is the only approach that affords access to the inferolateral region of the cone.

A lateral approach can therefore be used for lesions located in the lateral orbit, and possibly for those in the lateral part of the apex, but not for lesions which extend into the superior orbital fissure or the optic canal [3].

A superior approach is usually indicated for orbital tumors invading into the cranial region, cranial tumors which encroach on the orbit, and lesions in the medial part of the apex or the optic canal. For lesions located in the superomedial part of the cone or if the particularities of the case mean that the entire optic nerve has to be exposed, it will be necessary to pass medially between the superior muscles and the superior oblique muscle. The angle of vision afforded by this route is restricted by the medial edge of the craniotomy, which terminates at the median line.

Passage between the superior muscles and the lateral rectus muscle is the broadest one in the superior route with as broad an angle of view. It exposes the apical region, the superior orbital fissure and the part adjacent to the cavernous sinus. The only structure that can hinder access is the superior ophthalmic vein which can be displaced either medially (which restricts access to the apex) or—to better expose this region—it can be displaced laterally although there is a risk of damaging those neurological components that pass through the superior orbital fissure and the ciliary ganglion. In this case, the common tendon can be opened between the origin of the lateral and superior rectus muscles. This passage is equivalent to the superolateral route passing over the lateral rectus muscle.

The value of the passage between the levator muscle and the superior rectus muscle is somewhat limited because the space is so confined, meaning that displacement of the various structures or dissection procedures entail a significant risk of damage to nervous components. It can be used with care for certain lesions located in the middle third of the upper part of the cone although the superolateral approach is to be preferred.

In the superior approach the anterosuperior part can be hidden by the superior muscles. These can be put aside but if this displacement is to be avoided, the superolateral approach is preferable.

The superolateral approach is a hybrid of the preceding two routes which combines many of the advantages of each. Passage between the lateral rectus muscle and the superior muscles is the broadest and the least dangerous. It affords access to all regions of the orbit apart from the inferomedial region and the optic canal. In practice, the latter can only be opened via a superior approach route with extensive retraction of the frontal lobe.

In orbital surgery, it is better to reconstruct the orbital walls at the end of the operation, whenever possible [15]. This can be done using bone fragments or plastic surgery based on acrylic material. The advantage is primarily esthetic but it is also to prevent enophthalmos or transmission of beats from the brain to the eye. Plastic surgery will also make dissection easier in the event of the need to reoperate.

Conclusion

Intimate familiarity with the microanatomy of the region is indispensable before surgery is undertaken in the orbit. This type of surgery has become routine with the development of multiple approach routes. These can be divided into three groups, namely the superior, the lateral and the superolateral approaches. The superior approach which entails craniotomy and frontal displacement is the only one which allows opening of the optic canal and affords access to lesions which encroach into the cranial region. The lateral approach is supposed to be less disfiguring and does not involve craniotomy; however, it only affords access to the lateral lesions (including those in the apex, if the broader alternative is used). The superolateral approach that we have described is the broadest route that affords access to all regions of the orbit apart from the inferomedial region. The optic canal cannot be opened from this approach unless frontal craniotomy is performed with frontal displacement. It is the most suitable approach route to the region of the superior orbital fissure.

References

1. Al-Mefty O, Fox JL (1985) Superolateral orbital exposure and reconstruction. Surg Neurol 23: 609–613
2. Ammirati M, Spallone A, Ma J, Cheatham M, Becker D (1993) An anatomicosurgical study of the temporal branch of the facial nerve. Neurosurgery 33: 1038–1044
3. Arai H, Sato K, Katsuta T, Rhoton AL Jr (1996) Lateral approach to intraorbital lesions: anatomic and surgical considerations. Neurosurgery 39: 1157–1162

4. Bachs A (1962) Contribucion al tratamento de la exoftalmia unilateral por lesiones orbitarias y retroorbitarias. Ann Hospit Santa Cruz y San Pablo 22: 481–494
5. Berke RN (1953) A modified Krönlein operation. Trans Amer Ophtal Soc 51(89): 193–231
6. Berke RN (1954) A modified Krönlein operation. Arch Ophthal (Chicago) 51(5): 609–632
7. Blinkov SM, Gabibov GA, Tcherekayev VA (1986) Transcranial surgical approaches to the orbital part of the optic nerve: an anatomical study. J Neurosurg 65: 44–47
8. Brihaye J, Hoffman G, Francois J, Brihaye-Van Geertruyden M (1968) Les exophtalmies neurochirurgicales. Rapport à la société de neurochirurgie de langue française. Neuro-chirurgie 14: 188–486
9. Brihaye J (1976) Neurosurgical approaches to orbital tumors. In: Krayenbül M (ed) Advances and technical standards in neurosurgery. Springer, Wien New York, pp 103–121
10. Carta F, Siccardi D, Cossu M, Viola C, Maiello M (1998) Removal of tumors of the orbital apex via a postero-lateral orbitotomy. J Neurosurg Sci 42: 185–188
11. Colohan ART, Jane JA, Newman SA, Mggio WW (1985) Frontal sinus approach to the orbit. J Neurosurg 63: 811–813
12. Cophignon J, Clay C, Marchac D, Rey A (1974) Abord sous frontal élargi des tumeurs de l'orbite. Neurochirurgie 20: 161–167
13. Cossu M, Pau A, Viale GL (1995) Postero-lateral microsurgical approach to orbital tumors. Minim Invasive Neurosurg 38: 129–131
14. Dandy WE (1941) Results following the transcranial operative attack on orbital tumors. Arch Ophtal 25: 191–216
15. Delfini R, Raco A, Artico M, Salvati M, Ciappetta P (1992) A two-step supraorbital approach to lesions of the orbital apex. J Neurosurg 77: 959–961
16. Gonul E, Timurkaynak E (1998) Lateral approach to the orbit: an anatomical study. Neurosurg Rev 21: 111–116
17. Gonul E, Timurkaynak E (1999) Inferolateral approach to the orbit: an anatomical study. Minim Invasive Neurosurg 42: 137–141
18. Hamby WB (1964) Pterional approach to the orbits for decompression or tumor removal. J Neurosurg 21: 15–18
19. Housepian EM (1977) Intraorbital tumors. In Schmidek HH, Sweet WH (eds) Current techniques in operative neurosurgery. Grune & Stratton, New York, pp 143–160
20. Jane JA, Park TS, Pobereskin LH, Winn HR, Butler AB (1982) The supraorbital approach: technical note. Neurosurgery 11: 537–542
21. Karagezov L (1967) Transcranial operative approaches to the orbit (russian text). Moskva: Meditzina
22. Kennerdell JS, Maroon JC (1976) Microsurgical approach to intraorbital tumor. Technique and instrumentation. Arch Ophthalmol 94: 1333–1336
23. Kennerdell JS, Maroon JC, Malton ML (1998) Surgical approches to orbital tumors. Clin Plast Surg 15: 273–282

24. Koornneef L (1979) Orbital septa: anatomy and function. Ophthalmology 86: 876–880
25. Krönlein RU (1888) Zur pathologie und operativen behandlung der dermoid-cysten der orbita. Beitr Klin Chirg 4: 149–163
26. Maroon JC, Kennerdell JS (1976) Lateral microsurgical approach to intra-orbital tumors. J Neurosurg 44: 556–561
27. Maroon JC, Kennerdell JS (1979) Microsurgical approach to orbital tumors. Clin Neurosurg 27: 479–489
28. Maroon JC, Kennerdell JS (1984) Surgical approaches to the orbit. Indications and techniques. J Neurosurg 60: 1226–1235
29. Mourier KL, Cophignon J, D'Hermies F, Clay C, Lot G, George B (1994) Superolateral approach to orbital tumors. Minim Invasive Neurosurg 37: 9–11
30. Naffziger HC (1948) Exophthalmos. Some principles of surgical management from the neurosurgical aspect. Am J Surg 75: 25–41
31. Natori Y, Rhoton AL Jr (1994) Transcranial approach to the orbit: micro-surgical anatomy. J Neurosurg: 78–86
32. Oikawa S, Mizuno M, Muraoka S, Kobayashi S (1996) Retrograde dissection of the temporalis muscle preventing muscle atrophy for pterional craniotomy. J Neurosurg 84: 297–299
33. Yasargil MG, Reichman MV, Kubik S (1987) Preservation of the frontotem-poral branch of the facial nerve using the interfascial temporalis flap for pter-ional craniotomy. J Neurosurg 67: 463–466

Neurosurgical Concepts and Approaches for Orbital Tumours

J. C. Marchal and T. Civit

Department of Neurosurgery, Hôpital Central, Nancy University Hospital,
Nancy Cedex, France

With 16 Figures

Contents

Abstract

Orbital tumours are lesions that appear within the orbital craniofacial borders. To this end, treatment of these tumours is assured by teams of different specialists. Furthermore, these pathologies are different in adults and in children. We have endeavoured, in this chapter, to highlight the specifically neurosurgical features of orbital tumours or, to be more precise, tumours affecting the posterior two thirds of the orbit and tumours originating in or intruding into the optic canal. The list of aetiologies is long. After recapitulating the main types of tumour (as well as those of most concern), we have also studied the different stages of surgery, namely approaches and reconstructions which we have illustrated at each stage by a tumour that, in our view, seemed emblematic of the problem in question: the lateral eyebrow approach for schwannoma and cavernous angioma, the trans-orbital subfrontal approach for optic nerve glioma, the pterional and orbital approaches for spheno-orbital meningioma, problems with reconstruction and with plexiform neurofibroma affecting the orbit and fibrous dysplasia of bone.

Keywords: Orbital tumours; optic nerve glioma; plexiform neurofibroma; lymphangioma; rhabdomyosarcoma; orbito sphenoidal meningioma; optic sheath meningioma; orbital vascular lesions; fibrous dysplasia; orbital reconstruction; neurofibromatosis.

Introduction, Definition of Subject and Limitations

The mass of literature relating to orbital tumours testifies to the significance and complexity of the subject. Diagnosis and treatment of these tumours is often a major challenge that is undertaken by a wide variety of different specialists: ophthalmologists, plastic surgeons, otorhinolaryngologists, neurosurgeons, paediatricians, oncologists etc. The neurosurgeon deals primarily with lesions to the posterior third of the orbit [4], or those that overlap onto the posterior third of the orbit and the contents of the skull. This anatomical description must be completed by the following, more restrictive information: Orbital Tumours (OT's) are neoplasms of the bony orbit and its content with the exception of the eyeball. Certain lesions that we have encountered include occupying processes that are not strictly tumours (for example, lymphangioma). We have therefore used the term 'neoplasm' intentionally in this context. To this end, we have excluded secondary tumours: those that stem from the face or from the eyeball and which affect the orbit at a later stage such as retinoblastoma in children.

Surgical anatomy is not the same as that of anatomy books. However, even if the latter are essential to understand and carry out this surgery, they are not particularly pertinent in the domain of neurosurgery since access to the orbital cone, the nerve and the optic canal is strictly reserved to the domain of neurosurgery. This study is neither a thesaurus nor an exhaustive description of orbital tumours that presents all the questions that all specialists might ask. Therefore, in order to avoid repeating fastidious lists of aetiologies our preferred approach was to proceed, step by step, from biopsy to surgical excision and examine OT's that correspond to the field of neurosurgery and which are representative of an indication, and of surgical techniques, that we will describe here.

Historical Perspective

The history of tumourous proptosis is often confused with the history of medicine and ophthalmology, but the first efforts at coherent classification are much more recent [8]. Antonio Scarpa in 1816, [67] offered the first clinical description of optic nerve (ON) tumours. Jean Cruveilhier (1835) [14] thought that meningioma should be considered as clinical entities that are entirely separate from tumours of the central nervous system. Von Graefe in 1854 [79] considered ON tumours to be entirely separate and attributed to them the first semiological criteria of diagnosis. It was not before Hudson discovered it in 1912 [32] that a clear distinction was made between optic nerve glioma and meningioma of the optic nerve sheath. Harvey Cushing [15] made the first distinction between primitive meningioma of the optic nerve sheath and intracranial meningioma that spreads to the orbit. The idea of approaching the orbit via the endocranial approach was successfully described and practised by Durante 1887 [22]. Surgery to the posterior part of the orbit was at its peak when neurosurgery first came into being less than a century ago. The problem of retracting the frontal lobe, which had always been considered as the main obstacle for this approach, was partially solved as early as 1913 when Frazier [26] suggesting the deposing of the orbital margin. The first neurosurgical principles for this were described by Naffziger [55]. According to this perspective, a pterional approach of the orbit was proposed by Hamby in 1964 [29]. Simultaneously, during the 1960's the first developments in orbito-cranial surgery of cranio-facial malformations came about with the findings of Converse, Tessier and Mustarde. On the basis of these technical principles, numerous works appeared that aimed to define the principles of reconstruction following excision [45, 2, 75]. It is still difficult, to this day, to dissociate the history of orbital surgery from the recent history of imaging: ultrasonography, CT scan using X-rays and MRI. Although it is not strictly the history of neurosurgery, we cannot leave aside IT imaging and its clinical and therapeutic consequences. The first comparative description of orbital tumours examined by MRI and CT scan can be attributed to Li [42]. Today, imaging is not only a factor in diagnosis but also directly in the surgical procedures of intracranial neuronavigation as described by Shanno in 2001 [72] and possibly, in future, during surgery (intra operative MRI).

Aetiologies

Generalities

Aetiologies pertaining to orbital tumours are great in number, and often necessitate entirely different procedures. In order to avoid proposing surgery that may be, on occasions, useless (for example, rhabdomyosarcoma in children), or, inversely, to avoid deferring surgery that is imperative (in the event of orbital meningioma), it is necessary to detail the clinical and para-clinical characteristics of the principal aetiologies which relate to tumours. Adult and paediatric varieties of OTs constitute two clusters of distinctive histological entities (Table 1). Diagnosis of an orbital mass is based on clinical and evolutive elements as well as information obtained from modern imaging. Clinical examination has to be conducted with care: measurement, direction of proptosis, impairment of ocular motility, compressive optic neuropathy, age at the onset and unilateral or bilateral proptosis are important features. Benign tumours like dermoid cysts or haemangiomas grow slowly whereas rapid growth suggests a metastatic tumour or

Table 1. *Adult and Paediatric Varieties of OTs Constitute 2 Clusters of Distinctive Histological Entities*

OTs that occur primarily during childhood	OTs that occur primarily during adulthood
– Optic Nerve Glioma	– Orbital Meningioma
– Plexiform Neurofibroma	– Optic Sheath Meningioma
– Capillary Haemangioma	– Cavernous Haemangioma
– Lymphangioma	– Schwannoma (Neurilemoma)
– Rhabdomyosarcoma	– Fibrous Histiocytoma
– Dermoid Cyst	– Mucocele
– Teratoma	– Epithelial Tumours of the
– Granulocytic Sarcoma (Chloroma)	Lachrymal Gland
– Lymphoma	– Lymphoid Tumours
– Histiocytic Tumours	– Metastatic Tumours
– Metastatic Neuroblastoma	
– Fibrous dysplasia	

a rhabdomyosarcoma. Rapid bilateral proptosis associated with periorbital ecchymosis suggests a metastatic neuroblastoma. Pseudotumours of the orbit (idiopathic orbital inflammation) can go through growth spurts; however they manifest themselves with pain, impairment of ocular motility and hyperthermia. Fundi and visual acuity should be checked. General examination focuses especially on cutaneous abnormalities: café au lait spots in NF1, haemangioma that often involves the skin of the eyelid. Meningioma of the orbit is frequently associated with a palpable mass in the external temporal fossa. Ultrasonography is available in an office setting but is limited in its ability to accurately measure the size and infiltrating properties of the lesion. The depth to which it can penetrate in the orbit is limited (20 mm). This leaves the posterior third of the orbit inaccessible to this kind of imaging. Most of the time standard X-rays, CT and MRI permit diagnosis. In children, CT scans and MRI sometimes require the use of general anaesthesia in order to be properly performed.

Main Orbital Tumours Occurring in Adulthood

Orbitosphenoidal Meningioma (Fig. 1)

Orbito sphenoidal meningiomas represent 18% of intracranial meningiomas. They combine, with varying degrees, the following: invasion of the great wing of the sphenoid, invasion of the lesser wing of the sphenoid that may surround the optic canal and the anterior clinoid process, the adjacent frontotemporal dura, the dura of the superior orbital fissure with, on occasions a large bud in the orbit that is often in an extraconical position, and, sometimes, the cavernous sinus (Fig. 1 a, b). Proptosis (90% of cases) is clinically dominant. Other signs are: reduction of visual acuity (50% of cases), reduced field of vision, the oculomotor nerves may be affected but this is more rare, and the trigeminal nerve may be affected on very rare occasions. Analysis of the tumourous extension is conducted by means of a CT scan using the bone window level that is effective for visualising intraosseous extensions, and imaging by MRI to show intraorbital extensions, periorbital extensions and those affecting the area of the cavernous sinus. The inevitable progression towards blindness of the affected eye justifies a surgical indication [18]. The frontotemporal pterional approach is sometimes completed by deposing the orbito zygomatic arch. We never practise this according to this indication, but we do think it is necessary to systematically open the optic canal after having drilled the anterior clinoid process. The lateral wall of the orbit, opened wide, is not reconstructed in order to relieve and treat the proptosis. Attempts at excision within the cavernous sinus are deleterious for the oculomotor nerves and it seems

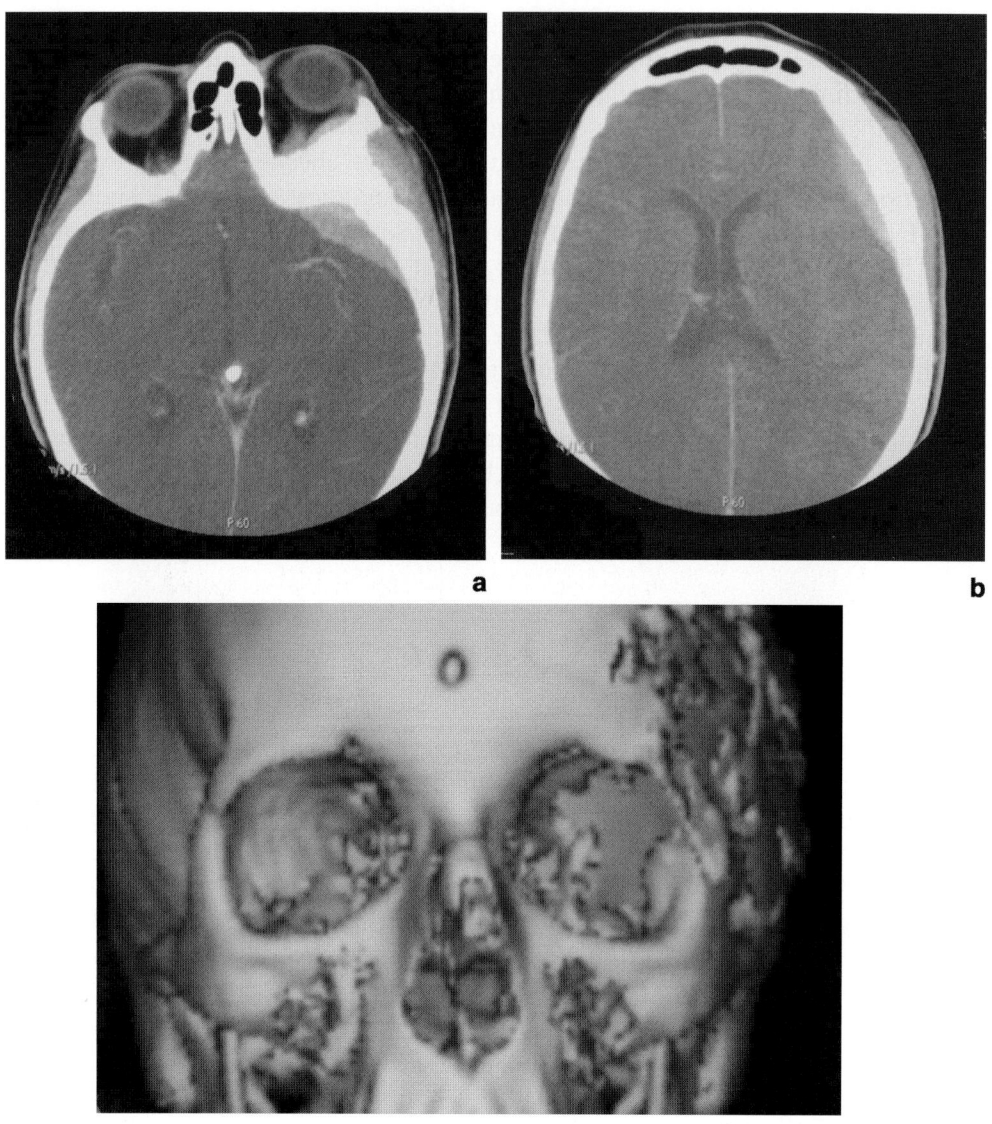

Fig. 1. Orbito Sphenoidal Meningioma: (a, b) Computed Tomography (*CT*) with iodine contrast: left sphéno-orbital meningioma involving the orbit, the vault and the soft tissues. (c–d) Frontal (c) and sagittal (d) views of 3D post-operative CT. Reconstruction with split ribs of the lateral orbital margin (d). Note the wide surgical opening of the superior orbital fissure (c)

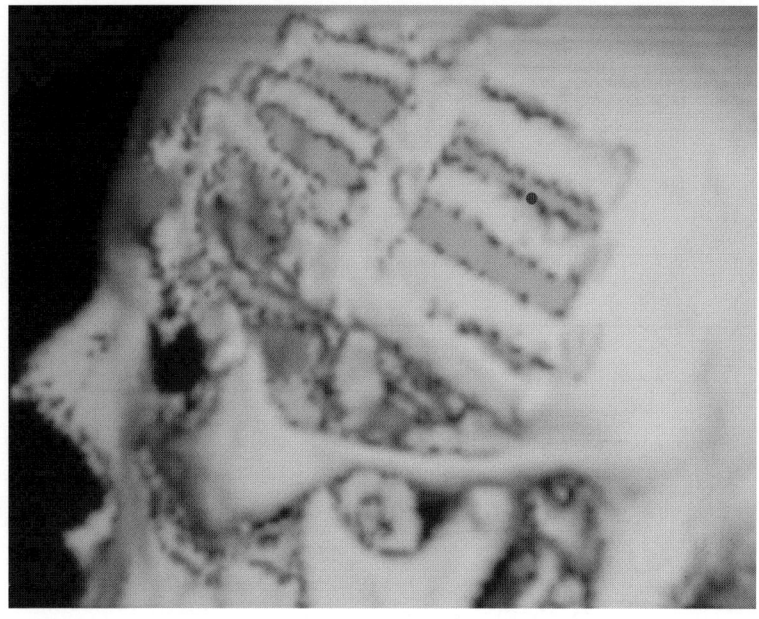

d

Fig. 1 (*continued*)

preferable to complete a partial excision at this level by radiosurgery or conformational radiotherapy. The results are often satisfying: reduction of proptosis and improvement in visual acuity [43]. Surgery can be complicated by the deterioration of visual acuity following surgery, especially if the optic nerve was previously affected [27]; post-surgical ptosis that is, in most cases, transient, is generally the result of manipulation of the levator palpebrae muscle/superior rectus muscle complex. Orbital-sphenoidal meningiomas tend to spread across the foramen, canals, fissures, periorbit, dura mater and the bone, thus giving these tumours a microscopic dimension that cannot be taken into account easily during the operation, and as a result the term 'complete excision' should be used with extreme care [49]. Partial excision alone can explain recurrences: from 34 to 54% of cases according to [51, 1], based on a follow-up study of 5 or 10 years. In the event of excision that is considered incomplete due to extension of the cavernous sinus, conformational radiotherapy or post-surgical radiosurgery is preferred instead of approaching the cavernous sinus directly: with a follow up of 8 years, 88% of patients undergoing surgery plus radiotherapy do not experience recurrence compared with 48% in the case of isolated surgery [49].

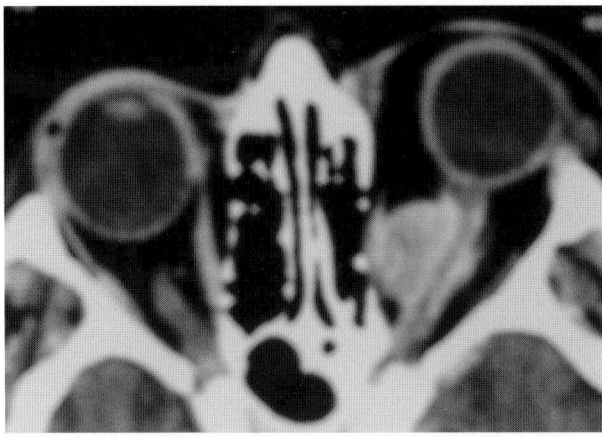

Fig. 2. Optic Nerve Sheath Meningioma: CT with iodine contrast: left optic sheath meningioma. This axial view shows the meningioma stems from the nerve sheath itself

Optic Sheath Meningioma (Fig. 2)

Second only to gliomas, this type of optic nerve tumour is the most frequent. They represent between 1 and 2% of intracranial meningiomas [23]. Efforts at excision of optic sheath meningioma are generally complicated by post-surgical amaurosis due to the problem of finding a suitable cleavage between the optic nerve and its tumour on the one hand, and vascular traumatism provoked by dissection around the nerve on the other [16, 71]. If visual acuity is preserved, radiotherapy that is fractionated under stereotactic conditions could, today constitute a good option for therapy: with an average period of 89 weeks of monitoring, Andrews *et al.* remarked that, for 92% of cases, visual acuity had been preserved compared with only 16% for the group with a single clinical follow-up [3]. When visual acuity is no longer functionally useful, a neurosurgical operation using the intraconical approach avoids the intracranial tumourous extension via the optic chiasm [56].

Cavernous Haemangioma

Cavernous haemangioma is mostly found in women aged between 40 and 50 [57]. It usually manifests itself as a proptosis and is distinct in that it appears progressively. It is indolent and reducible, thus revealing its vascular nature. When the cavernous haemangioma is in contact with the optic nerve, it can cause a reduction in visual acuity or diminish the field of vision. Echography is particularly useful here as it highlights the cystic nature of the lesion. Excision of cavernous haemangiomas is simple due to their non-infiltrating and encapsulating nature. The only surgical complication here would be adhesion to the optic nerve. The lateral orbital approach is the most suitable indication when the orbital cavernomas is located in the lateral part of the intraconal space. When the optic canal or the superior orbital fissure are concerned, an pterional extra/intradural approach should be preferred [69].

Schwannoma (Neurilemoma)

Schwannoma are nervous benign tumours that develop at the expense of the cells of the Schwann sheath. The name 'schwannoma' is attributed to Masson (1932) and the term neurilemoma to Stout (1935). It was not until 30 years later, when electronic microscopes were invented, that the original cell of this tumour was associated with Schwann cells. These tumours occur in young adults and are rarely associated with NF1. They represent 1% of orbital tumours [39, 65]. Orbital dystopia with proptosis is the revealing factor. They seem to develop more readily at the expense of the nervous fibres of the frontal nerve [25]. They develop mostly in the superior part of the orbit, touching the roof in an extra-conical or intra-conical position [60]. The tumour may be extirpated using a lateral intraorbital approach by debulking when it is large [70].

Main Orbital Tumours Occurring in Children [61, 68]

Optic Nerve Glioma (ONG) (Fig. 3)

Tumours of the optic nerve are rare, representing only 4% of all gliomas and 2% of all intracranial tumours. The peak frequency is between 2 and 8 years at a ratio of 3 girls to 2 boys. 75% of ONG's are diagnosed before the age of 10 years and 90% before the age of 20. They can be isolated or associated with NF1. They are pathognomonic of this disease when they are bilateral. 10% to 38% of patients suffering from an ONG have NF1, whereas 15% to 40% of those suffering from NF1's have optic pathway glioma. 50% to 85% of all optic pathway gliomas involve the optic chiasm or the hypothalamus. 80% of patients with an NF1 have an ONG that is limited to the optic nerve. ONG is the second most common imaging abnormality in patients with NF1 following the T2 hyperintensity seen in the basal ganglia, internal capsule, brain stem and cerebellum [31]. Tumours of the optic nerve can be classified according to their supposed point of origin: those situated on the optic disk which are intraocular, those situated in the orbit which are intraocular and constitute the subject of this study and those which invade the intracranial part of the optic pathways. In children, these tumours are usually quiescent and thus pose the problem of their

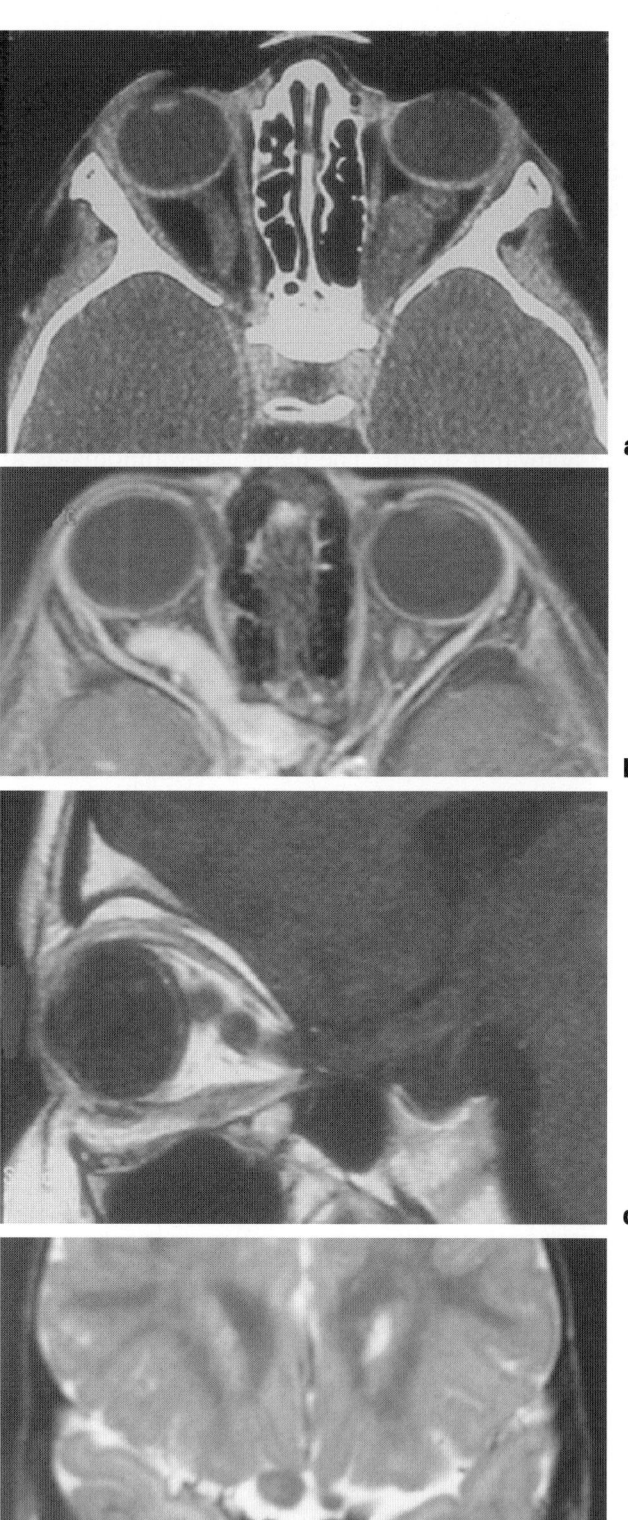

hypothetical tumourous nature. In rare cases these tumours may present evolutive growth spurts before the age of 2 or during adolescence. Spontaneous regressions are recognised. When they evolve in an aggressive way they can spread forwards and aggravate the proptosis, or backwards thus invading the optic chiasm via the optic canal (Fig. 3 b). Their histological nature likens them to pilocytic juvenile astrocytoma. Intraorbital gliomas are revealed by a reduction in visual acuity. There are no oculomotor palsies. Proptosis occurs later along with optic atrophy. Papille oedema is rare. A reduction in visual acuity in a patient suffering from NF1 is pathognomonic of an ONG. The clinical sign of an ONG in children is a reduction in visual acuity that cannot be corrected. When the visual field is affected, the ONG manifests itself as a central or paracentral field defect [31, 19, 21]. Before the invention of modern imaging techniques diagnosis was based on radiography of the orbit and optic canal that looked for a widening of the optic canal of more than 5 mm, and arteriography showing the widening of the loop of the ophthalmic artery as it passed around the optic nerve. Today CT scans permit effective analysis of osseous modifications but both with and without a contrasting agent, its use is limited in order to study the optic nerve in detail (Fig. 3). MRI is the most effective technique for studying modifications to the optic nerve (Fig. 3 d). The examination uses fat suppression techniques and a set of oblique sagittal images which are aligned parallel to the orbital course of the optic nerve (Fig. 3 c), from the globe to the chiasm [31]. These tumours have a low growth potential, therefore vision is affected at a later stage. Furthermore, the treatment is conservative and only requires annual evaluation by MRI. Tumours that grow rapidly are problematic for surgical indication. If visual acuity is preserved, surgical resection of the tumour would constitute mutilation. If the surgeon is confronted with a tumour that is clearly surgically, or radiologically evolutive, it is better to propose either chemotherapy or radiotherapy in the first instance rather than surgical excision. In this case, it is clear that repeated MR must be performed. In effect, surgical excision should take place before the tumourous extension spreads backwards in the direction of the optic chiasm. In the event of straightforward clinical and radiological surveillance, there is no question of a biopsy. The decision to carry out either radiotherapy or chemotherapy in the case of rapid growth may raise the question of surgical biopsy: on the one hand,

Fig. 3. Optic Nerve Glioma: (a) CT: left optic nerve glioma circumscribed into the orbit. (b) T1 weighted MRI: right optic nerve glioma spreading backward to the chiasm. (c) T1 weighted MRI: optic nerve glioma on an oblique sagittal image which is aligned parallel to the orbital course of the optic nerve from the globe to the chiasm. (d) T2 weighted MRI: frontal view of a right optic nerve glioma

the appearance of the lesion when examined by MRI is characteristic, but on the other hand biopsy on the optic nerve constitutes a functional risk that is sufficiently great to dismiss this option. Indication of a surgical resection of an ONG is justified if the tumour remains anterior to the optic chiasm, and may be completely resected if it presents clinical or radiological signs of evolution that suggest possible extensions to the optic chiasm, and if visual acuity is severely impaired or not functional (amblyopia). The techniques for such a resection include an intraorbital and intracranial approach of the optic nerve.

Plexiform Neurofibromas (Fig. 4)

It is common practice to separate neurofibromas into 4 groups [11]: plexiform, diffuse, circumscribed and postamputation. Plexiform neurofibromas are tumours which occur in children, and the circumscribed types are schwannoma. This type of tumour, the description of which is attributed to Von Recklinghausen in his 1882 study, seems to have existed for much longer as Huson identified in 1994 [33] when he related an illustration by Aldrovandi that dated from 1642. Ophthalmic-orbital deformity, generally unilateral, can be associated with what is obviously dystopia. Such a location is quite rare, estimated at 1% to 7% of patients with an NF1 [34, 41, 52]. Plexiform neurofibromas affect the eyelids, and the initial clinical manifestation is usually a visible swelling of the superior eyelid, thus giving it a hypertrophic appearance (Fig. 4 a). On palpation these nodules give the impression of "a bag of worms". They provoke intraocular hypertension that is responsible for proptosis in children, or glaucoma. Nonetheless, the most evocative sign is pulsatile proptosis [52, 66]. This is secondary to an intraorbital hernia of the temporal or frontal lobe, made possible by agenesis of the great wing of the sphenoid, enlargement of the superior orbital fissure and defects of the orbital roof. The aspect of facial asymmetry and proptosis stems from the imbalance between the contents of the orbit and the orbital container [10]. In effect, if the tumourous growth affects either one or the other of the parts, it is the ocular globe that suffers the constraint. During adolescence the tumour extends to the forehead, temple and superior orbit leading to downward displacement of the globe. Osseous defects of the affected part of the orbit are the result of lysis of the posterior orbit, but also of the constrained growth of the anterior orbit by hyper pressure. They associate partial or total absence of the great wing of the sphenoid and malar hypoplasia (Fig. 4 b) [6, 41, 37, 38]. CT and MR imaging show nodular lesions with significant contrast enhancement, with irregular limits, thus giving it an infiltrating appearance. Treatment of plexiform neurofibromas is not satisfactory, as they infiltrate the normal anatomical structures such as the oculomotor muscles and the extraocular

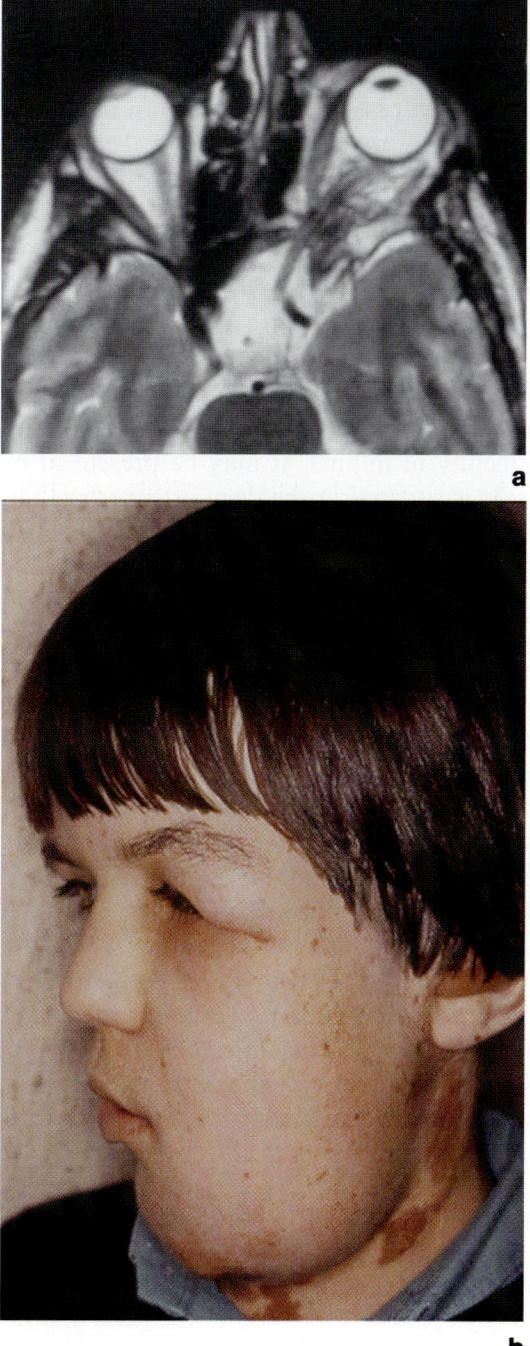

Fig. 4. Plexiform Neurofibroma: (a) plexiform neurofibroma in an adolescent: note the "S" shaped eyelid and the extension to the face. (b) CT (axial view): the dysplasia of the sphenoid wings is associated with a plexiform neurofibroma

muscles. They have a high rate of recurrence and may require numerous operations in order to obtain a satisfactory aesthetic result. In addition to difficult and often incomplete resections, due to their infiltrating nature, they pose the problem of osseous reconstruction of the orbital roof defect that has deteriorated due to orbital-sphenoid dysplasia. Bone grafts are often difficult in terms of finding a suitable resting point on an affected bone that has been destroyed or eroded by the plexiform neurofibroma. Surgery is therefore complex, multidisciplinary, staggered over time and is rarely neither aesthetically nor functionally satisfactory.

Orbital Vascular Lesions [5]

Capillary Haemangioma. Capillary haemangioma is the vascular lesion that occurs most frequently in infants. It may be present at birth or appear a short while after. It is situated in the deep or superficial areas of the orbit. There are, on occasions no cutaneous manifestations. The tumour tends to affect the superomedial part of the orbit. It can invade the entire orbit. They occur most often in females. They generally grow rapidly during the first 6 months of the infant's life and then begin to decrease in volume. Cutaneous localisations elsewhere on the body can be found in 30% of cases [21]. 60% disappear completely by the age of 4 years and 76% by the age of 7. The size of the tumourous mass can sometimes displace the ocular globe and hinder vision. Imaging is particularly important if the lesion is deep, and has no clinical cutaneous manifestation. Its internal architecture is characteristic: irregular mass, well-defined and enhancing with contrast. A differential diagnosis with a rhabdomyosarcoma, a metastatic neuroblastoma or an orbital lymphangioma is, in theory, simple. Their treatment depends essentially on their size and the way in which they hinder vision. The risk of amblyopia should be treated with corrective glasses, taking care to hide the unaffected eye. The treatments proposed are: corticosteroids by general administration or local injection or interferon (inhibitors of angiogenesis). Surgery must always take into account the natural evolution which is often spontaneously satisfying.

Orbital Lymphangiomas (OLs) or Venous Lymphatic Malformation (Fig. 5). It is likely that orbital lymphangiomas are of congenital origin [40]. These malformations have a malformed component that is a mix of venous and lymphatic. They appear at the level of the head and neck, spreading to the orbit. They are diagnosed at birth or in young infants (diagnosis is generally performed at the age of 6 years). They are slow to evolve but very invasive. Rapid growth spurts relate to hypertrophy of the lymphatic tissue that is provoked by infections of the ENT sphere, the upper respiratory

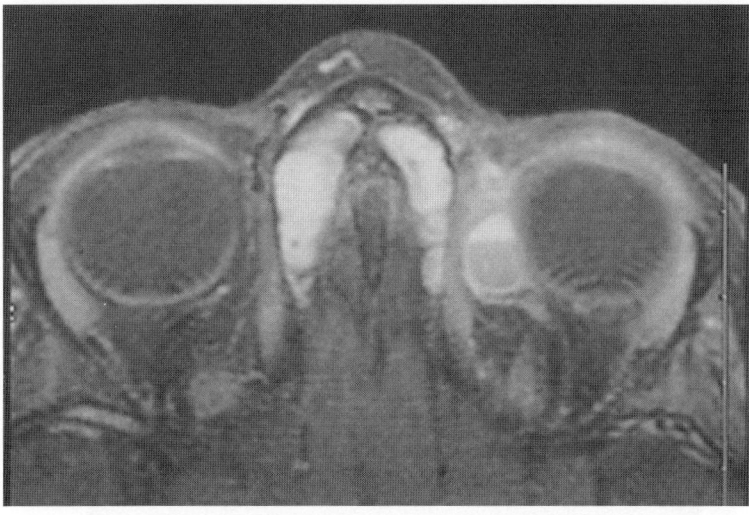

Fig. 5. Lymphangioma: Typical "chocolate cyst" in a left orbital lymphangioma

tracts or the face. In other locations, there are haemorrhagic phenomena that produce cysts, typically known as "chocolate cysts". Imaging allows for a differentiation between fleshy parts and cystic areas. OLs are not encapsulated and, when positioned at the level of the orbit, they infiltrate the fat, orbital septa, nerves and muscles. These characteristics make complete excision impossible. Surgery should therefore be limited to decompression in order to relieve troubles related to the optic nerve or oculomotor nerves, or reduce a proptosis that is of worrying proportions. Such decompression implies either aspiration of a cyst or debulking of the tumour (which explains why effective pre-surgical imaging is essential). It is on the basis of these principles that a limited approach can be proposed.

Primitive Bone Tumours

Fibrous Dysplasia (Fig. 6.a–c). Fibrous dysplasia is a rare, benign condition that causes the replacement of normal bone with fibro-cellular tissue. Islands of normal bone are dispersed within this fibro-cellular tissue. In children, however, there may be growth spurts that can provoke proptosis. Initially described by Von Recklinghausen in 1891 [80], fibrous dysplasia represents 2.5% of all bone tumours [24]. This pathology tends to affect the ribs, the tibia and the craniofacial bones. It may appear as one of two forms: a focal presentation which is more frequent but for which a craniofacial localisation is rare; 10% to 27% of cases according to Munro, [53, 54], and a multifocal form that is characterised by the impairment of at least two non-contiguous osseous structures. Albright's syndrome is a par-

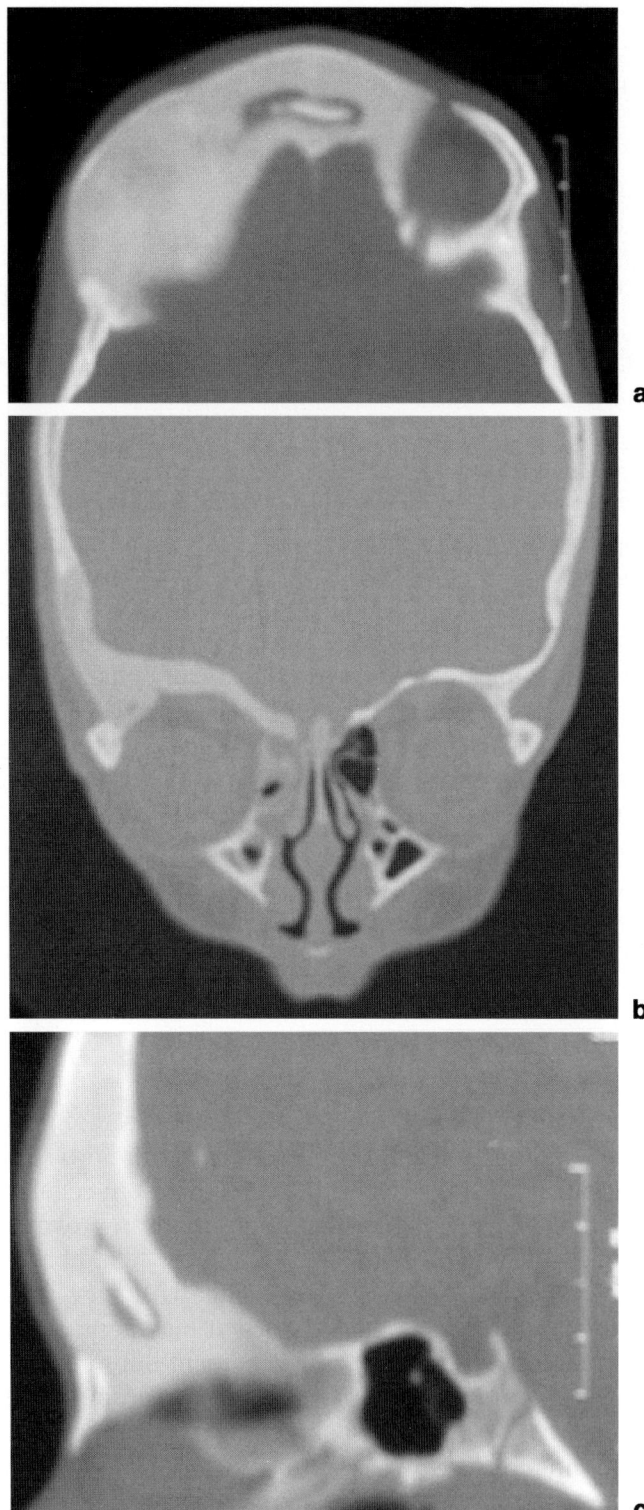

ticular type that combines fibrous dysplasia, sexual precocity, cutaneous café au lait spots, premature bone maturation and hyper parathyroidism. Orbital location is, in the majority of cases, isolated. Consequently the orbit is directly implicated at different levels, more readily in its superior hemi-circumference. In most cases, its initial manifestation is of aesthetic importance due to bone deformation. It appears as facial asymmetry. It may also lead to functional repercussions: reduction of visual acuity, epiphora caused by obstruction of the lachrymal canal in the event of ethmoidal involvement. From a vertical view, the ocular globe appears to be lowered and, from a horizontal view, the displacement can be lateral or medial. Imaging can be used for visualising a homogenous lysis or an osseous sclerosis, the two aspects mentioned can closely intermingle in various ways. Fibrous dysplasia of the orbit usually manifests itself during the first ten years of a person's life. Its growth during puberty is slow. Once this period has passed, evolution of dysplasia is extremely variable, ranging from a steady growth to stabilisation or, in exceptional cases, regression [76]. Its evolution is dominated by risk of damage to vision by compression of the optic nerve during its intra-osseous course. This justifies regular monitoring of acuity, both of the visual field and also the perception of colour. Risk of malignant transformation appears to be quite low and is estimated at between 0.5% and 4%. This potential evolution represents a greater risk for sufferers of Albright's syndrome. This evolution usually provokes pain and a rapid increase in the volume of the tumour. This has been observed after radiation of the dysplastic process [12, 13]. The surgical indication is entirely subject to the reduction in visual acuity, or to a severe narrowing of the optic canal that can be clearly radiologically demonstrated [12]. Aesthetic damage should justify the surgical indication on the condition that it does not infer a significant functional risk.

Osteosarcoma [19, 21]. Orbit osteosarcoma following radiation therapy for bilateral retinoblastoma is not rare. The occurrence of these tumours is due partly to radiation therapy in childhood, and partly to genetic abnormalities within chromosome 13. 30% of children who underwent radiation therapy for bilateral retinoblastoma harbouring chromosome 13 abnormalities will develop radio-induced sarcoma. Prognosis is poor and the five year survival rate does not exceed 30 to 40%.

Fibrosarcoma. Congenital, infantile, juvenile and adult forms of fibrosarcoma are recognized. Congenital and infantile fibrosarcomas are rare,

Fig. 6. Fibrous Dysplasia (*FD*) of the roof of the right orbit: (a–b) CT axial and frontal views of FD of the right orbit; (c) sagittal reconstruction

grow rapidly and have a low incidence of metastasis (less than 8%). Juvenile fibrosarcoma of the orbit has effective prognosis in children under ten years old when treated by radical local excision. Fibrosarcoma of the orbit is second only to osteosarcoma, following radiation therapy for retinoblastoma.

Histiocytic Tumours. Histiocytosis includes three groups of diseases of increasing gravity and juvenile xantho-granuloma.

Eosinophilic Granuloma: Eosinophilic granuloma has a rapid onset. It looks like an inflammatory pseudo-tumour of the orbit with pain and supero-temporal swelling. Proptosis is unilateral. Blood cell count reveals a hypereosinophily and biopsy a reticulo-histiocytic hyperplasia.

Hand-Shuller-Christian Disease: Is characterized by a polyuro, polydipsic syndrome (insipid diabetes), bilateral proptosis, and defects on plain films of the skull.

Letterer-Siwe Disease: Represents the acute presence of histiocytosis in infants. Patients suffer from generalized malaise, hyperthermia and axial purpura, which may affect the eyelids.

The first two diseases have a variable response to corticoids and chemotherapy whereas Letterer-Siwe disease evolves dramatically and prognosis is poor.

Juvenile Xantho-Granuloma: Is often classified with this group of histiocytic diseases. Only children under 1 year old are concerned. Clinical presentation combines cutaneous papules with ophthalmologic manifestations: anterior uveitis and secondary glaucoma. Biopsy shows an abnormal proliferation of histiocytes, lymphocytes, eosinophilic and Touton's giant cells, which are the hallmark of the disease.

Sarcomas [19, 21]

Rhabdomyosarcoma (Fig. 7). Rhabdomyosarcoma are tumours frequently found in children, an orbital position being the most common. They are the most frequent of all malignant tumours of the orbit in children. They stem from rectus muscles, although it is extremely difficult to ascertain their true origin. The average age for diagnosis is 7 years and upwards, and the tumours predominate in males (3 boys to 1 girl). Natural evolution is serious: local extension towards the nasal fossa, para-nasal and para-meningeal areas and general extension with pulmonary and hepatic metastasis. Tumourous growth and proptosis are rapid, thus necessitating emergency treatment. The proptosis often appears with inflammations. CT and

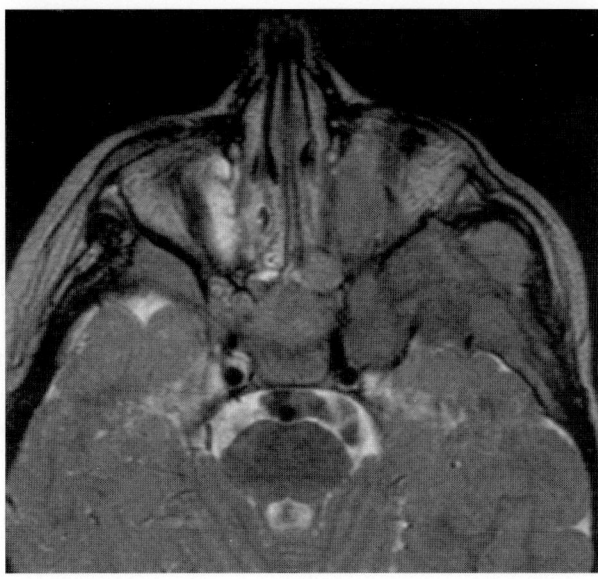

Fig. 7. Rhabdomyosacoma: Rhabdomyosarcoma of the left orbit. Dura and bone are concerned

MRI reveal the involvement of the intra-conical and extra-conical sections of the orbit. The most common starting point is at the supero-internal quadrant. The mass is ill-defined and has irregular limits. Contrast enhancement is low. Extension and invasion of the neighbouring osseous structures is very common, particularly the para-sinusal and para-meningeal areas. It is firstly necessary to confirm diagnosis by carrying out either surgical biopsy or biopsy using a needle. There are two kinds of histological rhabdomyosarcoma entities found in children: the embryonal variety and the alveolar variety. Prognosis is more positive for the alveolar variety than the embryonal variety. Currently, the preferred treatment is chemotherapy followed by radiotherapy. Surgery may be considered after the chemotherapy and radiotherapy, in order to confirm or invalidate the presence of tumourous residue on an image showing suspect residue after treatment. The global survival rate at the age of 5 years is 90%.

Fibromatosis. Is a benign but locally aggressive fibroblastic lesion that can be clinically and pathologically classified between fibrosis and fibrosarcoma. It affects mainly boys at the rate of 2 to 1 girl. It may be solitary (73% of cases), multicentric or generalized.

Nodular Fascitis. Is a common fibrous tumour found elsewhere in the limb and trunk in adults, but head and neck localization is more common among infants and children. The tumour is well circumscribed. It is best treated by complete excision.

Leukaemia and Orbital Lymphoproliferative Disorders

Leukaemia. Occurs at any age and represents one of the most common forms of malignancy in childhood. Acute Lymphoblastic Leukaemia (ALL) accounts for 80% of all cases in children. About 80% arise from B-cell lineage and 20% from T-cell. Acute Myeloid Leukaemia (AML) is more common in adults, although it represents 20% of cases in children. There are many manifestations of orbital and ocular leukaemia. Orbital leukaemia is rare in ALL, but more common in AML. Such involvement in AML occurs in the form of *granulocytic sarcoma*, characterized by an infiltration of immature cells of myeloid (granulocytic series). This tumour is also known under the term of *chloroma* because of the myeloperoxydase within the mass, producing the green hue observed during microscopic examination. This orbital mass may precede the blood and bone marrow findings of AML or develop after the diagnosis. The most common ocular manifestation is retinal haemorrhage, which occurs when the patient is anemic or thrombocytopenic. Leukaemic infiltration of the optic nerve is an emergency because vision may deteriorate rapidly. Radiation therapy posterior to the globe and orbit on an emergent basis may be necessary to preserve vision. Leukaemic cells may also infiltrate the iris and the anterior chamber as present as iritis. These infiltrations may cause a secondary glaucoma. *Granulocytic Sarcoma* is a chloroma with bone erosion in the absence of blood involvement. This may be the first manifestation of AML. This lesion affects the subperiosteal space, usually the lateral wall of the orbit with extension in the temporal fossa or medial wall of the orbit with extension to the ethmoidal air cells and cribriform plate.

Lymphomas [78]. Extranodal manifestations of Non Hodgkin Lymphoma are common, but the incidence of orbital localization is low. There is usually an insidious presentation. This characteristic helps to differentiate lymphoma from orbital pseudo-tumours that progress rapidly over days. Burkitt's lymphoma (BL) is a disease due to the Epstein-Barr virus. BL is much more frequent in Tropical Africa. In 60% of cases, BL appears as a tumour of the maxilla which then affects the orbit, leading to a huge proptosis.

Metastatic Tumours of the Orbit

Metastatic Neuroblastoma (NB). Is a malignant tumour of the sympathetic nervous system and the most frequent orbital metastasis to occur during childhood. Most of the time, primitive tumours are diagnosed prior to the metastasis. However, this diagnosis has to be kept in mind because orbital metastasis reveals the NB in 10% of cases and necessitates systematic abdominal palpation and echography. The primitive tumour usually appears in the medullo-suprarenal glands or other retroperitoneal structures. The sites of origin may be cervical, mediastinal and pelvic sympathetic system. These orbital metastases lead to unilateral or bilateral proptosis which is quite characteristic when they manifest ecchymosis of the eyelids (Hutchinson's syndrome). 90% of cases have an increased level of Vanilyl-Mandelic Acid in the urine because of cathecholamine secretion in the tumour.

Ewing's Sarcoma. Is a bone tumour, which may give rise to orbital metastasis.

Epidermoid Cysts (ECs) (Fig. 8)

ECs tend to appear either as asymptomatic superficial lesions in children or as complicated deep lesions in adolescents and adults. Superficial ECs are more frequently lateral than medial. An eyebrow location is common. Superficial lesions can be dealt with by a direct, uncomplicated surgical approach. Surgical procedure must include an *en bloc* excision to avoid rupturing the cyst and to avoid inflammatory reaction in the post-operative courses. The deep lesions, in contrast, are frequently extensive and difficult to remove (Fig. 8 d–f), requiring careful pre-operative planning (Fig. 8 a–c) [73].

Teratoma (Fig. 9)

Teratoma is a cystic, benign, and congenital tumour. It is visible at birth and appears as a huge, impressive proptosis. It is often difficult to know where the eyeball is. CT scans show a multilobulated mass involving the neighboring anatomical structures (brain, bony orbit, etc.). Histological findings show a cyst surrounding the epidermis, gastro-intestinal mucosa or respiratory epithelium. On rare occasions, certain cases are more distinct, making up parts of a foetal body. The growth of these tumours is very rapid, extensive, and life threatening. They must therefore be removed as completely as possible even if surgery in neonates may be hazardous.

Surgical Approaches

As a general rule, it is necessary to ask the question why, for orbital tumours an orbital approach is more readily practised in the domain of neurosurgery than in the domain of ophthalmology or plastic surgery.

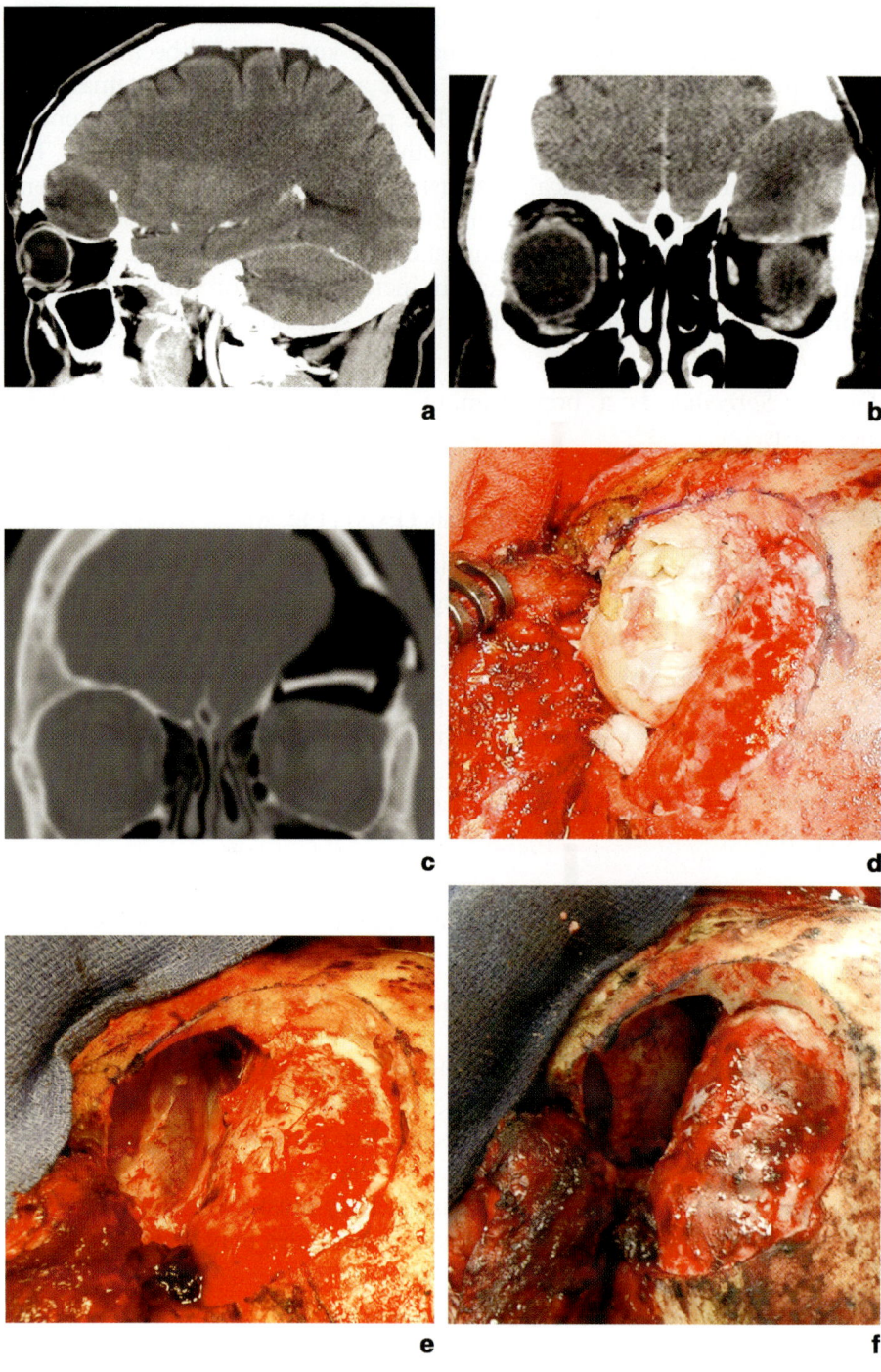

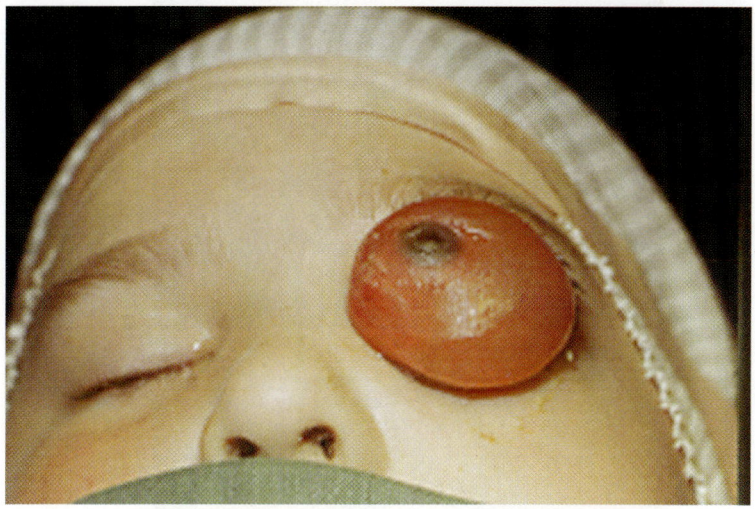

Fig. 9. Teratoma

There appear to be two circumstances that justify this decision: it is preferable to choose a neurosurgical approach in the case of lesions to the two
posterior thirds of the orbit and/or in the event that it is necessary to check
the excision of the tumour as well as that of reconstruction via an epidural
or intra-dural approach. There are 3 surgical approaches that correspond
to surgery of increasing difficulty which also suffers post-operative complications that become increasingly serious: the strictly intraorbital approach,
the epidural approach and the intra-dural approach. In order to guarantee
maximum safety during surgery, a difficult stage of the surgical approach
should give the opportunity to check the approach to which it is contigu-
ous: the intra-orbital approach should allow for epidural monitoring, and
the epidural approach should allow for intra-dural monitoring. The intra-
dural approach allows for monitoring all three but presents more deleterious side effects, and should therefore be reserved for accessing the anatomical structures such as the optic nerve, or for intra-dural tumourous
extensions. The different approaches described here take place in microsurgical conditions whenever the orbit or the interior of the dura are operated
upon.

◄──

Fig. 8. Epidermoid cyst of the left orbit (*EC*): (a–b) CT (a: sagittal reconstruction, b:
frontal view) of EC of the left orbit. (c) CT (frontal view) reconstruction of the roof of
the orbit with split bone flap. (d–f) Intra operative views (left sub frontal approach).
(d) EC becomes visible beneath the dura, through the orbital roof. (e) EC is excised,
note the large defect in the orbital roof. (f) the split bone flap is inlaid to replace the
orbital roof

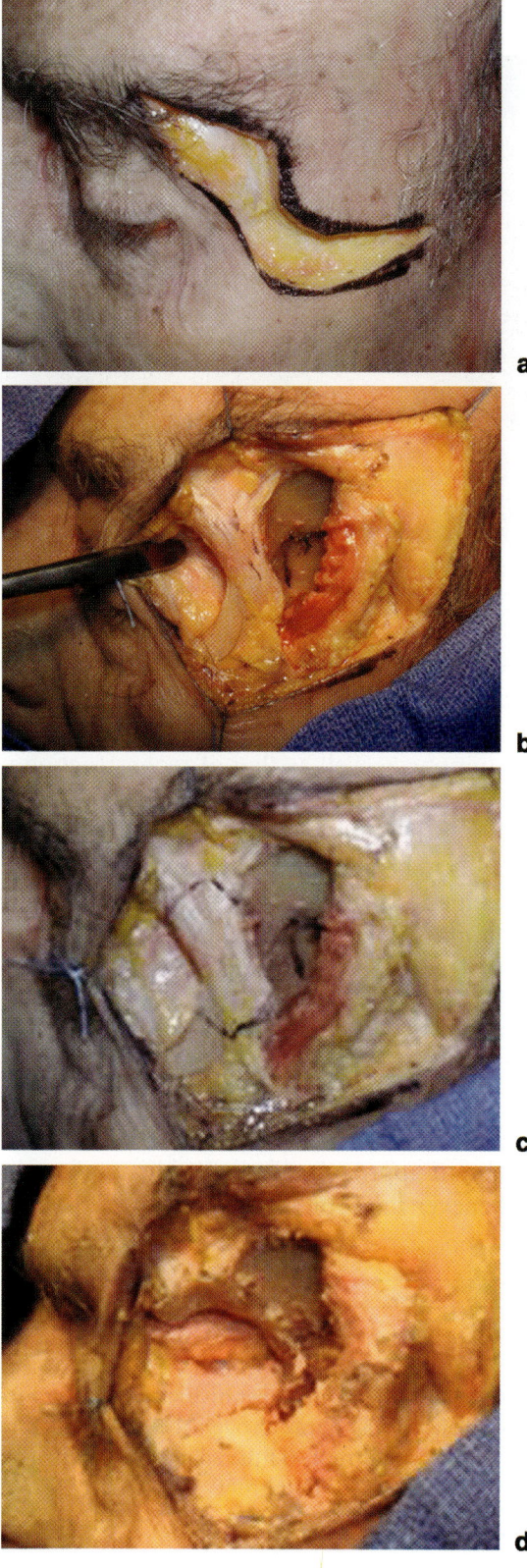

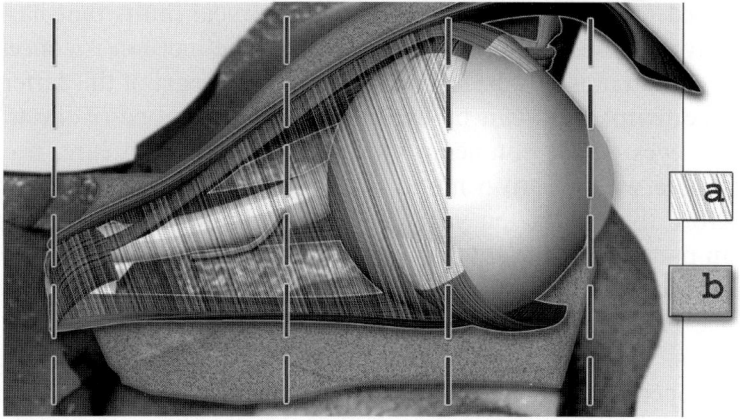

Fig. 11. The two posterior third of the orbit

The Lateral Intra-Orbital Approach (Fig. 10)

Surgical approach of the orbit and, in particular when concerning the two anterior thirds [4] can be adequately addressed by the classic ophthalmologic approaches, whatever variety or variant they may be. On the contrary, neurosurgical approaches that allow access to the posterior third of the orbit are more aggressive and can, on occasions represent a risk for the frontal lobe during retraction operations. Furthermore, the lateral approach as described by Krönlein is of particular interest since it associates the benefits of a trans browsal and those of an orbitotomy of the roof and the external margin of the orbit. It spares the frontal lobe since it does not necessitate cerebral retraction. This approach is useful for excision of accessible tumours from a frontal view by the superolateral and inferolateral quadrants, independently of their intra-conical or extra-conical location. As this approach permits access to the superior orbital fissure it concerns tumours situated, from a sagittal view in the two posterior thirds as defined by Benedict [4] (Fig. 11). Surgery should not include complex bone reconstruction after excision. In other words, it should be reserved for tumours of the soft intra-conical and extra-conical areas (for instance, neurilemomas, cavernous haemangiomas, dermoid cysts, etc.) according to the anatomical limits that we have identified and the biopsy or biopsy-excision with or without bone samples. The drawbacks of this approach are mini-

Fig. 10. The lateral intra-orbital approach (dissection): (a) The "S" shaped incision; (b) the periorbit is unfastened from the lateral wall of the orbit; (c) osteotomy takes the form of a dovetail; (d) the bone excision can be performed as far as the superior orbital fissure, towards the rear

mal provided the indication respects the anatomical limits that have been defined. From an aesthetic perspective, part of the incision is concealed in the eyelid (which must not be shaved); the posterior part of this latter bends towards the temple and can be usefully concealed in one of the wrinkles around the eyes. If the cutaneous closure is realised with care, it can become invisible within 3 to 6 months. The greatest obstacle is the frontal sinus. Before the age of ten years, it is little developed, if at all, and in adults can pneumatise the entire superior orbital margin as far as the external orbital process. Therefore, with this in mind, it is beneficial to use MR, and also CT, which provides more details of anatomical links with the frontal sinus. The frontal sinus may be opened accidentally or intentionally. If the opening is accidental, its size is generally small and the mucosa can be spared. There are more risks in ignoring the opening than there are difficulties in repairing it. In the case of accidental entry, the simplest method is to fill the opening with osseous powder and biological glue and close the gap from the exterior by applying a patch of temporal fascia stuck with acrylic-glue. When the frontal sinus has been opened intentionally it should be 'cranialised' by coagulating and resecting the mucosa of the sinus and filling the fronto-nasal canal with osseous powder and biological glue. The frontal sinus is the largest anatomical obstacle and brings many complications if these simple rules of exclusion are not respected.

The patient is in the supine position with their head turned 30 to 45 degrees towards the opposite side, in a simple headrest (a three pins holder cannot be safely used in children). The incision begins at the middle of the eyelid (it can be extended towards the interior, and only towards the interior if the objective is to gain access to roof of the orbit) and bends, in an 'S' shape towards the temple when the external orbital process meets the malar bone. At this stage, it runs alongside the superior rim of the zygomatic process (Fig. 10 a). The incision should go directly to the bone, and, if necessary the supra orbital nerve should be pushed aside from the supra orbital foramen. It is necessary to scrape the interior of the lateral orbital process and unfasten the periorbit from the lateral wall of the orbit (Fig. 10 b). The subcutaneous tissue is dissected from the fascia of the temporal muscle. Insertions of the temporal muscle are removed from the external temporal fossa and pulled towards the rear. This operation makes it possible to view the superior part of the external side of the great wing of the sphenoid. Osteotomy of the lateral orbital process may take the form of a step or, more simply, of a dovetail (Fig. 10 c). Once removed, the lateral wall of the orbit is gouged or drilled whilst keeping the periorbit at a distance. This bone excision can be performed as far as the superior orbital fissure, towards the rear (Fig. 10 d). Access to the intra-conical part of the orbit takes place between the superior and lateral rectus muscle. There is, therefore no interposition of important anatomical elements. By using this approach it is possible to check the dura and remove the lateral part of the orbital roof by drilling the bone immediately below the line of insertion of the muscle, as far as is necessary in order to reveal the dura. This hole should be situated underneath the ridge of insertion of the muscle in order to conceal it during closure. This approach, when compared with a trans browsal approach, has the advantage of being able to remain strictly epidural, to check the orbital roof using both the intra orbital and epidural approaches if necessary. Finally, it is always possible to carry out a supraorbital craniotomy by extending the cutaneous incision towards the interior of the eyebrow. However, this approach does not permit a satisfactory intra-dural opening or monitoring.

The Optic Nerve Approach

The optic nerve approach is justified for two kinds of tumours: meningioma of the optic nerve sheath and optic nerve glioma. In the first case, approaching the optic canal is the key point of the operation, and in the case of optic nerve glioma it is not absolutely necessary to open the canal although the optic nerve should be checked during its intra-dural course.

The Sub Frontal and Intra Conical Approach of the Orbit [7, 64]

The patient is in the supine position with their head towards the zenith or at a slight hyperextension of 10 degrees to distance the frontal lobe from the orbit, by gravity. After coronal incision the temporal muscle is pulled forwards en bloc with the periosteum and the skin flap. The notch or the supra orbital foramen is exposed with the supra orbital nerve. When it presents as a hole it should be open and the nerve individualized to be pulled forwards with the skin flap. A frontal craniotomy that is homolateral to the lesion is carried out. An accidental or intentional medial opening of the frontal sinus requires the same precautions to be taken as for a lateral approach of the orbit. In order to optimise the posterior field of vision and diminish cerebral retraction, it is possible to combine removal of the superior orbital margin after separating the supra orbital nerve from the orbital rim. It is possible, particularly with children, to carry out a combined approach in a single cranio-orbitotomy. The orbitotomy runs laterally as far as the lateral orbital process above the fronto-malar suture. It avoids opening the frontal sinus medially. The orbitotomy of the roof implies, in a first instance, perforating the roof with a one millimeter drill then detaching it like a postal stamp. The dura is separated from the orbital roof. This epidural separation does not, generally, allow the necessary space for exposing the whole of the orbital roof due to the impossibility of combining an epidural separation in the middle cerebral fossa. Under these conditions it is advisable to approach the orbit by an intra-dural approach. Careful retraction of the frontal lobe can be easily obtained after opening the optico-carotid cistern and the slow, progressive subtraction of the Cerebro Spinal Fluid (CSF). Exposure of the two posterior thirds of the orbit towards the rear of the eyeball and across the entire width is then complete. The axis of the optic canal permits easy identification of the internal part of the orbital apex. If necessary, the superior orbital fissure can be opened after drilling the anterior clinoid process. A wide dural flap, attached by a pedicle to the median line, is pushed back medially to expose the orbital roof. By opening the optic canal it is possible to determine the apical area, which would be exposed across its entire width, and complementing surgical exposure using either a drill or a thin bone rongeur. The frontal nerve appears beyond the transparent periorbit: it is an excellent landmark before opening the periorbit, a longitudinal incision is made at the base of the levator palpebrae superioris muscle and completed by two perpendicular contra-incisions, one of which is anterior at the level of the eyeball and the other posterior at the level of the orbital apex. During this stage, any lesions to the trochlear nerve should be avoided as the nerve is situated medial to the frontal nerve, almost on the optic axis, in front of the roof of the optic canal. Orbital fat is usually considered as a dominating element of muscular function. However its coagulation and retraction appear indispensable in order to obtain a satisfactory view of the intra-conical anatomical elements. From the opened periorbit, there are 3 ways in which to penetrate the orbital cone.

The Lateral Approach is the Most Frequent (Fig. 12 a, b). It permits visualisation of lateral lesions to the optic nerve as well as those that spread from the superior orbital fissure to the lateral and apical area of the optic nerve. By pulling up the muscular levator palpebrae superioris/superior rectus muscles complex en bloc and using a siliconed surgical loop, it is possible to access the external part of the optic nerve. Lesions to the superior branch of division of the oculomotor nerve must be avoided whilst innervating the superior rectus muscle, setting the surgical loop sufficiently to the anterior, underneath the muscles. Pulled medially along with the other elements, the superior ophthalmic vein hinders the view of the optic nerve in its apical segment. It can be sacrificed or pushed laterally. The ophthalmic artery and the nasociliary nerve must be identified: they cross the superior side of the optic nerve, laterally to medially. At its medial bend the ophthalmic artery gives rise to the ciliar arteries and the lachrymal artery which is pushed away laterally. Further laterally the abducens nerve runs along the internal side of the lateral rectus muscle. The branch of division of the oculomotor nerve destined for the inferior oblique muscle is situated at the level of the inferolateral side of the optic nerve/ophthalmic artery

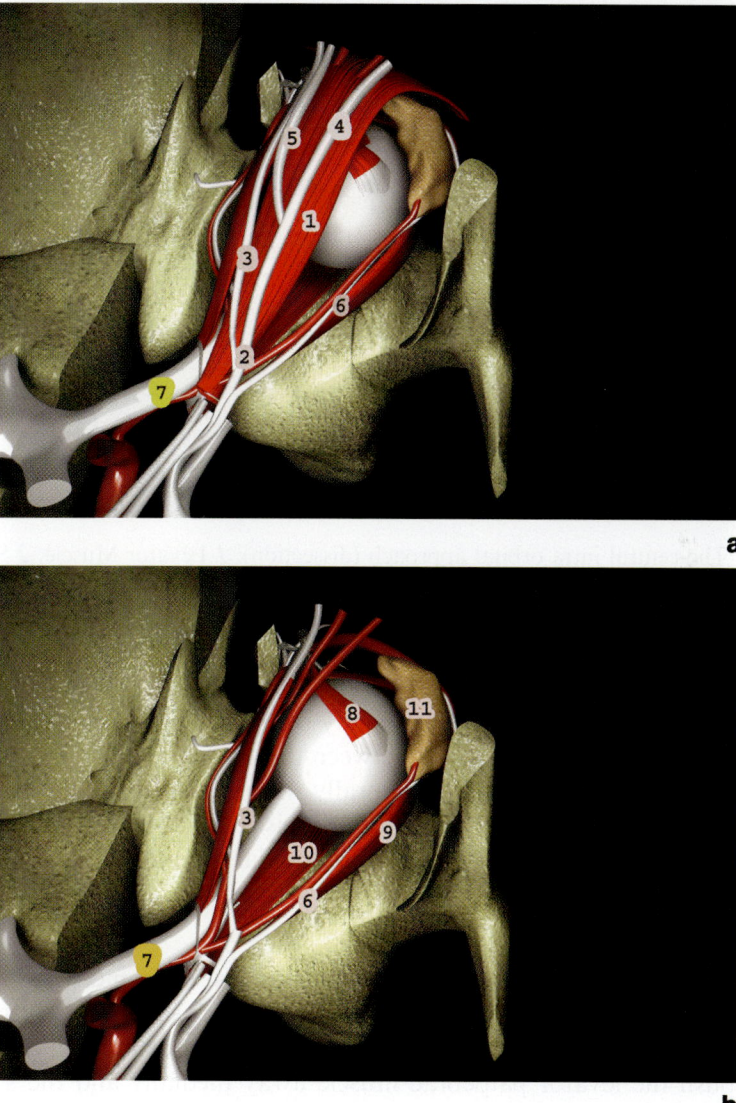

Fig. 12 (a–b): Superior aspect of the right orbit (unroofed) (a). Superior aspect of the orbital apex (The levator palpebrae and superior rectus muscles are reflected) (b). *1* Levator Muscle, *2* Frontal Nerve, *3* Trochlear Nerve, *4* Supra Orbital Nerve, *5* Supra Trochlear Nerve, *6* Lachrymal Nerve, *7* Optic Nerve, *8* Superior Oblique Muscle, *9* Lateral Rectus Muscle, *10* Inferior Rectus Muscle, *11* Lachrymal Gland

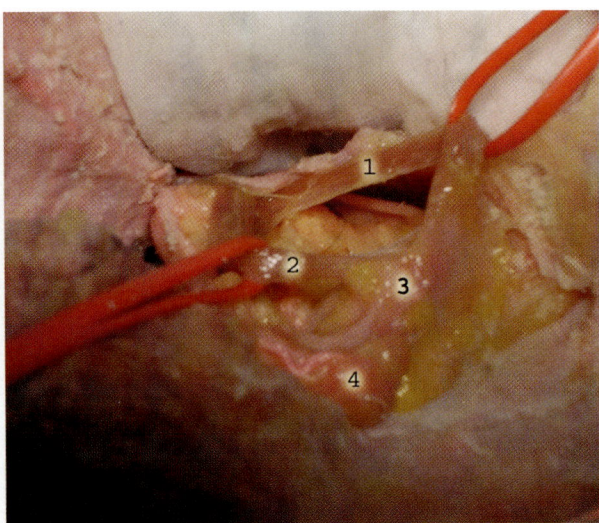

Fig. 13. The central intra orbital approach (dissection): *1* Levator Muscle, *2* Superior
Rectus Muscle, *3* Whitnall's Ligament, *4* Lachrymal Gland

complex. The superior orbital fissure may be opened by making an incision
to the annular tendon, backwards between the superior rectus muscle medially and the lateral rectus muscle laterally. By pushing apart the two muscular bodies at the level of the tendinous incision, the superior branches of
division of the oculomotor nerve, the nasociliary nerve and the abducens
nerve can be seen [20, 63, 64].

Central Approach (Fig. 13). This approach is most often indicated for limited surgery such as biopsy. As the levator palpebrae superioris muscle horizontally overlaps the medial side of the superior rectus muscle, it is possible to push the levator palpebrae muscle away medially and the superior
rectus muscle laterally. Each muscle is retracted by a surgical loop. The
frontal nerve can be kept in position and pushed medially with the levator
palpebrae, or dissected then pushed away laterally with the superior rectus
muscle. This permits a better view of the optic nerve at the level of the
annular tendon. The trochlear nerve, at this level, is in an extra-conical
position, directly in front of the optic nerve sheath and medial to the frontal nerve. It merits particularly careful consideration. This approach permits access to the entire width of the middle intra-orbital third of the optic
nerve. However, it can be deleterious for the branch of division of the oculomotor nerve that innervates the levator palpebrae superioris muscle and
should, therefore be undertaken with extreme care. Underneath this branch

of division that appears in the middle of the operating field of vision, the ophthalmic artery accompanied by the nasociliary nerve crosses the superior side of the optic nerve laterally to medially.

Medial Approach. This approach runs medially between the oblique superior muscle and the levator palpebrae superioris muscle and the superior rectus muscle laterally. The optic nerve is thus exposed across its entire length. The ophthalmic artery appears at the medial side of the optic nerve after crossing it. An incision may be made to the annular tendon towards the rear, between the levator palpebrae superioris muscle and the superior rectus muscle in such a way as to expose the optic nerve, at the level of the apex. During this incision, care must be taken to protect the trochlear nerve because it is in an extra-conical position.

Gliomas of the Optic Nerve

The slow or non-existent evolution of these tumours and the fact that vision is not affected make operations rare. The tumourous extension spreading backwards the chiasm is possible. The evolutive forms must therefore be operated on. After carrying out the sub frontal and intra-conical approach of the orbit the access to the intra-conical part of the orbit takes place between the superior and lateral rectus muscles. It is possible to access the optic nerve without removing the superior orbital margin and the adjacent part of the lateral orbital process. However, this orbitotomy increases the size of the area for using surgical instruments in the orbit and, to an equal degree, reduces the cerebral retraction which still remains epidural. An incision is made in the periorbit at the large axis of the orbit, parallel to the course of the optic nerve. The orbital fat will hinder access to the space between the lateral rectus muscle and the superior rectus muscle/ levator palpebrae superioris complex. No dissection of the superior right muscle from the levator palpebrae superioris muscle must be performed as there is a risk of lesions to the nerves destined to the levator palpebrae muscle. It is necessary to retract the muscular group as a whole medially. The fat is the only remaining obstacle between the periorbit and the optic nerve. The tumour can simply be identified using one's finger before it is dissected and visualised. The fatty orbital body is pushed back medially with the superior muscle group and downwards with the lateral rectus muscle. The optic nerve is cut as closely to the surface as possible to the posterior pole of the eyeball. Dissecting the tumour is then realised backwards, still in the space between the superior muscular group and the external group, as far as the summit of the orbital cone where the optic nerve is cut behind the posterior pole of the tumour, which can then be removed. At this point there is a risk of wounding the ophthalmic artery medial or lateral to the

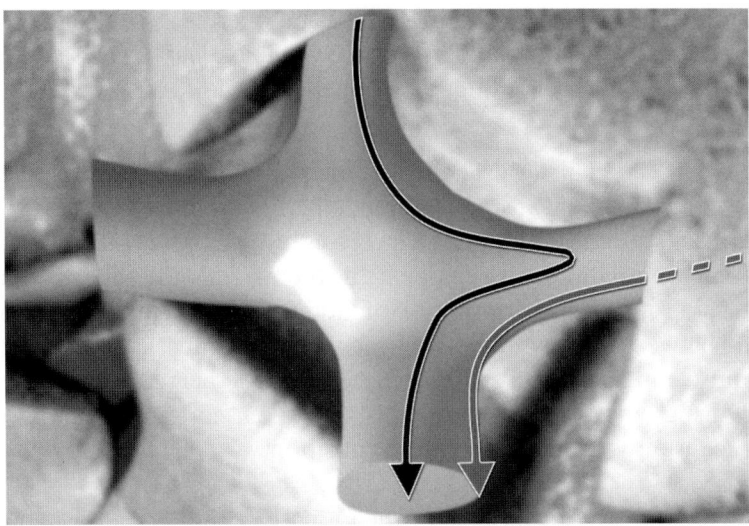

Fig. 14. Wilbrand's knee

optic nerve. This accidental injury is relatively harmless so long as the optic nerve is cut, but the difficulties of haemostasis in the fatty orbital body may lead to manoeuvres that are dangerous for the other nerve elements of the orbit. Furthermore, it is never certain that the incision to the optic nerve is in a safe area, since the incision is made as close to the outer cranial orifice of the optic canal as possible. Even though the tumour is strictly inside the orbit, it is preferable to complete the posterior incision of the intra-orbital optic nerve with a second intra-dural incision of the optic nerve, taking care not to open the optic canal. In this way it is possible to be sure that the second incision is made in a safe area. To do so, it is necessary to open the frontal dura. The frontal lobe must be carefully retracted half-way between a sub frontal and a pterional approach. The optic nerve is identified in the optochiasmatic cistern which is then opened. Resection of the optic nerve is carried out as far away as possible, forwards from the optic chiasm in order to avoid injuries to the nasal axons of the contra lateral optic nerve that have a short, recurrent passage in the optic nerve after crossing in the optic chiasm (Wilbrand's knee) [48, 59] (Fig. 14). The remaining part of the optic nerve in the optic canal is then simply co-agulated *in situ*. This second posterior incision protects the optic chiasm from a hypothetical posterior extension. If the tumourous extension directs itself backwards through the optic canal, it is approached in the same way as a meningioma of the sheath of the optic nerve, in order to resect the part of the tumour that lies in the optic canal Replacing the superior orbital margin and the part of the orbital roof which is adjacent to it, is generally

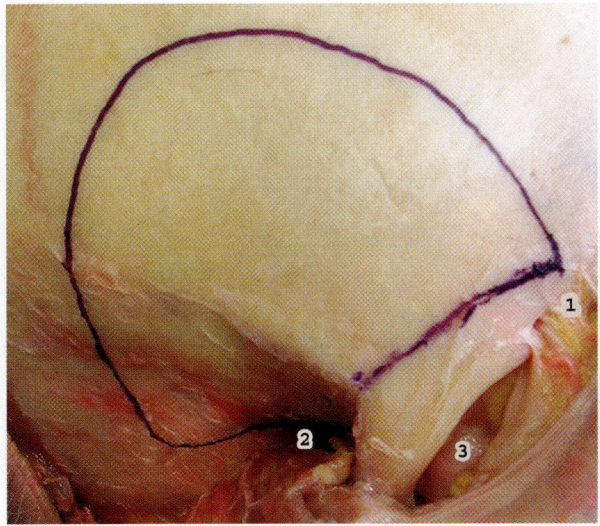

Fig. 15. Drawing of a fronto pterional craniotomy (dissection): *1* Supra Orbital Nerve,
2 Great Wing of the Sphenoid, *3* Lachrymal Gland

sufficient for reconstituting an osseous interface between the orbit and the
dura of the inferior side of the frontal lobe, and does not necessitate a bone
graft.

Frontopterional Approach of the Orbit

This surgical approach is mainly used for tumourous processes or pseudo-
tumourous processes that are particularly extensive, like orbito-sphenoidal
meningioma or fibrous dysplasia. It can be modulated according to the extension of the pathological process. The surgery is performed almost completely in an epidural situation after a wide separation of the dura above
the orbit and of the temporal fossa [18]. The patient is in the supine position, the head in a slight hyper flexion to extend the field of operating
vision at the level of the optic canal and the superior orbital fissure. The
cutaneous incision is performed upwards from the tragus and follows, at
an interval of 2 cm behind the hair implantation line until the frontal
midline. The cutaneous point that corresponds to the pterion should be situated on line which runs from the two extremities of this incision. The skin,
the pericranium and the temporal muscle are unfastened *en bloc* from the
skull in order to avoid injuries to the frontal branch of the facial nerve. The
fronto-pteriono-temporal area is thus exposed (Fig. 15). Some principles
must be kept in mind in order to properly perform the craniotomy: it
must be centred anteriorly by the pterion, and must be wide enough to

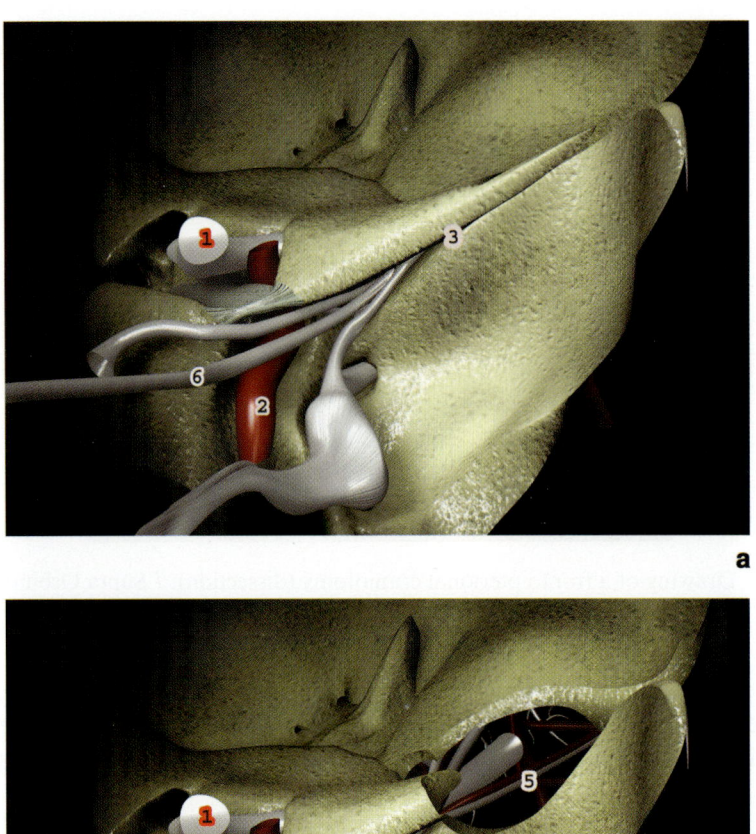

a

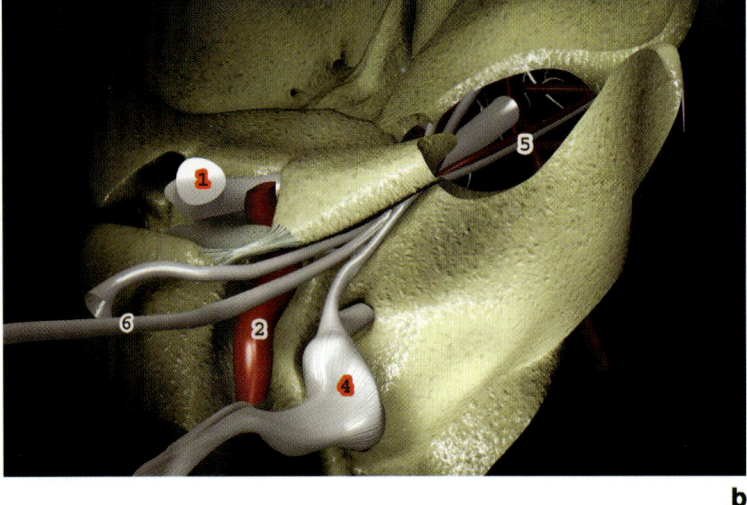

b

Fig. 16. Fronto ptérional approach of the right orbit (posterior and lateral aspects): (a) The pathological targets: the great wing of the sphenoid, the ridge of the lesser wing of the sphenoid, the superior orbital fissure; (b) the superior and lateral orbitotomy runs anteriorly to posteriorly from the pterion and the ridge of the lesser wing of the sphenoid; (c) the superior orbital fissure is opened; (d) the optic canal is opened and partially unroofed. *1* Optic Chiasm, *2* Intra Cavernous Carotid Artery, *3* Lesser Wing of the Sphenoid, *4* Trigeminal Ganglion, *5* Frontal Nerve, *6* Oculomotor Nerve

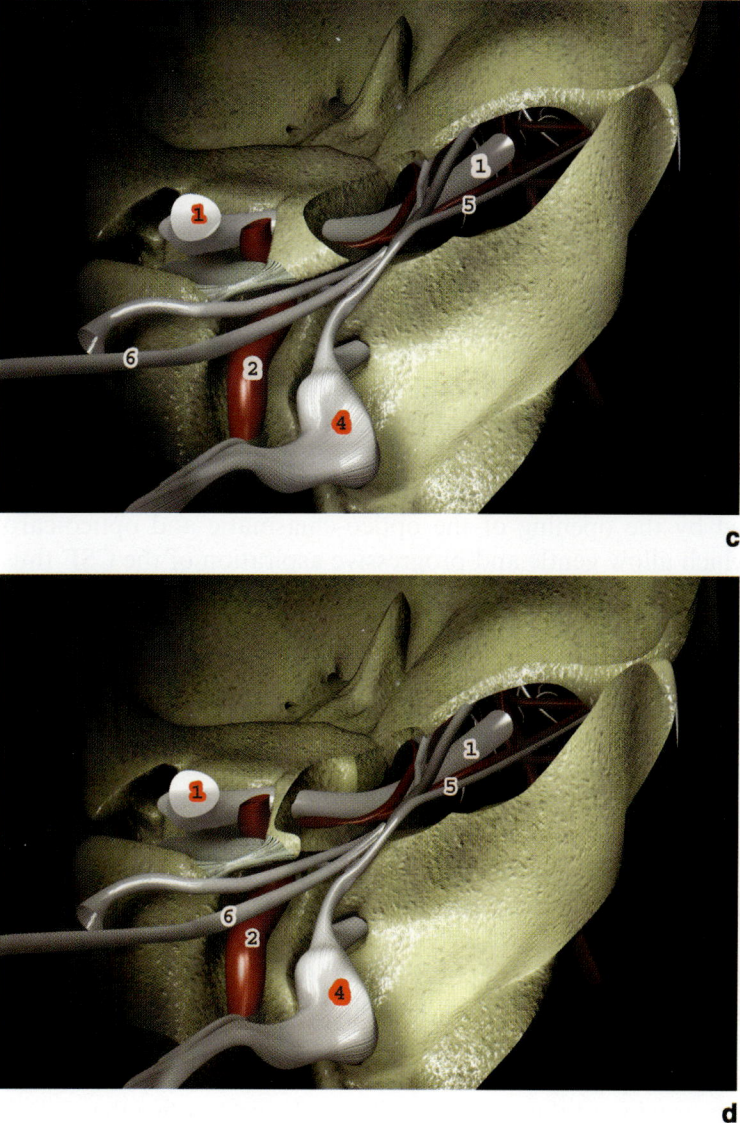

Fig. 16 (*continued*)

permit access to the pathological targets: the great wing of the sphenoid, the ridge of the lesser wing of the sphenoid, the superior orbital fissure (Fig. 16 a), the anterior clinoid process, the optic canal, the dura of the fronto-pterional-temporal area which is involved in the case of orbito-sphenoidal meningioma, opening the frontal sinus should be avoided unless it is particularly pneumatised and partially covers the orbital roof. This is a rare occurrence and the craniotomy in the frontal area has to follow an angle of approximately thirty degrees from the horizontal plane. It is sufficient to expose the orbital roof without opening the frontal sinus. Multiplying the burr holes generally implies aesthetic sequelae. A single posterior burr hole in the hypothetic area the lateral fissure is often sufficient. This necessitates a slow and careful epidural unfastening which can be done with the Gigli's saw guide that permits an adapted length of interposition between the bone and the dura. Dural suspension should only be carried out at the end of the operation in order to avoid interfering with the epidural unfastening on the skull base. Removal of the superior orbital margin and of the zygomatic arch which is often advised [44] does not improve access to the posterior third of the orbit itself. This orbital-zygomatic removal can only be justified in the case of tumourous extensions towards the middle area of the temporal fossa, the petrous portion of the temporal bone and the pterygo-maxillar fossa. This technique does not strictly apply to the orbital approaches as we have defined them. The dura is opened then followed by the opening of the optico-chiasmatic and optico-carotid cisterns which allow gentle and progressive aspiration of the CSF thus reducing the brain volume in order to facilitate the epidural unfastening on the skull base. Often the dura is involved by the meningioma and we prefer to excise the pathologic area and repair it with an epicranial graft before drilling the skull base. The superior and lateral orbitotomy runs anteriorly to posteriorly from the pterion and the ridge of the lesser wing of the sphenoid (Fig. 16 b). It is necessary to begin drilling the superior and lateral wall of the posterior half of the orbit as the periorbit is more fragile at its anterior point. If the periorbit is torn, the gap will cause periorbital fat to escape, thus hindering the posterior field of vision for the surgeon. After having opened the superior orbital fissure (Fig. 16 c) and the optic canal (Fig. 16 d), the superior and lateral orbitotomy is completed forwards and terminates at the frontal and temporal bones. A large orbitotomy allows a partial reduction of the proptosis and the excision of the intra-orbital bud which spreads to the periorbit in an extra-conical position. The superior orbital fissure is opened after having drilled its roof; in other words, the ridge of the lesser wing of the sphenoid. This is the first orifice that is individualised. The anterior clinoid process lies posteriorly, the inferior and medial part of which is supported by the posterior root of the lesser wing of the sphenoid that corresponds to the lateral wall of the optic canal. Removing the clinoid process allows, on the one hand completion of excision when the pathological process concerns the anterior clinoid process and, on the other hand allows access to the optic canal via its external part [9]. Opening the optic canal is recommended as visual acuity will be impaired. The drilling at this point must be carried out using drills of 1 or 2 millimetres. Monitoring the optic nerve is carried out after opening the sheath in the optic canal.

Drilling the lateral part of the orbit terminates at the floor of the orbit, thus exposing the inferior optic fissure that links the inferior and lateral part of the orbit with the infra-temporal fossa on the one hand, and the pterygo-maxillary fossa on the other. Drilling of the temporal fossa is carried out downwards from the inferior orbital fissure and the foramen rotundum and oval posteriorly. It is important to avoid any contact with the periorbit as the orbit's fatty body will infringe onto the area of surgery, thus complicating the drilling operation considerably. If the periorbit is affected, it must be resected as much as possible. The periorbit is opened to the exterior of the frontal nerve that goes above and along the superior muscular complex. At the same level as the superior orbital fissure is the trochlea nerve which is situated medial to the frontal nerve, and is particularly vulnerable at this stage. The frontal nerve, which is a branch of the ophthalmic nerve, partially carries frontal palpebral sensitivity and is not a physiologically determining factor. The pathological process can spread backwards to the cavernous sinus. Its external wall can be peeled and coagulated without risk. A tumourous extension in the lateral wall of the cavernous sinus can be easily removed by excision, without risk. This is quite different to excision that is carried out in the cavernous sinus itself. Intracavernous surgery of meningeal residue is deleterious and this tumourous residue is easily accessible via radiosurgery or conformational radiotherapy. The periorbit is reconstructed using a fragment of pericranial graft sewn onto the periorbit which is considered to be safe. Bone reconstruction of the orbit is increasingly abandoned. As far as we are concerned, the only reconstructive surgery required concerns the defect of the frontotemporal bone defect (Fig. 1 c, d) which can be replaced by acrylic or a bone graft, thus compensating the atrophy of the temporal muscle and not allowing it to herniate through this bone defect which is often the cause of inaesthetic sequelae. The pathological limits are: periorbital, dural or osseous, whether radiological or macroscopic, and often very difficult to determine with precision; under such conditions the term of complete excision should be employed with great care. Intracranial neuronavigation offers a more precise schedule of excision when pre-surgical imaging and planning is carried out, but does not permit the application of a concept of excision that goes beyond extensions to these tumours, whose exact limits are not precise. However, only excision in safe osseous areas can minimise the possibilities of recurrences for the patient [49].

Problems of Orbital Reconstruction

Reconstruction of the orbit after a neurosurgical approach depends on the pathology concerned by the surgery, the extension of the osseous resections and on the nature of the bone which is sometimes affected by dysplasia.

The Orbital Rim

For aesthetic reasons it is always necessary to reconstruct of the orbital rim. Reconstructing the superior orbital and the lateral orbital rims is generally simple and necessitates either steel wire or screwed plates that may or may not be resorptive. This depends on each individual's preferences. In the case of bone defects concerning the orbital margin the same aesthetic reasons necessitate reconstruction. The most reliable method is an autologous graft. The most malleable substance and most suited to the constraints upon reformation of the orbital structure is the ribs, whether split or not. Harvesting is generally carried out before surgery. The number of sites of costal harvesting varies between 1 and 3.

The Orbital Walls (Fig. 8)

Small defects of the orbital roof do not require attention. However, large losses of substances necessitate reconstruction in order to maintain a solid interface between the dura of the frontal lobe and the periorbit. This reconstruction involves the split ribs when their removal has been made necessary in order to reconstruct the orbital structure or splitting the bone flap when removal of the ribs is not required. The bone graft is then simply inlaid or fixed from the superior orbital margin. It is not generally useful to reconstruct the lateral wall of the orbit. Even after complete resection of the tumour, the proptosis does not regress on occasions. This is why removing its external wall from the orbit plays such an important role in orbital decompression, as described in the guidelines for surgery for proptosis of Basedow's disease.

Orbital Dysplasia of the NF1

In the case of an orbital location of Recklinghausen's disease, it manifests itself as a plexiform neuroma and radically alters the social life of affected patients due to its appearance and the functional difficulty that it entails. The objective of therapy is to re-establish the balance between orbital content and container, thus requiring simultaneous surgery to the osseous structures (for example, reducing the increased size of the orbital volume) and the soft tissues (resection of the tumour). It is imperative, in the case of facial symmetrisation, that the volume of the bony orbit is reduced. Excision or iterative resections of tumours that are carried out on the soft tissues appear to be factors of less importance in the short and long term results. On a frontal view, it is necessary to restore the inferior and superior orbital margins, which are thinner than usual and irregular, whereas on a sagittal view is necessary, on the one hand, to reduce the hernia of the frontal or temporal lobe and, on the other hand to set it by reconstructing the dehiscent orbital roof. The orbital roof, which is thinner and perhaps even absent, must also be reconstructed [28]. Indications for this depend on the size of the reduction in volume that is desired, and also on the preservation, or not, of visual function. Enucleation, despite allowing tri-dimensional control of the reconstruction operation, is always seen as a failure. Different approaches of reconstructive surgery have been suggested: the principle remains the same, which, in other words corresponds to a combination of separating the skull from the orbit, reducing the tumourous intra-orbital mass and remodelling the orbital skeleton. The osseous stage required for this operation associates to varying degrees, osteotomy and the on lay of organic material for grafts, or synthetic materials. Reconstruction of the orbit should begin with a transcranial approach [74], although the material to be used is still a subject of discussion [30]. Jackson [34] proceeds with the reduction of the hernia and contention of the temporal lobe by reconstructing the great wing of the sphenoid with the aid of split rib graft. Marchac [45, 46] proposes splitting the frontal bone flap to compensate for the bone dehiscence. Ulterior resorption of osseous grafts amongst this pathological tissue has caused certain authors to prefer this method to using acrylic [50] or titanium [74]. Reduction of the orbital content necessitates partial resection as total resection is rarely possible without inflicting serious functional impairment. Furthermore, in this instance the quality of resection depends entirely on the preservation of visual capacity. In many cases it is limited to tissular structures that are easily accessed by the superior defect. Remodelling the orbital skeleton aims to re-establish the balance between content and container by reducing the volume of the orbital whilst keeping the face symmetrical. This may necessitate osteotomy, but also appositional grafts or osseous substitutes. Associating osteotomy with autologous grafts would appear to yield better results in the long term. Jackson [34, 36] performs symmetrisation of the inferior orbital segment and the internal and lateral walls by means of osteotomy of the entire malar: this is possible by vertically cutting the internal wall of the orbit as far as the inferior orbital fissure. The zygomatic arch is sectioned and a horizontal maxillary osteotomy carried out below the infra-orbital nerve, in order complete the vertical osteotomy. Cutting the orbital floor facilitates the movement of the segment obtained in an upwards and medial direction. Any defect of maxillary bone can be compensated for with bone grafts, and the superior orbital margin benefits from split rib grafts. Munro [53] proposes inversed facial osteotomy in the event of maxillary asymmetry: this surgery combines a hemi-osteotomy of the type Lefort 1, and cranial-facial osteotomy of the type Lefort 3. The fragment that is obtained can be moved via posterior and superior rotation-translation. Facial symmetrisation is completed by bone grafts of iliac or costal origin Marchac [46, 47] completes an

osteotomy of the external orbital margin and repositions the canthal ligaments. He combines this with the installation of osseous grafts to the orbital roof. Remodelling of the soft tissues follows osteotomy because, if
the eyelid is affected this contributes to the imbalance between content
and container. As far as possible, surgery at this stage should respect its
function. This requires transfixing resection of the eyelid on horizontal
and vertical planes, whilst taking care to maintain the functioning of the
levator palpebrae superioris muscle. Marchac [45, 47] carries out resection
of the cutaneous excedent, on the whole thickness by comparing the unaffected side. The levator palpebrae superioris muscle is re-inserted after reduction of its movement on the tarsus. An external canthopexy would conclude surgery. This pathology demands delicate surgery both to the osseous
structures but also to the soft tissues. This is all the more true as surgery
may still engender post-surgical complications [62]. However, it does offer
aesthetic and functional solutions that may be partial and imperfect, but
are desired by patients.

Surgery in the Case of Fibrous Dysplasia (FD)

Modelling and conserving resection in fibrous dysplasia reduces the unde-
sirable aesthetic element, but does not control the potential growth of
the remaining process and the recurrence of deformation. Total or partial
resection of the pathological tissue followed by reconstruction during the
same operating stage would seem preferable [54, 17]. This method is based
on ablation of the entire pathological process by an approach that may be
intracranial when the FD is at the base of the skull, or facial when the FD
has a lower extension. The surgical approach is endo cranial, facial or
combined according to the size of the FD extension at the base of the skull.
Reconstruction necessitates autologous bone grafts [17, 24, 35, 53, 54, 77]
proposes reinstating the pathological tissue after remodelling. Decompression of the optic nerve in the optic canal follows the same principles as for
the epidural approach of the optic canal in the case of orbito-sphenoidal
meningioma. The indication of decompressing the optic canal is given
when there is a reduction in visual acuity due to the optic canal being affected by the FD or when surgery is justified on the grounds of aesthetic
consideration and there is a threat of compression of the optic nerve by
its surrounding bone structures [58]. The epidural and intra-orbital endo
cranial approach yields the best results as it allows decompression of the
whole course of the optic nerve. Moreover, accurate identification of the
different anatomical elements makes this surgical approach safer and more
secure [13]. The main difficulty in this type of surgery is linked to the thickness of the dysplastic bone since the roof of the orbit can, on occasions
reach a thickness of 1 cm.

Conclusions

Neurosurgical approach of orbital tumours concern the posterior two thirds of the orbit. Whether concerning intra- or extra-conical pathologies, it is their direct or indirect repercussion on the optic nerve that, essentially justifies the neurosurgical indication. Any lesion that necessitates an intra-operative access to of the orbit, its contents and its peri-cerebral spaces should be approached by using one of the neurosurgical techniques that we have discussed. The current methods of diagnosis by imaging should make it possible to foresee the degree, and succession of surgery required. As a result of these findings, a pluridisciplinary approach is organized and the specific therapeutic orientations of each tumour can be considered.

Acknowledgements

We thank Professor Marc Braun of the University of Nancy (France) for allowing realisation of the dissections displayed in this chapter.

References

1. Adegbite AB, Kahn MI, Paine KWE, Tan LK (1983) The recurrence of intracranial meningiomas after surgical treatment. J Neurosurg 58: 51–56
2. Al Mefty O, Fox JL (1985) Supero-lateral orbital exposure and reconstruction. Surg Neurol 23: 609–613
3. Andrews DW, Faroozan R, Yang BP, Hudes RS, Werner-Wasik M, Kim SM, Sergott RC, Savino PJ, Shields J, Shields C, Downes MB, Simeone FA, Goldman HW, Curran WJ (2002) Fractionated stereotactic radiotherapy for the treatment of optic nerve sheath meningiomas: preliminary observations of 33 optic nerves in 30 patients with historical comparison to observation with or without prior surgery. Neurosurgery 51: 890–904
4. Benedict WL (1950) Diseases of the orbit. Amer J Ophthal 33: 1–10
5. Bilaniuk LT (1999) Orbital vascular lesions. The radiologic clinics of North-America (Imaging in Ophtalmology II) 37: 169–183
6. Binet EF, Kieffer SA, Martin SH, Peterson HO (1969) Orbital dysplasia in neurofibromatosis. Radiology 93:829–833
7. Blinkov SM, Gabibov GA, Tcherekayev VA (1986) Transcranial surgical approaches to the orbital part of the optic nerve: an anatomical study. J Neurosurg 65: 44–47
8. Brihaye J, Hoffmann GR, François J (1968) Les exophtalmies neurochirurgicales. Neurochirurgie 3: 13–485 (in French)
9. Brotchi J, Bonnal JP (1991) Lateral and middle sphenoid wing meningiomas. In: Al Mefty O (ed) Meningiomas. Raven Press, New York
10. Burrows EH (1978) Orbitocranial asymetry. Br J Radiol 51:771–781
11. Carroll GS (1999) Peripheral Nerve Tumors of the orbit. The radiologic clinics of north-america (Imaging in Ophtalmology II) 37: 195–202

12. Chen YR, Breidahl A, Chang CN (1997) Optic nerve decompression in fibrous dysplasia: indications, efficacy, and safety. Plast Reconstr Surg 99: 22–30
13. Chen YR, Kao CC (2000) Craniofacial Tumors and Fibrous Dysplasia. Mosby, St Louis
14. Cruveilhier J (1835) Anatomie Pathologique du Corps Humain. JB Balliere, Paris, pp 252, 464, 566 (in French)
15. Cushing H, Eisenhard TL (1938) Meningiomas: their classification, regional behaviour, life history and surgical end results. Charles C Thomas, Springfield, pp 250–282
16. Delfini R, Missori P, Tarantino R, Ciappetta P, Cantore R (1996) Primary benign tumors of the orbital cavity: comparative data in a series of patients with optic nerve glioma, sheath meningioma, or neurinoma. Surg Neurol 45: 147–154
17. Derome P (1972) Spheno-ethmoidal tumors. Possibilities for exeresis and surgical repair. Neurochirurgie 18: 1–164 (in French)
18. Derome P, Visot A (1991) Bony reaction and invasion in meningiomas. In: Al Mefty O (ed) Meningiomas. Raven Press, New York
19. Desjardins L (2000) Ophtalmological tumours in children: diagnosis and therapeutic strategy. J Fr Ophtalm 23: 926–939 (in French)
20. Ducasse A (1999) Anatomy of the Orbit. In: Adenis JP, Morax S (Ed) Pathology of the orbit and the eyelids. Masson. Paris, pp 38–68 (in French)
21. Duffier JL and Al (1986) Orbital tumours in children. Arch Fr Pediatr 43: 133–139 (in French)
22. Durante F (1887) Contribution to endocranial surgery. Lancet 654–655
23. Dutton JJ (1992) Optic nerve sheath meningiomas. Surv Ophthalmol 37: 167–183
24. Edgerton MT, Persing JA, Jane JA (1985) The surgical treatment of fibrous dysplasia. With emphasis on recent contributions from cranio-maxillo-facial surgery. Ann Surg 202: 459–479
25. Enringen FM, Weiss FW (1995) Soft tissue tumors, 3rd edn. Mosby, St Louis, pp 821–881
26. Frazier CH (1913) An approach to the hypophysis through the anterior canial fossa. Ann Surg 145–152
27. Gaillard S, Lejeune JP, Pellerin P, Pertuzon B, Dhellemmes P, Christiaens JL (1995) Résultats à long terme du traitement chirurgical des osteo-méningiomes sphéno-orbitaires. Neurochirurgie 41:391–397 (in French)
28. Gurland JE, Tenner M, Hornblass A, Wolintz AH (1976) Orbital neurofibromatosis: involvement of the orbital floor. Arch Ophthalmol 94: 1723–1725
29. Hamby WB (1964) Pterional approach to the orbit for decompression or tumor removal. J Neurosurg 21: 15–18
30. Havlik RJ, Boaz J (1998) Cranio-orbital-temporal neurofibromatosis: are we treating the whole problem? J Craniofac Surg 9:529–535
31. Hollander MD et al (1999) Optic gliomas. The radiologic clinics of North-America (Imaging in Ophtalmology II) 37: 59–71
32. Hudson AC (1912) Primary tumors of the optic nerve. Rev Ophthalmo Hosp Rep 18: 317–439

33. Huson SM, Hughes RAC (1994) The neurofibromatoses. A pathogenetic and clinical overview. Chapman & Hall, London
34. Jackson IT, Laws ER, Jr, Martin RD (1983) The surgical management of orbital neurofibromatosis. Plast Reconstr Surg 71:751–758
35. Jackson IT, Webster HR (1994) Craniofacial tumors. Clin Plast Surg 21: 633–648
36. Jackson IT (2001) Management of craniofacial neurofibromatosis. Facial Plast Surg Clin North Am 9: 59–75
37. Jacquemin C, Bosley TM, Liu D, Svedberg H, Buhaliqa A (2002) Reassessment of sphenoid dysplasia associated with neurofibromatosis type 1. Am J Neuroradiol 23:644–648
38. Jacquemin C, Bosley TM, Svedberg H (2003) Orbit deformities in craniofacial neurofibromatosis type 1. Am J Neuroradiol 24:1678–1682
39. Jakobiec FA, Font RL (1986) Ophthalmic pathology. WB Saunders, Philadelphia, pp 2603–2632
40. Kazim M (1992) Orbital Lymphangioma. Ophthalmology 99: 1588–1594
41. Krohel GB, Rosenberg PN, Wright JE, Smith RS (1985) Localized orbital neurofibromas. Am J Ophthalmol 100: 458–464
42. Li KC, Poon PY, Hinton et al (1984) MR imaging of orbital tumors with CT and ultrasounds correlations. J Comput Assist Tomogr 8: 1039–1047
43. Macarez R, Bazin S, Civit T, Grubain S, De La Marnierre E, Tran Huu D, Guigon B (2003) Récupération fonctionnelle après chirurgie d'exérèse d'un méningiome sphéno-orbitaire: à propos d'une observation. J Fr Ophtalmol 26: 375–380 (in French)
44. Mac Dermott MW, Durity FA, Rootman J, Woodhurst WB (1990) Combined fronto-temporal-orbitozygomatic approach for tumors of the sphenoid wing and orbit. Neurosurgery 26: 107–116
45. Marchac D, Cophignon J, Clay C (1975) Subtotal reconstruction of the orbit after destruction by benign tumors. Mod Probl Ophtalmol 14: 541–544
46. Marchac D (1984) Intracranial enlargement of the orbital cavity and palpebral remodeling for orbitopalpebral neurofibromatosis. Plast Reconstr Surg 73: 534–543
47. Marchac D, Renier D, Dufier JL, Desjardins L (1984) Orbitopalpebral neurofibromatosis: orbital enlargement by an intracranial approach and palpebral correction. J Fr Ophtalmol 7: 469–478
48. Martin HM (1996) Neuroanatomy. Mc Graw-Hill, p 170
49. Maroon JC, Kennerdell JS, Vidovich DV, Abla A, Sternau L (1994) Recurrent spheno-orbital meningioma. J Neurosurg 80: 202–208
50. Mayer R, Brihaye J, Brihaye-van Geertruyden M, Noterman J, Schrooyen A (1975) Reconstruction of the orbital roof by acrylic prosthesis. Mod Probl Ophtalmol 14: 506–509
51. Mirimanoff RO, Dosoretz DE, Linggood RM, Ojemann RG, Martuza RL (1985) Meningioma: analysis of recurrence and progression following neurosurgical resection. J Neurosurg 62: 18–24
52. Mortada A (1977) Neurofibromatosis of lid and orbit in early childhood. J Pediatr Ophthalmol 14: 148–150

53. Munro IR, Martin RD (1980) The management of gigantic benign cranio-facial tumors: the reverse facial osteotomy. Plast Reconstr Surg 65: 776–785
54. Munro IR, Chen YR (1981) Radical treatment for fronto-orbital fibrous dysplasia: the chain-link fence. Plast Reconstr Surg 67: 719–730
55. Naffziger HC (1948) Exophtalmos: some principles of surgical management from the neurosurgical aspect. Am J Surg 75: 25–31
56. Newman SA, Jane JA (1991) Meningiomas of the optic nerve, orbit, and anterior visual pathways. In: Al Mefty O (ed) Meningiomas. Raven Press, New York
57. Ngohou S, Herdan ML (1998) In: Adenis JP, Morax S (ed) Pathology of the orbit and the eyelids. Masson, Paris, pp 591–597 (in French)
58. Papay FA, Morales L, Flaharty P, Smith SJ, Anderson R, JM WA, et al (1995) Optic nerve decompression in cranial base fibrous dysplasia. J Craniofac Surg 6: 5–10
59. Patten H (1977) Neurological differential diagnosis. Springer, Berlin Heidelberg New York Tokyo
60. Perez Moreiras JV, Prada Sanchez MC (1998) In: Adenis JP, Morax S (ed) Pathology of the Orbit and the Eyelids. Masson, Paris, pp 523–525 (in French)
61. Peter JC (1999) In: Choux M, Di Rocco, Hockley A, Walker M (eds) Pediatric Neurosurgery. Churchill, Livingstone, pp 561–580
62. Poole MD (1989) Experiences in the surgical treatment of cranio-orbital neurofibromatosis. Br J Plast Surg 42: 155–162
63. Rhoton AL, Natori Y (1996) The orbit and sellar region. Microsurgical anatomy and operative approaches. Thieme, New-York Stuttgart, pp 4–121
64. Rhoton AL, Natori Y (1996) The orbit and sellar region. Microsurgical anatomy and operative approaches. Thieme, New-York Stuttgart, pp 208–236
65. Rootman J, Robertson WD (1988) In: Rootman J (ed) Diseases of the orbit. Lippincott, Philadelphia, pp 310–325
66. Rovit RL, Sosman MC (1960) Hemicranial aplasia with pulsating exophthalmos. An unusual manifestation of von Recklinghausen's disease. J Neurosurg 17: 104–121
67. Scarpa A (1816) Trattato delle Principali Maletties degli Occhi. Edn Quinta Pavia Pietro Bizzoni, pp 507–509
68. Schick U, Hassler W (2003) Pediatric tumors of the orbit and optic pathway. Pediatr Neurosurg 38: 113–121
69. Schick U, Dott U, Hassler W (2003) Surgical treatment of orbital cavernomas. Surg Neurol 60: 234–244
70. Schick U, Blayen J, Hassler W (2003) Treatment of orbital schwannomas and neurofibromas. Br J Neurosurg 17: 541–545
71. Schick U, Dott U, Hassler W (2004) Surgical management of meningiomas involving the optic nerve sheath. J Neurosurg 101: 951–959
72. Shanno G, Maus M, Bilyk et al (2001) Image-guided transorbital roof craniotomy via a suprabrow approach: a surgical serie of 72 patients. Neurosurgery 48: 559–567

73. Shermann RP (1984) Orbital dermoids: clinical presentation and management. Br J Ophthalmol 68: 642–652
74. Snyder BJ, Hanieh A, Trott JA, David DJ (1998) Transcranial correction of orbital neurofibromatosis. Plast Reconstr Surg 102: 633–642
75. Stricker M, Van Der Meulen J, Raphael B, Mazzola R (1990) Craniofacial malformations. Livingstone, Edinburgh
76. Tanaka Y, Tajima S, Maejima S, Umebayashi M (1993) Craniofacial fibrous dysplasia showing marked involution postoperatively. Ann Plast Surg 30: 71–76
77. Tessier P (1977) Reconstitution de l'Etage Antérieur de la Base du Crâne dans les tumeurs Bénignes Orbito-Craniennes. In: Rougier J, Tessier P, Hervouet F, Woillez M, Lekieffre M, Derome P (eds) Chirurgie Plastique Orbito-Palpebrale. Masson, Paris (in French)
78. Valvassori GE *et al* (1999) Imaging of the lymphoproliferative disorders. The radiologic clinics of North-America (Imaging in Ophtalmology II) 37: 135–150
79. Von Graefe A (1864) Zur Casuistik der Tumoren. Arc Ophtalmol 10:176–200 (in German)
80. Von Recklinghausen FD (1891) Die fibröse oder deformirende Osteitis, die Ostomalacie und die osteoplastische Carcinose in ihren gegenseitigen Beziehungen. Virchow Berlin (in German)

Endoscopic Third Ventriculostomy in the Treatment of Hydrocephalus in Pediatric Patients

C. Di Rocco[1], G. Cinalli[2], L. Massimi[1], P. Spennato[2], E. Cianciulli[3], and G. Tamburrini[1]

[1]Pediatric Neurosurgical Unit, Catholic University Medical School, Rome, Italy
[2]Neuroendoscopy Unit, Department of Pediatric Neurosurgery, Santobono-Pausilipon Children's Hospital, Naples, Italy
[3]Department of Pediatric Neuroradiology, Santobono-Pausilipon Children's Hospital, Naples, Italy

With 28 Figures

Contents

Abstract

Advances in surgical instrumentation and technique have lead to an extensive use of endoscopic third ventriculostomy in the management of pediatric hydrocephalus. The aim of this work was to point out the leading aspects related to this technique. After a review of the history, which is now almost one century last, the analysis of the endoscopic ventricular anatomy is aimed to detail normal findings and possible anatomic variations which might influence the correct conclusion of the procedure. The overview of modern endoscopic instrumentation helps to understand the technical improvements that have contributed to significantly reduce the operative invasiveness. Indications are analysed from a pathogenetic standpoint with the intent to better understand the results reported in the literature. A further part of the paper is dedicated to the neuroradiological and clinical means of outcome evaluation, which are still a matter of debate. Finally a review of transient and permanent surgical complications is performed looking at their occurrence in different hydrocephalus etiologies.

Keywords: Hydrocephalus; endoscopic third ventriculostomy; pediatric age.

Historical Background

The increasing diffusion of endoscopy in the surgical practice is one of the most impressive results of the continuous search of methodologies aimed at decreasing patient's discomfort while maintaining a therapeutic efficacy comparable to that of traditional approaches requiring more invasive procedures. Its progressive wider use is supported by a constantly developing technology. Actually, endoscopic third-ventriculostomy (ETV) has become the routine treatment for obstructive hydrocephalus in many neurosurgical centers and neuroendoscopic procedures are more and more employed for both diagnostic and therapeutic purposes in various pathological conditions. In spite of its wide diffusion in recent years, neuroendoscopy is, however, an old technique.

The concept of internal visualization of the human cavities through natural orifices or small wounds was introduced by Bozzini in 1806 (Bozzini 1806). He carried out the first endoscopic procedure with directed light by using candlelight and a series of mirrors placed at an angle of 45°. Bozzini applied the methodology to the study of the urethra and the rectum. His experience was inherited by other authors during the century who attempted to improve the technique, especially the quality of the light source. The year 1879 was fundamental for endoscopy: Thomas Edison invented the electricity bulb and Nitze created a cystoscope to remove bladder stones (Nitze 1879). This last instrument was widely used and modified for the different human cavities and, thanks to the advances in anesthesiology and in the medical care, actually opened the era of endoscopy. The development of the computer chip TV camera during the Eighties finally ratified the beginning of the modern endoscopic surgery (Stellato 1992).

Neuroendoscopy originated during the first years of the last century with the aim to find an effective treatment for hydrocephalus. Its history started in 1910 when Lespinasse, an urologist, used a cystoscope to explore the lateral ventricles of two hydrocephalic children in order to coagulate their choroid plexus (Lespinasse 1910). After a few years, Dandy (Dandy 1918) reported the avulsion of the choroid plexus in five hydrocephalic children (four of them died during the operation) under direct cystoscopic visualization. He made use of rigid Kelly's cystoscope and alligator forceps, besides headlight and transillumination of the heads of his patients as light source; he called his instrumentation "ventriculoscope". In 1922 he proposed also a subfrontal approach to open the floor of the third ventricle by sacrificing an optic nerve (Dandy 1922). One year later, Mixter (Mixter 1923) synthesized the ideas of Dandy in a single procedure and performed a third-ventriculo-cisternostomy using an urethroscope introduced through the anterior fontanel and a flexible probe in a 9-months-old child with non-communicating hydrocephalus. The presence of contrast dye (previously injected in the lateral ventricle) in the lumbar subarachnoid space, demonstrated the success of the first ETV ever realized. In the same year, Fay and Grant (Fay and Grante 1923) were able to take the first endoscopic photographs of the cerebral ventricles. In the following years the efforts of the scientists were addressed towards the improvement of the endoscopic techniques (Putnan 1943). In fact, although the procedures were correctly carried out and based on a sound theory, the long-term results were not rewarding yet and the morbidity and mortality rate unacceptable. Poorly design of the instruments and optical apparatus were the main causes of the disappointing outcomes. For such a reason, Dandy, the "father of neuroendoscopy", was compelled to abandon ventriculoscopy after his first attempts and devoted himself to the development of ventriculographic techniques and direct craniotomic approaches. Other authors, on the other hand, looked for different surgical approaches (for example, stereotaxy) for the treatment of hydrocephalus, so that endoscopic neurosurgery never achieved a widespread popularity. The interest in neuroendoscopy further declined during the second half of the 20th century after the introduction of low morbidity/mortality procedures for the implantation of CSF shunting devices.

Several factors have contributed to the renaissance of the neuroendoscopy in the last two decades. Among them, the pioneer work of Bosma, who applied a 8-mm film registration in his interventions, and the introduction by Harold Hopkins of a solid-rod lens system during 1960s. The system, which still represents the base of the current nonflexible endoscopes, was further enhanced by Guiot who introduced the solid quartz rod lenses, with their internal reflective properties. These innovations were the base for the following development of modern instruments, as, for example, the ductile Fukishima's ventriculofiberscope introduced in 1973 (Fukushima et al. 1973). The advent of microsurgery initially diverted neurosurgeons' interest from endoscopic neurosurgery, but afterwards made the neurosurgeons more confident with the neuro-microanatomy and more conscious about the potential applications of the neuroendoscopy. The development of the computerized technology, either for diagnostic or therapeutic purposes, stimulated a renewed interest towards a technique which could represent a valid alternative to the CSF shunt devices and their excessively high risk of infective complications and mechanical malfunctions. The first modern and important clinical experience with ETV in the management of hydrocephalus was reported by Vries in 1978 (Vries 1978). In 1990, Jones et al. (Jones et al. 1990) reported about the possibility to manage different types of noncommunicating hydrocephalus. Their work became a milestone for the indications and the evaluation of the results after ETV. An increasing number of reports on large series of hydrocephalic patients treated by ETV, which appeared in the literature in recent years, justifies the increasingly wider use of the technique throughout the world. Many innovative clinical and experimental studies concerning the procedure itself, the instrumentation and the possible integration with other techniques represents the last steps in the diffusion of ETV (Burtscher et al. 2002, Broggi et al. 2000, Decq et al. 2000, Foroutan et al. 1998, Horowitz et al. 2003, Vandertop et al. 1998).

Thanks to the improvement in the video imaging and in the endoscopic instrumentation, the current use of the neuroendoscope is not limited only to the treatment of hydrocephalus. Though intraventricular surgery still remains its main field of application (third-ventriculostomy, aqueductal plasty, septostomy of the septum pellucidum, choroid plexectomy, tumor biopsy, arachnoid or colloid cysts marsupialization,...), endoscopy or the endoscopic-assisted microsurgery is more and more used in the management of intra-axial lesions and for skull base, spinal and peripheral nerves surgery. It is likely that neuroendoscopy will be further enriched by the advances of three-dimensional imaging, image fusion, surgical armamentarium, and telepresence surgery in the very near future (Frazee and Shah 1998).

Ventricular Anatomy

A correct preoperative study of individual ventricular anatomy and the endoscopic recognition of ventricular structures is mandatory to increase the success rate of ETV and to decrease the percentage of operative complications. Anatomic anomalies and morphological variants of the cerebral ventricular system have been indeed extensively described, most of them in relation with the etiology of the hydrocephalus and the patient's age; their potentially adverse effects on the surgical procedure have been pointed out (Rohde and Gilsbach 2000, Rohde *et al.* 2001). Preoperatively, modern Magnetic Resonance (MR) apparatuses allow the 3D anatomic reconstruction and measurement of linear distances in any chosen image plane, with the possibility to evaluate anatomical characteristics in the single patient (Duffner *et al.* 2003, Rohde *et al.* 2001). Virtual neuroendoscopic planning, based on MR, 3D ultrasonography and neuronavigation systems have been also recently claimed to improve the safety of the endoscopic procedures (Duffner *et al.* 2003, Riegel *et al.* 2001). Furthermore a wealth of information comes from the improvement of intraoperative digital imaging (Decq 2004, Grant 1998, Kamikawa *et al.* 2001c, Lang 1992, Longatti 2003, Oka *et al.* 1993a, Pavez Salinas 2004, Riegel *et al.* 2001, Rohde and Gilsbach 2000, van Aalst *et al.* 2002). More precise endoscopic view reports are indeed extremely helpful, as they increase our knowledge of the anatomical variations that can be found in hydrocephalus of different etiologies.

Preoperative Evaluation of Ventricular Anatomy

It is almost universally accepted that whenever an ETV procedure is planned a MR study of the brain should be performed. Basic MR studies allow to obtain an overall view of ventricular structures and their anatomical relationships, width of the cerebral mantle, size of the basal subarachnoid spaces, and basilar artery position. Three-D reconstructions and distances measurements can be subsequently obtained and used in the preoperative planning.

In such a regard, the recently published study of Duffner *et al.* (Duffner *et al.* 2003) is of particular importance. The authors compared preoperative MR findings of thirty patients with a diagnosis of obstructive hydrocephalus and thirty healthy volunteers. After acquisition the images were analyzed with a software that enabled the visualization of the three scanning planes (sagittal, axial and coronal) through any free chosen point in the image volume, the definition of oblique planes displaying two structures of interest a time, and the measurement of angles and distances within each of the selected planes. Significant anatomic differences were found between the two groups; in particular lateral ventricles height was 2.08 times higher in the hydrocephalic patients so as third ventricle width (4.39 times larger in the hydrocephalic group). The mean distance between anterior and posterior commissures was 1.19 times longer in patients than in volunteers and the distance between the ventricular system and the cortical surface was significantly higher in this latter group; moreover the mean size of the Monro foramina was about 20 times the size in hydrocephalic patients if compared with normal individuals and it was larger than 5×5 mm in 24 of them. The position of an optimally located burr hole for third ventriculostomy was also calculated and was found to vary significantly between different patients. In an anterior-posterior direction it varied between 16.1 mm in front of and 46.5 mm behind the coronal suture, with a mean value of 8.2 mm behind the coronal suture, that is about 2 cm posterior to the point suggested by most authors (1 cm anterior to the coronal suture). The study provided two further interesting conclusions: the first is that a rigid endoscope used for ETV should not exceed an external diameter of 5 mm and the second is that they should be longer than 120 mm to allow a safe access to the floor of the third ventricle.

Another useful application of 3D MR reconstructions is related with virtual neuroendoscopic procedures. Virtual MR neuroendoscopy has been introduced in the clinical practice in late '90s. Brutscher and co-workers (Brutscher *et al.* 1999) produced virtual endoscopic images of 5 non-hydrocephalic brain specimens and compared the obtained images with intraventricular endoscopic views. The foramen of Monro, fornix, choroid plexus, clivus, mammillary bodies and basilar artery could be virtually visualized and the images obtained were comparable to the actual views. Similar results were obtained by Auer and Auer (Auer and Auer 1998), who simulated several approaches to the ventricular system by virtual MR endoscopy in healthy volunteers as well as in patients with hydrocephalus. Rohde *et al.* (Rohde *et al.* 2001) analyzed the sensitivity of virtual MRI endoscopy in detecting anatomic variations that could be found at surgery. Seven anomalies of the normal ventricular anatomy were encountered during ETV in 5 of 18 patients; five of the seven anomalies had been already identified by virtual MR with an overall sensitivity of 71%. All the missed information concerned anatomical variations of the third ventricular floor. This anatomic structure invariably appeared as a defect when it was translucent so as when it was thick and opaque or steeply inclined. According to the authors, the advantage of this virtual endoscopy of the third ventricle region is that the surgeon can "look through" the third ventricular floor onto the first segment of the posterior cerebral arteries and onto the basilar artery tip and relate these data with the planned surgical approach. However, other authors have underlined the difficulties to study the anatomic relationships of the basilar bifurcation with the designated site of ventriculostomy due to the lack of segmentation between cerebral vessels and brain tissue (Jodicke *et al.* 2003). On these grounds alternative devices for virtual neuroendoscopy have been proposed such as, for example, 3D ultrasonography (3D-US). This examination has the advantage of an high resolution for ventricular pathologies and can be performed on a routine basis without the need of sedation. Jodicke *et al.* (Jodicke *et al.* 2003) recently evaluated the sensitivity of 3D US-based virtual neuroendoscopy in the identification of parenchymal and vascular anatomical landmarks of the third ventricle. A software able to reconstruct sequential 2D images in a 3D mathematical model was used and a power-doppler mode was employed to depict vessels in relation to ventricular walls. In the authors experience, the definition of ventricular and vascular structures position was comparable to virtual MR neuroendoscopy. One main advantage of 3D US-based over MR virtual neuroendoscopy was the coregistration of parenchymal ventricular and vascular anatomy with one single image acquisition due to flow detection and coding using the sensitive Doppler properties of ultrasonography. On the other side, the non vascular anatomical arrangement of the basal cisterns cannot be studied on ultrasonography images, due to reflexion artifacts from the bony clivus and the narrow space of the basal cisterns with multiple anatomical borders; for this reason the same authors maintained as essential a preoperative MR study.

Neuroendoscopic Ventricular Anatomy

The majority of ETV procedures are performed in patients with noticeable dilatation of the ventricular system. It should be therefore considered that the endoscopic anatomic views correspond to this kind of situation, the anatomic structures being often displaced and separated from each other. Regarding the cranial surface parameters most of the authors agree that the burr hole for ETV should be placed immediately (up to 1 cm) anterior to the coronal suture, 2–3 cm. from the midline, on the mid-pupillary line (Decq 2004). The distance between the cortical surface and the frontal horn of the lateral ventricle is extremely variable and should be individually calculated. In the previously quoted paper by Duffner *et al.* (Duffner *et al.* 2003), this distance varied from 5.4 to 34.6 mm.

Anatomy of the Frontal Horn of the Lateral Ventricles and of the Foramen of Monro; Key-Points for Endoscopic Orientation

The frontal horn of the lateral ventricle is delimited by a medial wall, formed by the septum pellucidum, an anterior wall and roof, formed by the genu of the corpus callosum, a lateral wall, composed of the head of the caudate nucleus, and a narrow floor formed by the rostrum of the corpus callosum (Rhoton 2002). Most of the endoscopic orientation inside the frontal horn is based on the visualization of the Monro foramen (MF). This structure is bound anteriorly and superiorly by the column of the fornix and medially by the interventricular septum. The anterior septal vein is visible along the septum and crosses the fornix column. The thalamus appears as a bulging on the posterolateral margin of the MF. The thalamostriate vein passes backward at its boundary with the caudate nucleus, turns medially and empties into the internal cerebral vein. The choroid plexus constitutes the posterior wall of the MF; it is attached medially to the fornix by the tenia fornix and laterally to the thalamus by the tenia thalami. It travels over the superior surface of the thalamus, either in a straight line or with a sinuous course. Reflecting posteriorly it contributes to the formation of the roof of the third ventricle. The posterior and medial margin of the MF is also composed of the angle of anastomosis of the anterior septal vein, choroidal veins (rarely visible within the choroid plexus) and thalamostriatal veins (Rhoton 2002). The Y-shaped angle between these veins may be extremely variable, usually being 80–90°. The veins usually have the same diameter, but in some case either the thalamostriatal vein, or the anterior septal vein is much larger. Looking laterally to the thalamostriate vein several affluent venous branches may be recognized, draining the anterior part of the caudate nucleus. The consequent striped appearance has lead to the name of "striatum" for this anatomic structure (Decq 2004).

Anatomy and Endoscopic View of the Third Ventricle

Once separated from cerebral hemispheres and viewed from inside the lateral ventricles, the third ventricle has nearly a prismatic shape in which we can distinguish a roof, a floor, an anterior, a posterior and two lateral walls. The roof forms a gentle upward arch, extending from the foramen of Monro anteriorly to the suprapineal recess posteriorly. It is composed by four layers: one neural layer formed by the fornix, two thin membranous layers of tela choroidea and a layer of blood vessels between the sheets of tela choroidea. During endoscopic procedures the roof of the third ventricle can almost only be seen from above when there is a partial or complete agenesia of the septum pellucidum; it appears as a thin vascularized triangular membrane peripherically bounded by the two columns of the fornix (Decq 2004, Grant 1998, Rhoton 2002). The floor of the third ventricle extends from the optic chiasm anteriorly to the orifice of the aqueduct of Sylvius posteriorly. When viewed from inferiorly, the structures forming the floor include, from anterior to posterior the optic chiasm, the infundibulum of the hypothalamus, the tuber cinereum, the mammillary bodies, the posterior perforated substance and most posteriorly the part of the tegmentum of the midbrain located above the medial aspect of the cerebral peduncles. The optic chiasm is located at the junction of the floor and the anterior wall of the third ventricle, the inferior surface forming the anterior part of the floor, the superior surface constituting the lower part of the anterior wall. During endoscopic procedures the optic chiasm is viewed as a prominence at the anterior margin of the floor. Immediately behind it and inferiorly a graysh hole, circumscribed by a pink anular ring represent the infundibular recess. The thin whitish parenchimatous structure which can be visible at the base of the infundibulum is the tuber cinereum (Vinas *et al.* 1996). One of the most important reference points inside the third ventricle are the mammillary bodies, which form paired prominences on the inner surface of the floor. Commonly, they form a narrow angle, but they can be widely separated from each other and occasionally they are not clearly recognizable (Decq 2004). Just anteriorly to the mammillary bodies and behind the tuber cinereum lies the premammillary recess, which appears as an almost constantly translucent area; it can sometimes be very small or, on the contrary appear very large and even deep. Its anterior margin is considered as the safest area to perform the orifice for third ventriculostomy (Vinas *et al.* 1996). The termination of the basilar artery and its branches, posterior cerebral arteries, or even the superior cerebellar arteries may be visible through under this recess, particularly in case of extreme hydrocephalus. The part of the third ventricular floor between the mammillary bodies and the aqueduct of Sylvius has a smooth surface and is concave from side to side. This smooth surface lies above the anterior perforated substance, a triangular area of gray matter which has a punctated appearence due to multiple branches of the posterior cerebral arteries passing through it and directed to the brainstem. The subarachnoid space under the floor of the third ventricle is the interpeduncular cistern. It has a conic shape and is bound posteriorly by the cerebral peduncles and anteriorly by the Liliequist's membrane. Laterally, the interpeduncular cistern is surrounded by the oculomotor, crural and ambient cisterns. A membrane arising from the posterior edge of Liliequist's membrane separates this cistern into two compartments: anterior and posterior. The anterior compartment contains the bifurcation of the basilar artery, the origin of both posterior cerebral arteries and both medial and lateral posterior choroidal arteries. The posterior compartment contains posterior thalamo-perforating branches, arising from the basilar and posterior cerebral arteries.

The anterior wall of the third ventricle extends from above the foramina of Monro to the optic chiasm below. During endoscopic procedures only its lower two-thirds can be seen; indeed, the upper third is hidden posterior to the rostrum of the corpus callosum. The part of the anterior wall that can be endoscopically viewed is formed by the optic chiasm and the lamina terminalis. This last appears as a thin sheet of gray matter and pia mater that attaches to the upper surface of the chiasm and stretches upward to fill the interval between the optic chiasm and the rostrum of the corpus callosum (Rhoton 2002). Two arachnoid cisterns can be found underneath the anterior third ventricle: the chiasmatic cistern and the lamina terminalis cistern. The first one is bordered by the superior surface of the optic nerves, chiasm, lamina terminalis cistern and Liliequist's membrane. It commonly has extensions through the diafragma sellae and optic foramen and contains the optic nerves, pituitary stalk, branches of the internal carotid artery and ophtalmic artery. The lamina terminalis cistern lies above the chiasmatic cistern. It is delimited by the superior surface of the optic chiasm, the lamina terminalis, the rostrum of the corpus callosum, gyrus cinguli, interhemispheric fissure and gyrus recti; laterally it is bordered by the olfactory gyrus and the anterior cerebral membrane. The lamina terminalis cistern contains both anterior cerebral artery and veins, the recurrent artery of Heubner, the anterior communicating arteries and veins, the fronto-orbital arteries and the most proximal A2 segments of the anterior cerebral arteries (Lang 1992, Oka *et al.* 1993a, Vinas *et al.* 1994, Vinas *et al.* 1996). The lateral walls of the third ventricle are formed by the hypothalamus inferiorly and the thalamus superiorly. Endoscopically they have an outline like the lateral silhouette of a bird's head with an open beak. The head is formed by the oval medial surface of the thalamus and the two beaks are respectively formed: the upper by the optic recess and the inferior by the infundibular recess. The columns of the fornix form distinct prominences in the lateral walls of the third ventricle, just below the foramina of Monro, but inferiorly they disappear under the surface of the floor (Kamikawa *et al.* 2001c, Oka *et al.* 1993a, Rhoton 2002).

Endoscopic Ventricular Anatomy Variations

Variations in the anatomy of the lateral and third ventricles can be found in more than one third of the cases (Decq 2004, Vinas *et al.* 1996). Some of them might hamper the correct conclusion of the endoscopic procedures, be responsible of a longer operating time and increase the complications rate. The most frequent anatomic variations regard the thickness of the floor of the third ventricle and its position. In cases of acute hydrocephalus (shunt malfunction; hydrocephalus associated with posterior fossa tumors) the floor may be undistended and extremely thick, with the mammillary bodies hardly recognizable. On the contrary, in children with long-standing hydrocephalus (i.e. hydrocephalus associated with aqueductal stenosis) the floor of the third ventricle can be extremely distended and bulge downward into the interpeduncular cistern because of the pressure gradient between the third ventricle and the subarachnoid space. According to van Aalst *et al.* (van Aalst *et al.* 2002) this finding may complicate ETV procedures. In a recent paper these authors reported four cases of triventricular hydrocephalus with deeply located third ventricular floor. The stretching of the floor in front of the basilar artery increased the risks of damaging this vascular structure; moreover in all four cases an immediate upward ballooning of the third ventricular floor was observed soon after the performance of the stoma, completely filling the third ventricle, obscuring the vision of the fenestration site with consequent problems of orientation.

Two other common findings in cases of extreme and long-standing hydrocephalus are the "loss" of the posterior margin of the MF and the partial or complete absence of the septum pellucidum, which may appear as a spiderweb structure; the only recognizable structure in the lateral ventricles may be the choroid plexus, which appears adherent to the outline of the thalamus laterally; through the dehiscence of the septum pellucidum the controlateral anatomic ventricular structures may be seen (Decq 2004).

Rohde and Gilsbach (Rohde and Gilsbach 2000) recently compared the above mentioned anatomical variations with a critical review of their personal experience. The video recordings, operative reports and preoperative MR images of 25 patients who underwent third ventriculostomy at their institution were analyzed. All the patients were affected by long standing hydrocephalus due to aqueductal stenosis in 18 cases, to obstruction of the fourth ventricle outlets in 4 patients and to presumed CSF malresorption mechanisms in the last three patients. Overall 10 anatomic variants were identified in 9 cases. Six anatomic variants were identified at the floor of the third ventricle. In four patients the floor of the third ventricle was not a thin transparent membrane, as could be expected in long-standing hydrocephalus, but a firm opaque structure. Blunt perforation was more time-consuming and the stretching of the hypothalamic structures seemed to be higher than usual. The identification of the basilar artery was not possible in two of these patients and repeated minor bleeding occurred in three of them, leading to the abandoning of the surgical procedure in one case. The lack of sufficient third ventricle dilatation lead to the abandoning of the surgical procedure in a further case. In two children the floor of the third ventricle atypically inclined steeply from the mammillary bodies to the infundibular recess. This finding increased the operative time, because of the slipping off of the tip of the perforation catheter and lead to a functionally insufficient stoma in one case. The other anatomic variations described by Rohde *et al.* regarded the Monro foramen; it was reduced in size in spite of chronic hydrocephalus in two cases with an associated agenesia of the corpus callosum and septum and division of the body of the fornix in one case. Differently from what occurred for the third ventricle anatomic variations the ones related to the MF did not lengthened the operative time nor influenced the correct conclusion of the surgical procedure.

Different anatomic variations have been described in hydrocephalus of different etiologies. Kamikawa *et al.* (Kamikawa *et al.* 2001c) analyzed the ventriculoscopic findings of four neonates with posthemorrhagic hydrocephalus who underwent ETV. The choroid plexus was atrophic with hematoma clots attached to it as well as at the orifice of the aqueduct of Sylvius in all cases. The septum pellucidum was widely fenestrated with a number of small varices at the level of the septal veins in two cases. In all patients fragments of old hematomas were scattered on the ventricular walls and on the floor of the third ventricle, appearing brown because of the presence of hemosiderin, an aspect defined by the authors as "leopard-like" (Fig. 1). At time distance from the hemorrhage thickening of the ventricular walls, due to fibrous scarring is common and septations can be found in the ventricular cavities as well as at the inlet of the aqueduct of Sylvius (Fig. 2). This could explain the occurrence of an acquired aqueductal stenosis in some of these patients (Scavarda *et al.* 2003). Septations as well as a thickened ventricular floor are common findings also in children with previous CSF infections. An extensive arachnoid sepimentation under the third ventricular floor might further complicate the correct conclusion of an ETV procedure in this subset of patients (Riegel *et al.* 2001).

Hydrocephalus associated with spinal dysraphism is another clinical condition that is frequently associated with ventricular anatomic anomalies. In a review of ten personally performed endoscopic procedures, Pavez Salinas (Pavez Salinas 2004) described significant anatomic variations in this kind of patients if compared with other hydrocephalus etiologies. In particular: a small Monro foramen and unrecognizable infundibulum, mammillary bodies and basilar artery occurred in 40% of the cases; septations inside the third ventricle and third ventricular floor umbilications occurred in 50% of the patients; 60% of the children presented atypical veins inside the third ventricle and 70% of them arachnoid adherences under the floor of the third ventricle (Figs. 3, 4). Most of these alterations are consistent with previous histopathological findings. In particular the presence of septations, and abnormal veins inside the third ventricle, or a

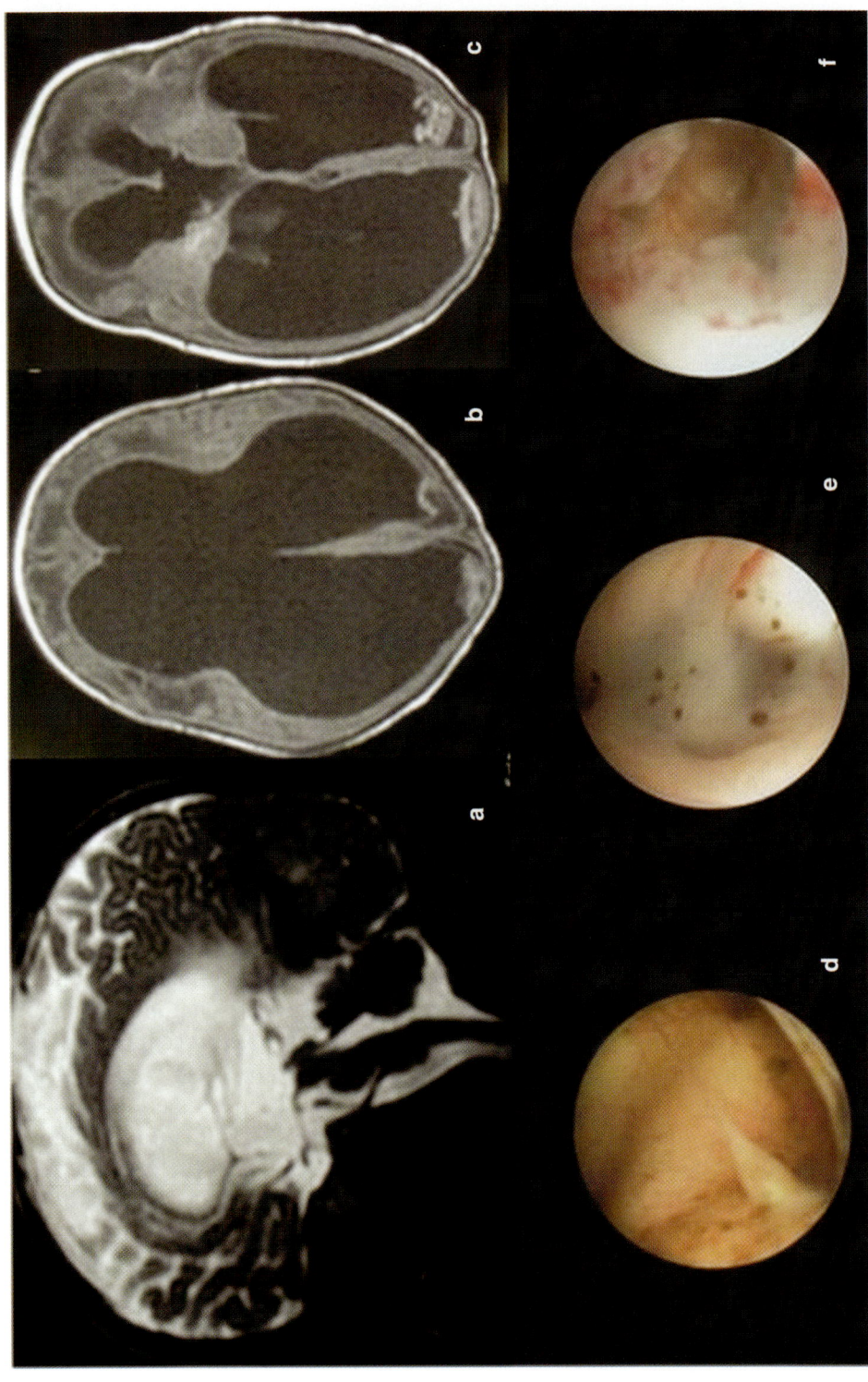

second arachnoidal membrane under the floor of the third ventricle, might be in relation with embryonic development alterations with abnormal degrees of hypothalamus fusion.

Table 1 summarizes the possible variations of ventricular anatomy in different types of hydrocephalus.

Modern Neuroendoscopic Instrumentation

Optic Devices

The neuroendoscope is the only instrument that allows access to deep anatomic structures in a minimally invasive way. Pre-requisites are the ability to bring the illumination and the subsequent ability to transmit the images accurately, clearly, and brightly to the eyes of the neurosurgeon.

To understand the functioning of an endoscope, one should know some principles of optical physics. These, focused on the neuroendoscope, have been recently reviewed by Liu *et al.* (Liu *et al.* 2004). More detailed treatments of principles of physics can be found in the optical engineering literature (Fisher and Biljana 2000, Laikin 2001).

The endoscopes specifically designed for neuroendoscopy can be classified into four types:
– Flexible fiberscopes
– Steerable fiberscopes
– Rigid fiberscopes
– Rigid rod lens endoscopes
These endoscopes varies for diameter, optical quality and number, and diameter of working channels.

Flexible Fiberscopes

These endoscopes are not useful for third ventriculostomy. They have a very small diameter (<2 mm), which allow their use inside the lumen of ventricular catheters for optimal ventricular catheter positioning during ventriculoperitoneal shunting. Their limitations are poor quality of vision and the absence of a working channel.

◄——————————————————————————————

Fig. 1. Ventricular anatomy variations in children with posthemorrhagic hydrocephalus. (a, b, c) T2 sagittal (a) and T1 axial (b, c) MRI images showing an irregular ventricular walls signal (a) and the appearance of intraventricular clots at long distance from the hemorrhage (b, c). (d, e, f) endoscopic views of the same patient documenting a "leopard-like" appearance of the lateral ventricle walls (d) and of the floor of the third ventricle (e). After the endoscopic opening of the third ventricle floor extensive arachnoid adhesions in the interpeduncular cistern are detected (f)

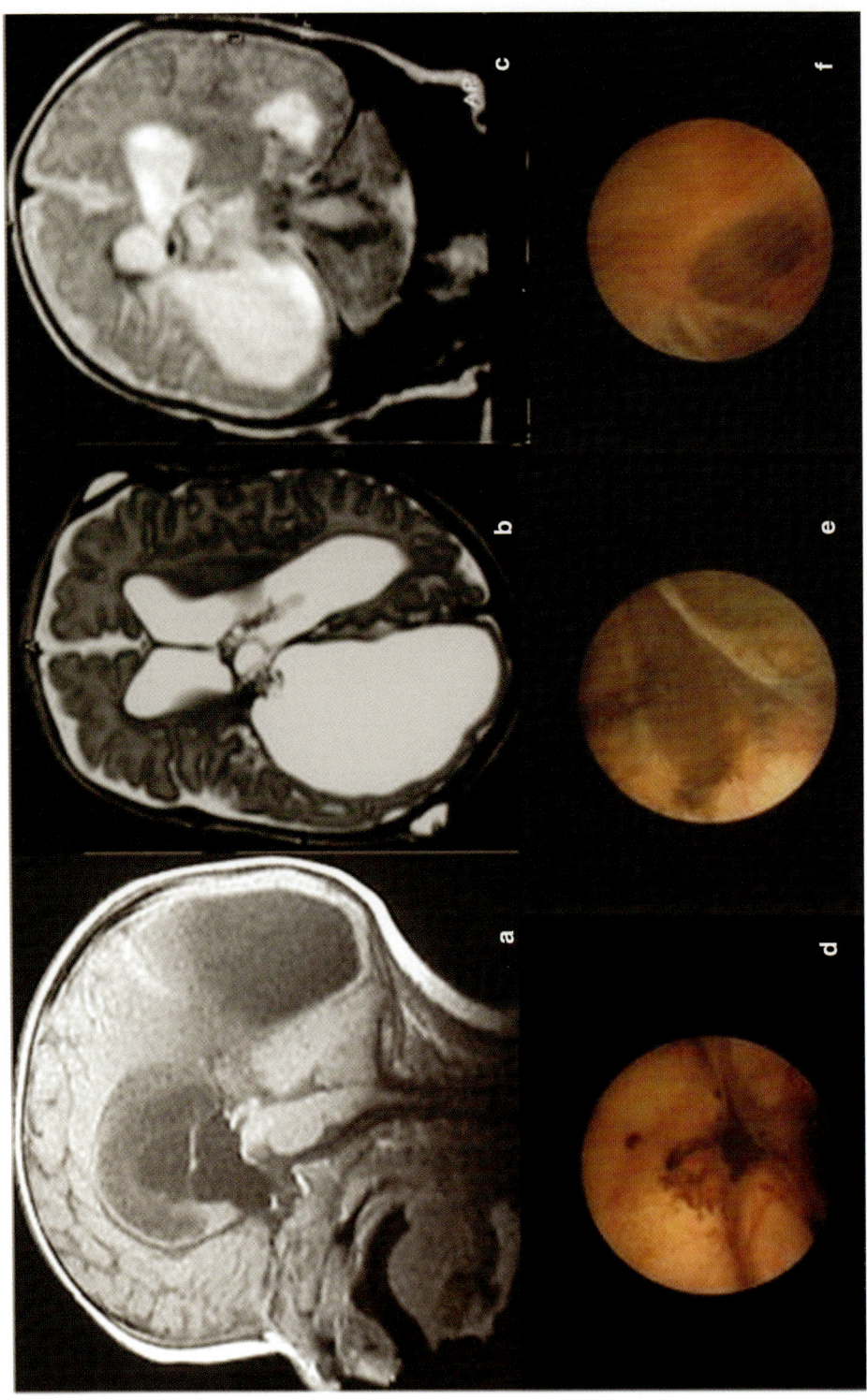

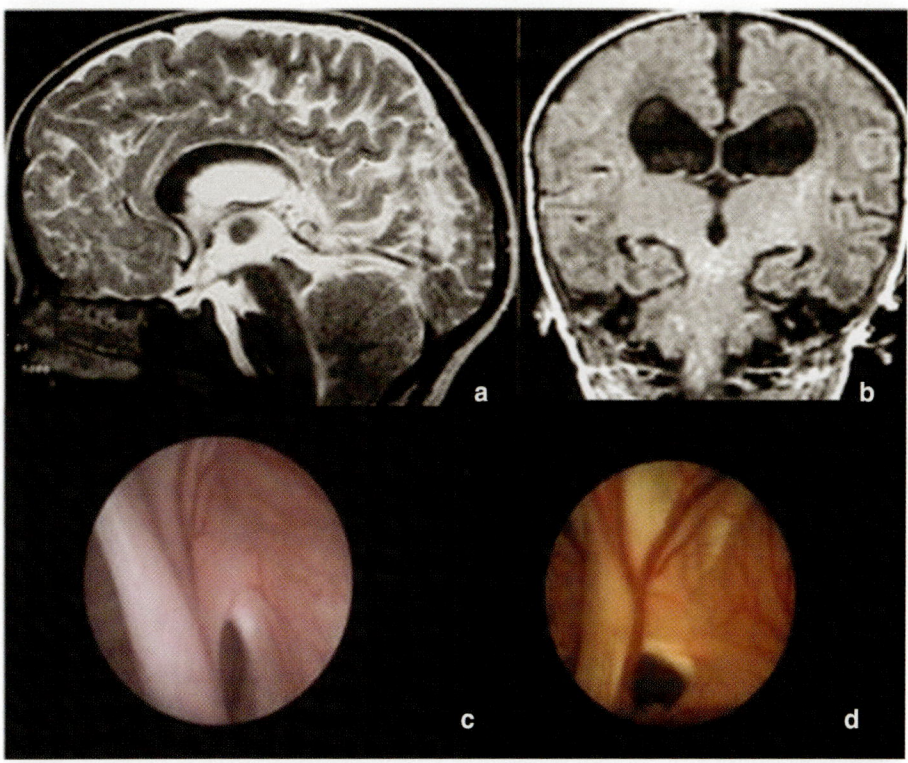

Fig. 3. Ventricular anatomy variations in children with hydrocephalus and myelome-ningocele. (a, b) T2 sagittal (a) and proton density coronal (b) MRI images of a 2 years old child: the foramen of Monro is narrow and steeply oriented; the interventricular septum is apparently preserved. (c, d) endoscopic views of the same patient confirming the reduced size and the oblique orientation of the foramen of Monro; differently from what documented by the MRI images the inteventricular septum is widely opened

Steerable Fiberscopes

Flexible-Steerable endoscopes became a reality with the development of fiber optic technology (Fukushima *et al.* 1973, Hecht 1999). They are constructed of silica glass (which can be flexed without breaking) (Nobles

Fig. 2. Ventricular anatomy variations in children with posthemorrhagic hydrocepha-lus. (a, b, c) T1 sagittal (a) and T2 axial (b) and coronal (c) images of an infant with posthemorrhagic hydrocephalus. Intraventricular septa (a, b, c) and clots (b, c) lead to a compartimentalization of the right lateral ventricle. The interventricular septum seems to be preserved. (d, e, f) endoscopic views of the same patient showing the aspect of intraventricular clots (d) and of the right ventricle cella media septation (e); the interventricular septum is opened and has a "spider-web" appearance (f)

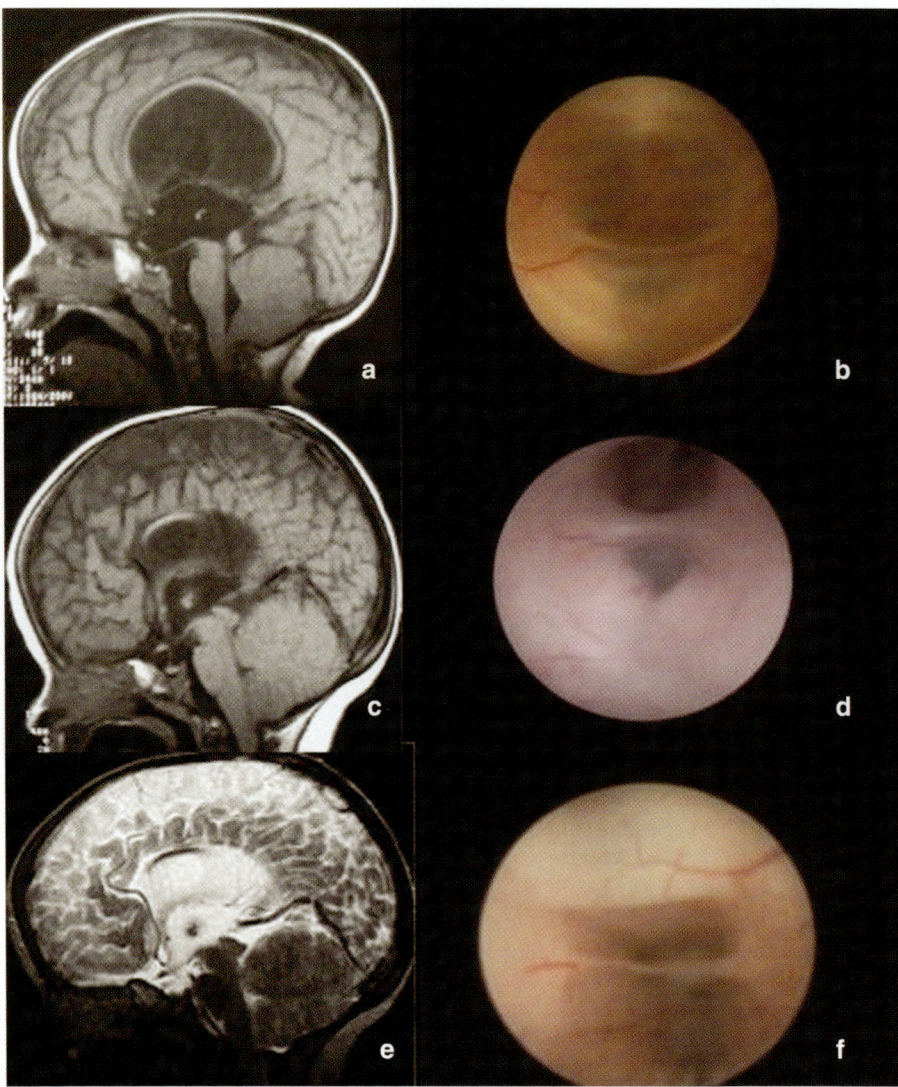

Fig. 4. Ventricular anatomy variations in children with hydrocephalus and myelomeningocele. (a, c, e) T1 (a, c) and T2 (e) sagittal images of three different patients. The
third ventricle does not seem to have significant anatomic variations, as in the patient
of Fig. 1 it appears oblique and steeply oriented. (b, d, f) endoscopic views of the same
patients showing different third ventricle anatomic variations which were not detected
by the MRI. (b) the mammillary bodies are not recognizable; the premammillary recess
is crossed by a septum and an overlapping vein. (d, f) the route to the premammillary
recess is crossed by and interthalamic adhesion, which can be found at different distan
ces from the floor of the third ventricle

Table 1. *Possible Variations of Ventricular Endoscopic Anatomy in Different Types of Hydrocephalus*

Diagnosis	Foramen of Monro	Ventricular walls	III ventricle floor	Various	Associated risk
Aqueductal stenosis	loss of the posterior margin	absence of the septum pellucidum	1 – distended and bulging downward 2 – more rarely thick and opaque		1 – basilar artery damage 2 – stretching of the hypothalamus
Posterior cranial fossa tumors			1 – undistended, thick and opaque	2 – mammillary bodies hardly recognizable	1 – stretching of the hypothalamus 2 – orientation
Post-hemorrhagic	1 – atrophic choroid plexus, with attached clots	fenestrated septum pellucidum; hemosiderin deposition ("leopard like" appearance)	2 – (at long distance) thickened, with fibrous scarring		1 – orientation 2 – stretching of the hypothalamus
Myelomeningocele			unrecognizable infundibulum, basilar artery and mammillary bodies; septations and umbilications; atypical veins; arachnoid adherences under the floor		orientation; intraoperative hemorrhage; insufficient CSF flow through the soma (arachnoid adherences)

Table 2. *List of main steerable neuroendoscopes with fiberoptic system suitable for endoscopic third ventriculostomy*

Manifacturer	Model	Outer diameter	Channels*	Working channel diameter	Bending
Aesculap[1]	flexible – steerable endoscope	4.3 mm	1 W, 1 I/A	1.4 mm	280°
Codman & Shurtleff[2]	steerable neuroendoscope	4.0 mm	1 W	1.0 mm	260°
Olympus	flexible – steerable endoscope	4.2 mm	1 W + 2 I/A	2.0 mm	180°
Storz[3]	neurofiberscope	2.9 mm	1 W + 2 I/A	1.2 mm	290°
		3.7 mm	1 W + 2 I/A	1.5 mm	

[1] Aesculap – Tuttlingen, Germany; Wolf, Knittlingen, Germany.
[2] Codman & Shurtleff (Johnson & Johnson) New Brunswick, NJ.
[3] Karl Storz GmbH & Co., Tuttlingen, Germany.
* *W* Working channel; *I/A* Irrigation/Aspiration channel.

1998). The image formed by the objective lens is relayed to the eye lens by multiple fibers contained in a very small package. Image fibers are formed in a coherent bundle that allows the image to be properly reconstructed on the proximal end of the fiber. By contrast, light fibers are not constructed in a coherent fashion.

The main characteristic of the flexible and steerable fiberscope is that the last 4 cm can be oriented 100° upward and 160° downward. This is the only system that makes looking and working around a corner possible. The diameter of the scope ranges from 2.3 to 4.6 mm (Table 2) allowing to work also in small ventricles and through a small foramen of Monro. The size of the device usually determines the number of individual fibers, consequently the number of pixels: the smaller the size, the fewer the number of pixels. For each endoscope, designated fiberoptic cables and lighting equipment are used in combination with a standard camera and television monitor. The diameter of the working channel ranges from 1.0 mm to 1.5 mm, allowing the introduction of 3-French (1 mm) flexible instruments (micro scissors, micro grasping forceps, micro biopsy forceps, monopolar electrodes, Fogarty balloon). Distinct irrigation channel and the outflow channel are usually available. Access to the ventricle is possible using a dedicated peel-away sheath. Some steerable scopes with only one working channel present the limit of very slow irrigation when the instrument occupies the channel. For this reason it is impossible to work under continuous irriga-

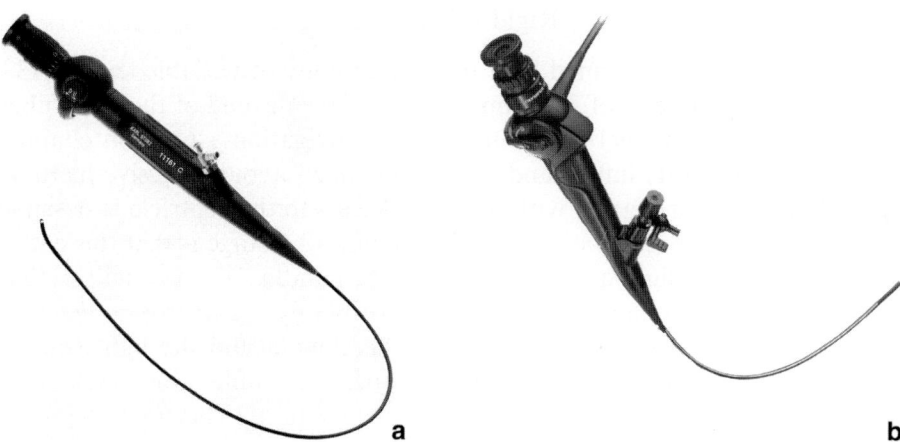

Fig. 5. (a) Storz steerable neuroendoscope. The largest part is usually kept by a holder, the proximal steerable tip is inserted into the ventricle through a peel-away sheath and oriented with the left hand, while the right hand manipulates the instruments (image courtesy of Karl Storz GmbH & Co., Tuttlingen, Germany). (b) Olympus steerable endoscope. A short fiberscope, easier to manipulate for neuroendoscopic procedures (Courtesy of Olympus opt, Tokyo, Japan)

tion and this makes more complex operations (tumor biopsies or removal) more complicated. When an instrument is introduced into the working channel, the steering properties are decreased, sometimes significantly, according to the stiffness of the instrument introduced. The steerable scope modifies the orientation of the optical fibers but also of the working channel, allowing the instruments to reach all the structures visualized. A holder is necessary to maintain the rigid part of the scope (Fig. 5a,b). The distance from the target should be carefully and precisely evaluated. Being too close to the target obliges the surgeon to work with a curved endoscope, while being too far obliges the surgeon to release the holder, with possible rough movements in the proximity of potentially delicate anatomical structures. When the endoscope is in the position of maximum flexion, the flexible instruments can have some problems in progressing through the working channel. Finally, the scope must be in the neutral position before backing out the ventricular system, to avoid the risk of damage to the fornices and other intraventricular structures. Some steerable scopes cannot be sterilized according to the protocols used in some countries (i.e. France) in order to prevent prion transmission. These protocols include decontamination with alkaline medium and sterilization for 20 minutes at 134°C. Therefore, these devices should be considered as disposable, that makes their use prohibitive in these countries.

Rigid Fiberscopes

These fiberscopes are formed by a main rigid body of variable length (13–27 cm) with a diameter of 3–4 mm. This contains the end of the optic fiber tract, a working channel (1–2 mm), and the irrigation-aspiration channel (Fig. 6a–c). Separate inflow and outflow channels avoid excessive increase of ICP balancing irrigation with outflow. Access to the ventricle is possible using a dedicated peel-away sheath. The major advantage is that this endoscope is extremely light and short, and can be handled like a pencil, so that it is easier to manipulate. This is made possible by the fiberoptic technology, which allows remote placement of the camera and the light source: they can be placed 40 cm away on the operating table. One single soft, fiberoptic cable is the only link to the fiberscope itself. The small diameter of the scope allows its use in neonates and in case of small ventricles –

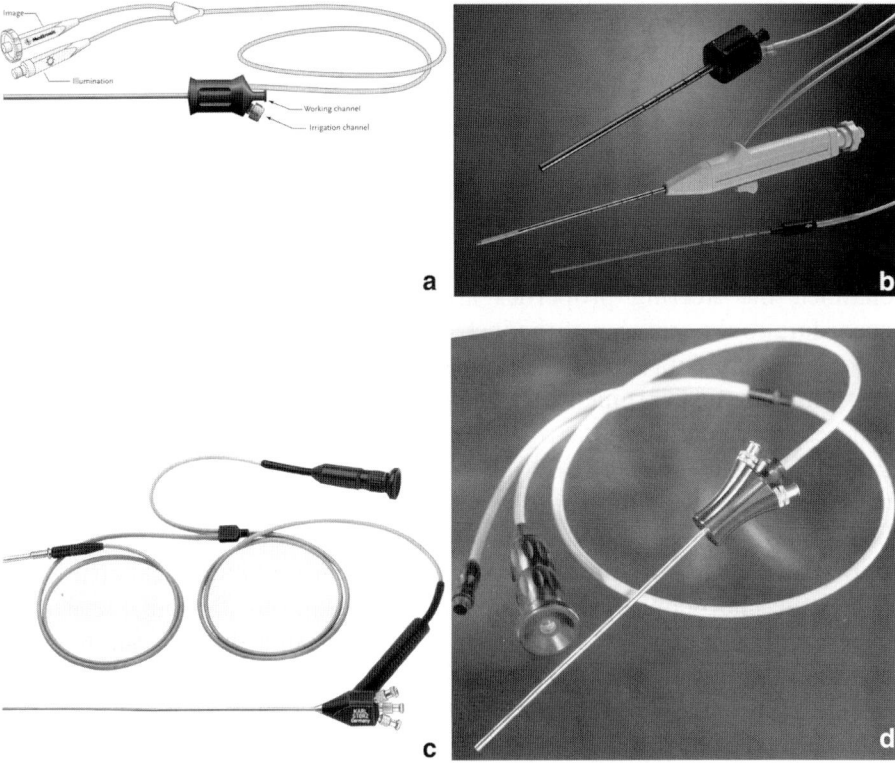

Fig. 6. Extremely light and easy to handle, their quality of vision is higher than steerable fiberscopes but lower than rod lenses endoscopes. They can be disposable like the Medtronic (a) or the Integra Neuroview (b) or re-sterilizable for a limited number of operations like the Storz (c) or the Paediscope Aesculap (d) (images courtesy of manufacturers)

Table 3. *List of main rigid straight neuroendoscopes with fiberoptic system suitable for endoscopic third ventriculostomy*

Manifacturer	Model	Outer diameter	Channels*	Working channel diameter
Aesculap	paediscope (autoclavable)	3 mm	1 W, 2 I/A	1.2 mm
Medtronic	channel neuroendoscope (disposable)	3.5 mm (10000 pixel fibers)	1 W, 1 I/A	2.13 mm
		4.2 mm (30000 pixel fibers)	1 W, 1 I/A	2.13 mm
		4.5 mm (10000 pixel fibers)	1 W, 1 I/A	3.12 mm
Integra neuroview	neuroview	2.3 mm	1 W	1.0 mm
	disposable rigid/ semirigid scope	4.6 mm	1 W + 1 I/A	2.4 mm
Storz	Gaab miniature neuroscope (autoclavable)	3.2 mm	1 W, 2 I/A$^+$	1.3 mm

* *W* Working channel; *I/A* Irrigation/Aspiration channel.
$^+$ Lateral irrigation channel allows use of additional instruments with a diameter of 1 mm parallel to the instrument channel.

small foramen of Monro (Table 3). The absence of a rigid rod lens system allows a very wide working channel and a wide irrigating channel, making it possible to use virtually all endoscopically designed surgical instruments, of any length. However, the fiberscopes with smaller diameter have smaller working channel as well, allowing the use of 1 mm diameter instruments.

Vision is superior to that with the steerable fiberscopes because the number of optic fibers can be higher since there is no need for tip orientation. More modern fiberscopes provide higher quality images, thanks to the presence of as high as 30000 pixel fibers (Medtronic, Aesculap). Nevertheless, the quality of vision cannot be compared to that offered by rigid rod lens endoscopes.

Rigid Rod Lens Endoscopes

The quality of vision is the main advantage that makes the rigid rod lens scope an indispensable item in the armamentarium of any neuroendoscopist. In the sixties, Hopkins described a series of glass rods with small air gaps, which is the exact opposite of the design since then used which was

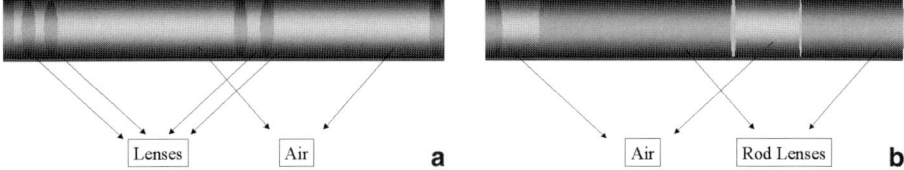

Fig. 7. Traditional (a) and Rod lenses optic systems (b). In the rod lenses optical system the air gap between lenses is significantly reduced, decreasing light dispersion and increasing the stability of the system

composed by a series of small glass lenses interspersed with large air spaces (Fig. 7a,b). This technique forms the basis of most modern endoscopic systems and bear his name (Nobles 1998, Siomin and Constantini 2004). Further reduction in light loss is achieved through the coating of the glass surfaces with an ultrathin layer of magnesium fluoride. This layer markedly decreases the reflection and improves the optic characteristics of endoscopes and cameras (Nobles 1998, Shiau and King 1998, Siomin and Constantini 2004). With this technology the quality of vision is extremely sharp, and allows easier and more precise identification of the anatomical structures encountered, with an excellent visual definition of details. The rod lens system requires the presence of the camera and of the fiberoptic cable for the cold light attached to the proximal extremity of the endoscope (Figs. 8–11). The whole system requires good surgical training to be manipulated freehand during navigation and throughout the whole surgical procedure. A holder may be useful. The rigid lens system only allows targets to be reached that are located on a straight line from the burr

Fig. 8. Wolf Neuroendoscope (image courtesy of Richard Wolf, Henke-Sass, Tuttlingen, Germany)

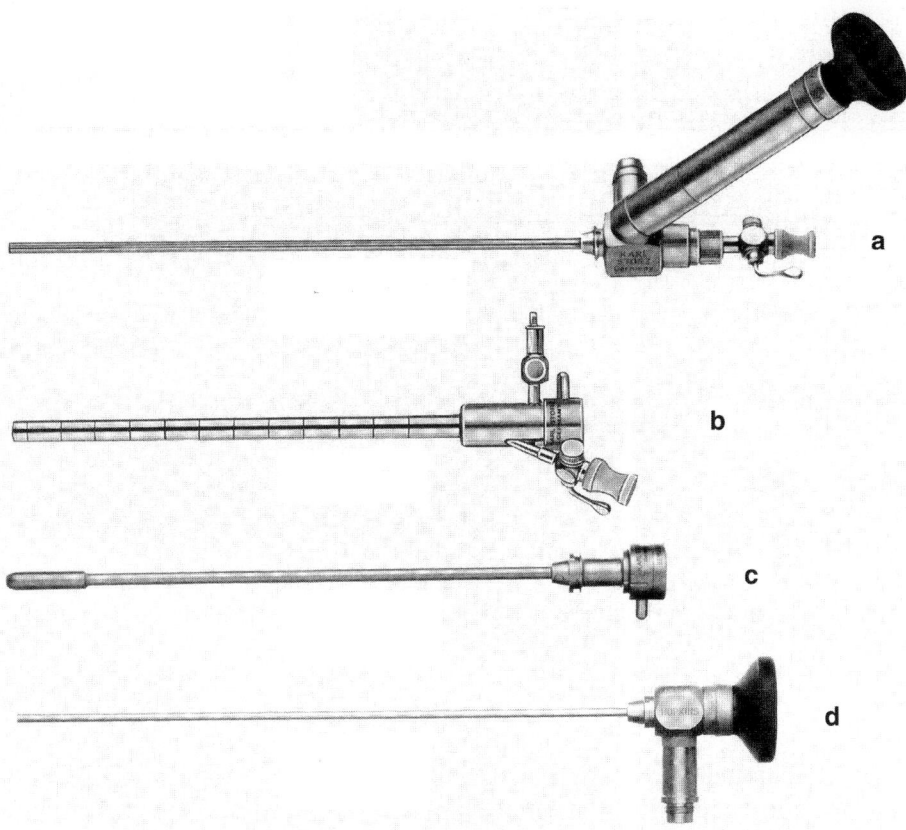

Fig. 9. Elements of the Gaab endoscope: (A) optic, (B) trochar, (C) stylet (D) optic
(image courtesy of Karl Storz GmbH & Co., Tuttlingen, Germany)

hole. Unlike rigid fiberscope, that incorporates an instrument channel and
two separate channels for irrigation in- and outflow, the rigid lens system
requires the presence of an endoscope trochar with multiple parallel operative channels: the optic is inserted in the "optic channel" while the instruments are inserted in different operating channels. The operating sheath
allows one to change scopes intraoperatively without reinserting them
through brain tissue, thus avoiding unnecessary damage to the surrounding
brain.

The trochars are of different diameters, according to the number of
channels available (Table 4). For third ventriculostomy 4 channels should
be available: an optic channel (in which the optic is inserted), a working
channel for surgical instruments, an irrigation channel and an overflow
channel. The diameter of the whole system ranges from 3.2 to 6 mm, the

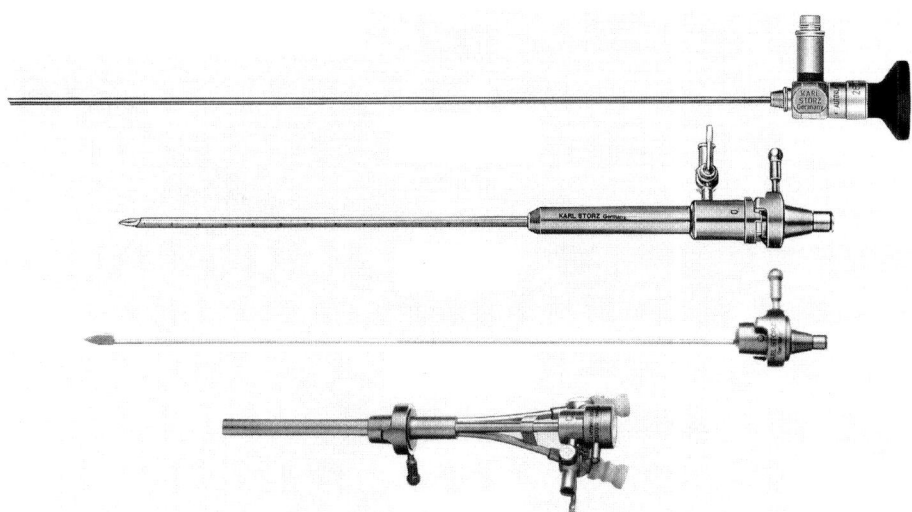

Fig. 10. Elements of the Decq endoscope (image courtesy of Karl Storz GmbH & Co., Tuttlingen, Germany). When used with a holder, two hands work is possible through the two symmetric working channels (image courtesy of Dr. Philippe Decq, Hopital Henri Mondor, Paris, France)

working channel allows the use of 2 mm instruments. However, also "miniature endoscope" (3.2-mm outer diameter) has been developed for use in pediatric patients (4.0-mm Gaab II miniature system, Karl Storz GmbH & Co., Tuttlingen, Germany). In some endoscopes (Minop System, Aesculap, Tuttlingen, Germany), 1 mm instruments can be introduced short term through the irrigation channel for bi-instrumental operation. Decq endoscope (Karl Storz GmbH & Co., Tuttlingen, Germany), is the only system provided by 2 working channels of 2 mm. To allow the presence of the two working channels, the section of the endoscope is oval-shaped with the two diameters ranging from 3.5 and 4.0 mm × 5.2 and 7.0 mm. This is specifically designed for tumor or cyst surgery allowing simultaneous grasping and fenestration (or aspiration). However it is also a useful tool in third ventriculostomy in case of large, redundant, highly pulsating third ventricle floor (as occur in long standing hydrocephalus). The endoscope trochars are available in "short version" (150–160 mm) for "freehand" use (or with the aid of a holder) or "long version" (250 mm) to be used with stereotactic frame.

Angle of view of the endoscope can range from 0° to 120°, according to the objective lens used. The most used are 0° and 30°. The 0° objective portray only what it is viewing head-on, minimizing the risk of disorientation (Siomin and Constantini 2004). The 30° objective offers some advantages:

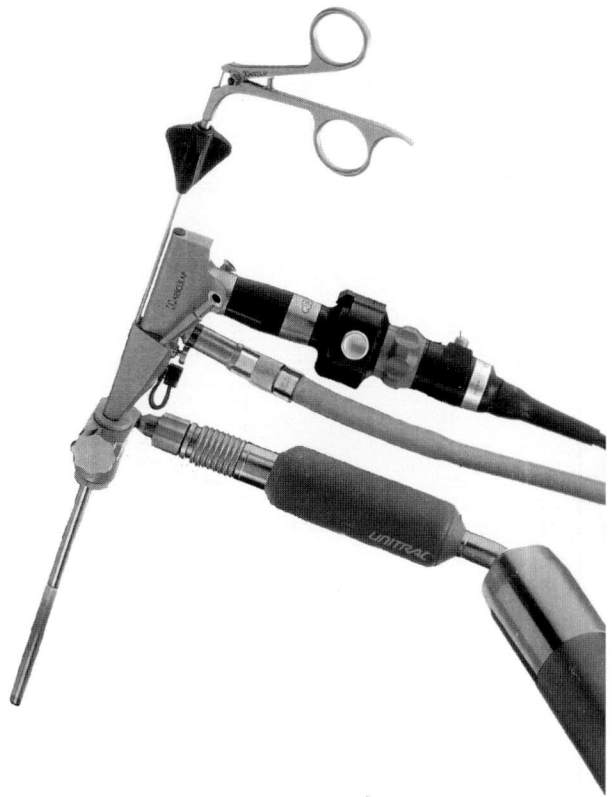

Fig. 11. The Minop endoscope (image courtesy of Aesculap – Tuttlingen, Germany)

by simple rotation it provides an angle of view with a surface area twice as large as that obtained with 0° objective and it allows a better control of the instruments because with the 30° objective the instruments introduced in the working channel (parallel to the endoscope) converge towards the center of the image (directed at 30°), while with the 0° objective the instruments remain in the periphery of the image. Major disadvantage of angled scopes is that the indirect image may cause the surgeon to become disoriented. More than 30° angled objective are useful only to "look around the corner" (Decq 2004).

The Future: The Videoscope

The future development of the technology of the endoscopes will allow the wide diffusion of the group of Videoscopes. A videoscope is characterized by a 1 CCD chip camera positioned at the proximal tip of the endoscope.

Table 4. *List of main rod lenses neuroendoscopes suitable for endoscopic third ventriculostomy*

Manufacturer	Model	Outer diameter of the trochar	Channels*	Working channel diameter	Optic tip orientation
Aesculap[1]	neuroendoscope	6.2 mm	1 O, 1 W, 2 I/A	2.2 mm	0°, 30°
Aesculap[1]	Minop system	3.2 mm	1 O/W	2.8 mm (optic/work)	0°, 30°
		4.6 mm	1 O/W, 2 I/A		
		6.0 mm	1 O, 1 W, 2 I/A	2.8 mm (optic/work) 2.2 mm	
Codman[2]	Gaab neuroendoscope system	5.8 mm	1 O, 1 W, 2 I/A	1.6 mm	0°, 30°, 70°, 120°
Integra neuroscience[3]	neuroview 700R	5.6 mm	1 O, 1 W, 2 I/A	2.0 mm	0°, 30°, 70°
Storz[4]	Gaab neuroendoscope	6.5 mm	1 O, 1 W, 2 I/A	3.0 mm	0°, 30°, 70°, 120°
Storz[4]	Decq	Oval 3.5 mm × 5.2 mm	1 O, 2 W, 2 I/A	1.7 mm	30°
		Oval 4.0 mm × 7.0 mm	1 O, 2 W, 2 I/A	3.0 mm	
Storz[4]	Oi-Samii Handy Pro®	Oval 3.5 × 2.5	Single space lumen	1.3 mm	0°–12°
Wolf[5]	pediatric neuroendoscope system	3.3 mm × 4.5 mm	1 O, 1 or 2 W, 2 I/A	1.6 mm	0°, 25°, 70°
		4.8 mm × 5.8 mm			

[1] Aesculap – Tuttlingen, Germany; Wolf, Knittlingen, Germany.
[2] Codman & Shurtleff (Johnson & Johnson) New Brunswick, NJ.
[3] NeuroNavigational (Integra NeuroSciences) Plainsboro, NJ.
[4] Karl Storz GmbH & Co., Tuttlingen, Germany.
[5] Richard Wolf, Henke-Sass, Tuttlingen, Germany.
* O Optic channel; W Working channel; O/W Optic and Work channel; I/A Irrigation/Aspiration channel.

This allows for the extreme simplification of the optical system, without the need for complex and long rod lenses systems improving at the same time significatively the quality of vision if compared to the fiberscopes. Although the 1 CCD camera is somehow less performant than the 3 CCD camera (see below) that can be used with rigid rod lenses systems and fiberscopes, the proximity of the camera to the target of vision allows for excellent magnification and sharpness of the image. Moreover, the steerable properties of the device are preserved because of the lack of rigid lens systems. Some videoscope are already commercially available for ENT use in some countries, with an outer diameter of 6–8 mm; prototypes with smaller outer diameter for neuroendoscopic use are under clinical validation studies and should be available commercially in the next 2–3 years (Kamikawa *et al.* 2001a, Kamikawa *et al.* 2001b). The real, significant advantage of this device is the excellent quality of vision (1 CCD camera like), comparable to the 1 CCD rod lens systems, associated with the steerable properties, allowing for perfect fusion of rigid and steerable systems (Fig. 12a–d).

Camera and Monitor

Two basic cameras are available: a single chip charged coupled device (CCD) and a three chip CCD. A good resolution for neuroendoscopy is available with 0.5 inches single chip cameras (resolution of 500 lines) (Schroeder *et al.* 2001). If the resolution is poor, the image needs computer-enhancement. The three chip CCD produces images of better quality (more than 800 horizontal lines) but is more expensive and heavier. So, most endoscopic system use single chip cameras. Some manifacturers produce both the type of digital cameras (David 1 and David 3, Aesculap, Tuttlingen, Germany; Image1 and Image3, Karl Storz GmbH & Co., Tuttlingen, Germany). In some models all the function can be controlled by the surgeon in the operative field. Zoom endo-lens is useful to enlarge the image section.

To achieve good quality images, a monitor with the highest possible resolution should be selected. However, the resolution of the monitor should not greatly exceed that of the camera. The size of the screen is limited by the loss of quality when an image is enlarged. In fact, one should remember that the images of the camera are displayed over an area larger than cross section of the optic cable. In monitor larger than 13 inches, the picture is enlarged too much and decreases in quality. This is especially true in case of fiberoptic endoscopes, where the spaces among the pixels may become evident. Larger monitors (19 or 20 inches) are useful for displaying multiple images (Cinalli 2004, Nobles 1998, Schroeder and Gaab 1999, Siomin and Constantini 2004).

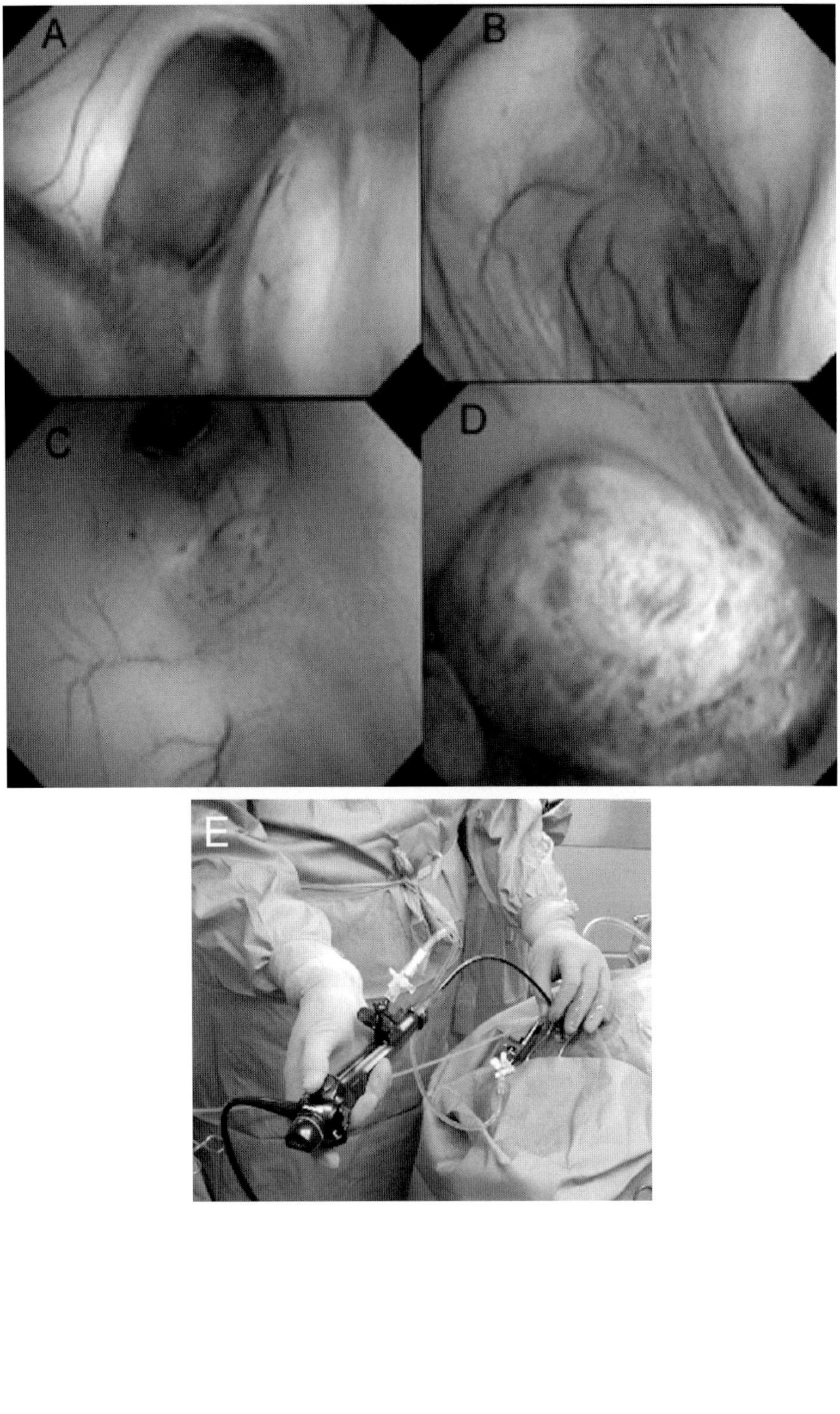

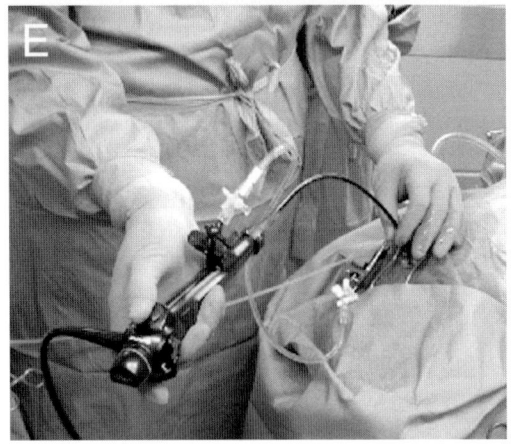

Illumination

Xenon light sources provide the best illumination for neuroendoscopic procedures. The light is transmitted via fiber bundles from the light fountain to the endoscope. Setting the light source to between 300 and 500 W provides a superior picture quality. Other types of light source, such as halogen, are not able to generate a light bright enough for neuroendoscopy. Siomin and Constantini (Siomin and Constantini 2004) have calculated that, due to the significant light loss in the fiberoptic system, only 30% of the light generated within the light source reaches the distal tip of the endoscope.

Accessories (Irrigation, Holders)

Imaging with an endoscope requires the clearest possible medium for optimum light with the lowest diffraction. So, irrigation is important to assure good visualization. It should be balanced by the egress of fluid. Care should be given to avoid entrapment of fluid inside the ventricle: it may lead to disastrous sequels (Cinalli 2004, Teo 2004). Irrigation can be performed simply by hand with a catheter connected to the irrigation channel of the endoscope. It can be also provided with the use of a pump irrigator for which the flow is easily controlled using a foot switch (The Malis CMS-II Irrigation Module, Codman and Shurtleff, Inc., Randolph, MA; Endoscopy Pump, Medtronic, Minneapolis, USA).

The use of a holder is sometimes advised when using a rigid rod lens endoscope. During these procedures it allows the surgeon to use both hands and two instruments through two different working channels (Fig. 13). The disadvantages of use of holders is the minor freedom of movements, especially when configuration needs to be frequently changed. However, holders with pneumatic (Fig. 14a) or electromagnetic (Fig. 14b) brakes offer a significant improvement over the mechanical systems, combining the advantages of freehand movements with the possibility of very secure and firm positioning, and are certainly the gold standard for both beginners and expert surgeons (Fig. 14c). With traditional holding devices, a precise steering of the neuroendoscope is not possible, but only a rough positioning. A new device has been developed (NeuroPilot, Aesculap, Tuttlingen, Germany) that used in combination with a pneumatic holder

---

Fig. 12. Images of a Neuroendoscopic third ventriculostomy and pineal tumor biopsy obtained with a prototype of videoscope (Olympus opt, Tokyo, Japan). (a) foramen of Monro. (b) ventricular trigone with choroids plexus. (c) from up to down, stoma of the ETV, mammillary bodies, mesencephalic roof. (d) pineal tumor. (e) prototype of videoscope during manipulation (images courtesy of Professor Shuji Kamikawa, Isesaki Sawa Medical Association Hospital, Japan)

Fig. 13. The use of a holder allows working with both hands if two working channels are available (image courtesy of Dr. Philippe Decq, Hopital Henri Mondor, Paris, France)

(Unitrac, Aesculap, Tuttlingen, Germany) allows, after positioning of the neuroendoscope, fine, sub-millimetric adjustment in the three dimensional space by three screws.

Neuronavigation and Stereotaxy

Stereotactic guidance was used before the advent of neuroendoscopy to perform third ventriculostomy (Hoffman et al. 1980, Kelly 1991) and was used in association with neuroendoscopy by several authors at the beginning of their experience (Grunert et al. 1994, Hellwig et al. 1998b, Hopf 1999a). In fact, stereotactic guidance can be of some value only in choosing the correct entry point and entering the lateral ventricle in a small-sized ventricular system. The limit of the technique is that the stereotactic frames are bulky and sometimes interfere with the endoscopic procedures and most importantly, frame-based stereotactic systems do not provide an on-going intraoperative feedback to the surgeon about anatomical structures encountered in the surgical field (Tirakotai et al. 2004).

A good alternative to traditional stereotactic frames can be the combination with frameless neuronavigation (Alberti et al. 2001, Broggi et al. 2000, Hopf et al. 1999a, Riegel et al. 2000, Riegel et al. 2002, Schroeder et al. 2001, Tirakotai et al. 2004). Unlike based stereotaxis, frame-less navigation is still useful for intraoperative orientation, especially in cases of impaired visualization, distorted anatomy or narrowed ventricles. In endoscopic third ventriculostomy, the use of neuronavigation may not be

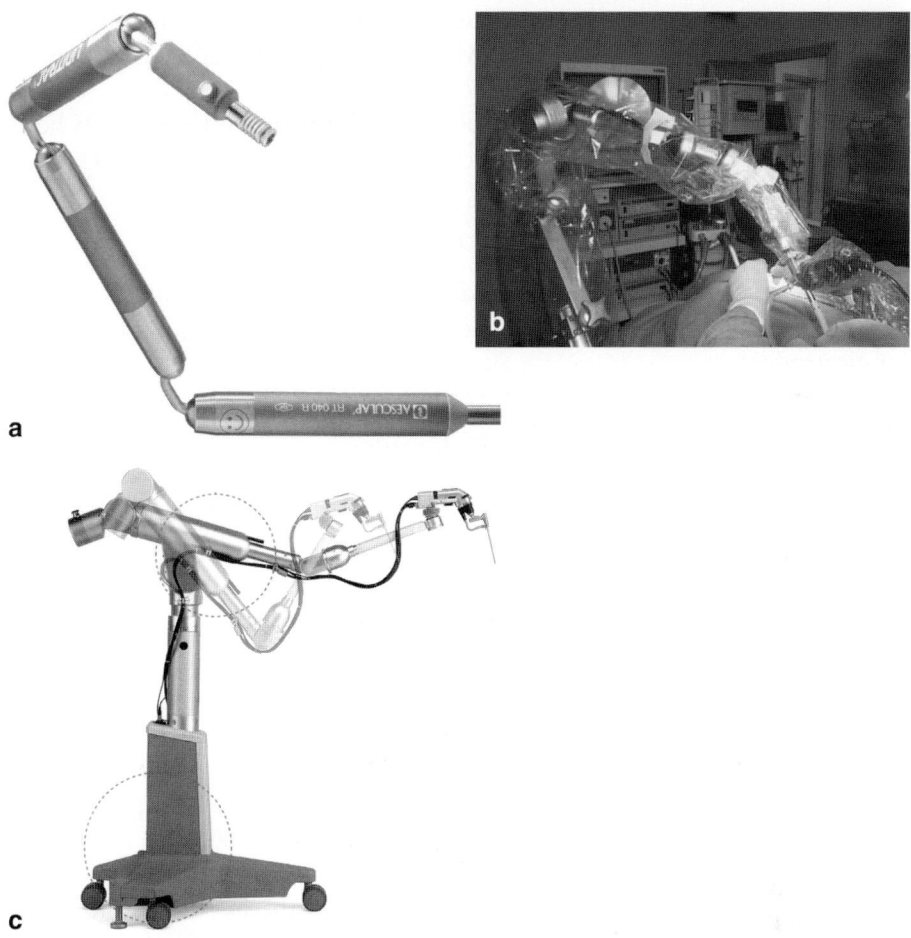

Fig. 14. (a) Holder Unitrac Aesculap: Pneumatically assisted holder (image courtesy of Aesculap – Tuttlingen, Germany). (b) the Storz-Mitaka arm. (c) the Endo-Arm from Olympus

necessary (Schroeder *et al.* 2001); however, in cases with thickened, non-translucent third ventricular floors, neuronavigation is useful for anatomical orientation (Alberti *et al.* 2001, Tirakotai *et al.* 2004). Brain shift can be a major factor in influencing the accuracy of the target localization. This problem occurs less often if some precautions are taken to prevent the abrupt change of CSF compartments or cystic lesion. The position of the burr hole should be at the highest point in order to minimize CSF loss. Moreover, brain distortion occurs rarely in midline structures and most endoscopic procedures use midline structures as anatomical landmarks.

Neuronavigation requires a rigid three-pin head fixation, difficult to obtain in case of younger babies. Moreover the neuronavigation can be coupled only with rigid endoscopes.

Equipment for neuronavigation coupled with neuroendoscopy has been discussed in a recent paper by Tirakotai *et al.* (Tirakotai *et al.* 2004).

Operative Instruments

Operative instruments for neuroendoscopy include sharp micro scissor, blunt micro scissor, biopsy forceps, grasping forceps, monopolar and bipolar electrodes.

Floor Perforation

The perforation of the floor can be achieved mechanically (by either a sharp instrument or, in combination with more force, a blunt instrument), electrically or with the aid of a laser.

Perforation With the Endoscope Itself

The endoscope can be gently pushed through the floor behind the clivus, stretching the fibers of the floor progressively until complete perforation is achieved and entry into the subarachnoid spaces is ensured by the sudden, direct visualization of the anatomical structures of the interpeduncular cistern (El-Dawlatly *et al.* 1999, El-Dawlatly *et al.* 2000, Teo and Jones 1996). This technique has several inconveniences: the traction on the floor can be significant and it is directly transmitted to the hypothalamic structures situated above. Until perforation is achieved this is a blind procedure, with no visual control of the depth reached by the endoscope or of the space remaining behind the membrane to be perforated.

Monopolar or Bipolar Coagulation

It is the most widely used technique (Cinalli 2004, Hellwig *et al.* 1999, Sainte-Rose and Chumas 1996). The advantages are evident. The point at which to perforate can be precisely chosen: if the floor is translucent, the tip of the coagulating wire can be positioned where the interpeduncular cistern is wider, as far as possible from the basilar bifurcation. Without applying cautery current, the tip of the wire can be used as a probe to "palpate" the floor of the third ventricle or to pierce it (Siomin and Constantini 2004). The coagulation is especially useful when the floor is very large and floating in the lumen of the ventricle: it allows the catheter tip to adhere to the chosen point, avoiding the natural tendency of the tip to slide. Coagulation should be used at the lowest effective energy to bring about coagulation of the floor. In most cases it is not necessary to maintain the coagulation until the perforation is achieved. A very short coagulation (<1 s) is usually sufficient to weaken the floor enough to allow perforation easily and atraumatically with the inactive probe, avoiding the risk of entering the interpeduncular cistern with an electric device on.

Both monopolar and bipolar coagulation are useful in this regard. Coagulation is also useful to achieve hemostasis. Most bleedings are venous with a slow flow, and can be managed only with irrigation. Sometimes a Fogarty balloon can be used to tamponade a bleeding vessel or the margins of a cutting (i.e. the stoma in the floor of the third ventricle). However, in some instances, neurosurgeon must appeal to coagulation to achieve hemostasis. Monopolar cautery can be used in both cutting and coagulation modes to achieve fenestration, dissection or cauterisation. The use of electrical current can be associated with some problems (Vandertop *et al.* 1998). The pathway of the currents flowing out of a tip cannot be controlled because the fluid in the ventricles is conductive and the current flows along the way of least resistance. Moreover energy losses caused by resistance in the leads makes them less efficient, so that very high currents could be necessary. Thus, tissue adherence and thermal damage of surrounding neural tissue are the major limitations of these instruments. However, the thermal damage to the hypothalamic region following coagulation has never been accurately studied. It may perhaps explain the fever sometimes observed after third ventriculostomy (Decq *et al.* 2000, Decq 2004, Sainte-Rose and Chumas 1996). Because of these problems, Heilman and Cohen (Heilman and Cohen 1991) invented a "saline torch": a device that sends a jet of saline past a monopolar wire. The saline acts as a conductor and coagulation can be achieved without direct contact with the probe (Shiau and King 1998).

Bipolar cautery may represent a more controlled method of coagulation: it has demonstrated minimal current spread; it permits sharply demarcated coagulation fields and precise cuts; damage to lateral or underlying structures is kept to a minimum. Therefore, bipolar coagulator should be preferred (Shiau and King 1998). The simplest way to achieve bipolar coagulation is through a fork electrode (Aesculap 2.1-mm fork electrode). The use of grasping bipolar forceps (2.5 mm – Codman & Shurtleff, Johnson & Johnson, Raynham, MA) allows the surgeon to pick up tissue for dissection and fenestration, and to coagulate vessels of more than 2 mm in diameter. Riegel *et al.* (Riegel *et al.* 2002) developed a new microbipolar forceps (ERBE Elektromedizin GmbH, Tübingen, Germany) that can be used for grasping, dissection, dilation, shrinkage of tissue, and precise coagulation even of larger vessels. The branches of the forceps are moved via elastic deformation of the metal without the use of a mechanical joint of

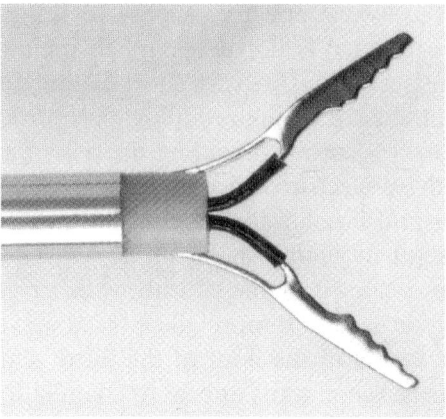

Fig. 15. The lack of mechanical joint allows very slow and delicate movements in this bipolar forceps produced by ERBE (Courtesy of ERBE Elektromedizin GmbH, Tübingen, Germany)

any kind (Fig. 15). The instrument has a outer diameter of 1.5 mm along its entire length and is compatible with most working channels of neuro-endoscopes. It can be opened up to a width of 6 mm. So, it can be used either to perforate or to enlarge the stoma. Bipolar electrodes of different shape are also available and are extremely effective for coagulation and perforation (Fig. 16).

Decq Forceps

Also this instrument can perforate the floor and enlarge the stoma (Decq *et al.* 2000). It is a modified endoscopic flexible grasping forcep with an outer diameter less than 1 mm, allowing it to be used with working channels of virtually all endoscopes. The tip is thin enough to allow easy perforation of the floor of the ventricle by the application of gentle pressure, while its pointed but blunt shape do not damage structures like vessels as a needle could.

The peculiarity of this forcep is that the inner surface is smooth whereas the outer surface presents indentations: this avoids accidental catching of vessels during closure and slipping of the edges of the stoma during opening, allowing easy dilatation with one single movement and avoiding the repeated manoeuvers that are often necessary to enlarge the first opening and that are potentially hazardous (Fig. 17a–c). The opening is approximately 4 mm in diameter. The advantage of this forcep is that it combines a thin, almost pointed tip with the potential for performing a gentle dissection by opening the jaws, especially when the floor is thick and difficult to puncture (Cinalli 2004, Decq *et al.* 2000).

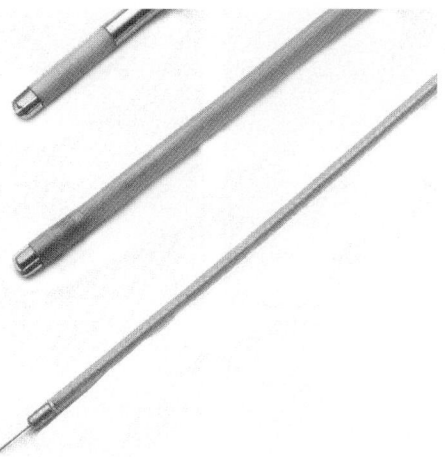

Fig. 16. Single, smooth tip bipolar electrodes allow atraumatic and small coagulation and perforation (Courtesy of ERBE Elektromedizin GmbH, Tübingen, Germany)

Laser

The physics of lasers in medicine has been reviewed by Nobles (Nobles 1998) and Siomin and Constantini (Siomin and Constantini 2004). Application of laser to tissue causes an instantaneous increase in temperature, leading to vaporization of the cells. Most lasers cannot be used in neuroendosopy, because they are not able to work through water and transmit through a miniature fiberoptic cables (600 μm and 400 μm). The only lasers suitable for neuroendoscopy are the neodymium:yttrium aluminum garnet (Nd:YAG) laser, the argon laser and the potassium-tetanyl-phosphate (KTP) laser (Nobles 1998, Shiau and King 1998, Siomin and Constantini 2004, Wharen *et al.* 1984, Wong and Lee 1996). A sharper-edge tip maximizes the cutting ability, but diminishes its coagulative properties (Shiau and King 1998). The lasers can be used in a contact or non-contact mode. Free laser light would be rapidly absorbed by the CSF with scattering and possible thermal injury to the surrounding structures (Vandertop *et al.* 1998). The contact probe offers more controlled tissue vaporization and requires less energy (Shiau and King 1998). In this case, the tip of the probe can become heat. The tip may remain hot even when the laser is off, so that care should be given to do not inadvertently damage neural tissue. Some authors have proposed the use of contact tipped Nd:YAG laser (1064 nm) (Miller 1992, Oka and Tomonaga 1992, Ymakawa 1995), although this only offers a partial solution to the problem. Vandertop *et al.* (Vandertop *et al.* 1998) propose the use of specially designed laser probes with atraumatic ball-shaped fiber tips coated with a layer of carbon particles. This allows 90% absorption of the laser light that is converted into heat,

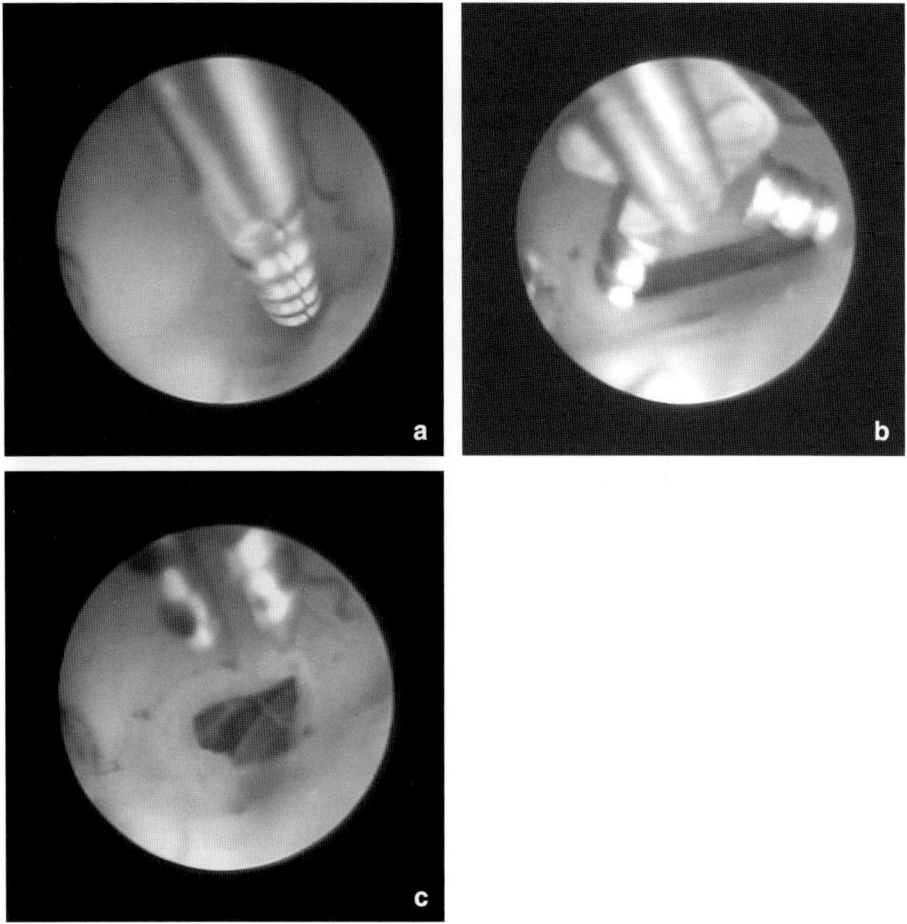

Fig. 17. The forceps described by Decq (Karl Storz GmbH & Co., Tuttlingen, Germany) present a smooth tip when closed, suitable for floor perforation (a), and indentation on the outer surface, allowing opening of the perforation (b) to a satisfying diameter (c) with one single movement (image courtesy of Dr. Philippe Decq, Hopital Henri Mondor, Creteil, France)

allowing both the amount of laser light used and the length of exposure to be reduced (Willems *et al.* 2001). Other authors (Büki *et al.* 1999) have proposed combined pulsed holmium (Ho)-Nd-YAG laser and claim that it is superior to mechanical cutting of the tissues, both for third ventriculostomy and for cyst fenestration. According to Siomin and Constantini (Siomin and Constantini 2004), the KTP laser offers some advantages on Nd:YAG laser: the emission of a visible light, that makes it easier to manipulate; an inferior tissue penetration, that makes it safer, and less dependency on tissue pigmentation, that makes it more versatile.

Nd:YAG laser and KTP laser are more useful in case of tumor removal and cyst fenestration than in case of third ventriculostomy. In fact, great care should be given when using a laser to perform a third ventriculostomy, since a case of injury of the vessels of the interpeduncolar cistern has been reported (McLaughlin *et al.* 1997).

Suction-Cutting (Grotenhuis) Device

The suction-cutting device (Synergetics), developed recently by Grotenhuis, is composed of a thin suction cannula that can be introduced through an operative channel at least 2 mm in diameter. The outer surface and the edges of the inlet of the cannula are smooth, whereas small blades are inserted into the lumen of the cannula. When the tip of the cannula comes into contact with the floor of the third ventricle, the suction hole on the handle is closed and the membrane is sucked into the lumen of the cannula. Rotation of the cannula allows section of only the tissue aspirated into the lumen, limiting the risk of accidental injury to vascular structures.

"Semisharp" Instruments

The cautious blunt perforation is usually safe, but in case of more resistant floor of the third ventricle a forceful pushing of the instruments is necessary and might be dangerous. Surgical tools specifically designed for safe perforation of a resistant and/or thick floor have a semisharp, slightly angulated tip that, directed anteriorly and pushed inferiorly along the clivus, would allow safe perforation minimizing the risk of injury of the basilar artery (Kehler *et al.* 1998).

Ultrasound Microprobes

Ultrasound microprobes have been specifically designed for use through the working channel of the endoscope (6 French) or paraendoscopically (8 French). These probes offer the major advantage of direct visualization of the anatomical structures of the interpeduncolar and prepontine cisterns and can be used for blunt perforation of the floor. This allows safer perforation under the double control of the floor of the third ventricle (endoscopic) and of the anatomical and vascular structures hidden behind the floor membrane (ultrasonographic) (Paladino *et al.* 2000, Resch and Reisch 1997, Resch and Perneczky 1998, Resch 2003) (Fig. 18a–c).

Forceps and Scissors

Instruments of various design are commercially available, suitable for rigid or flexible endoscopes. These include: grasping forceps, biopsy forceps

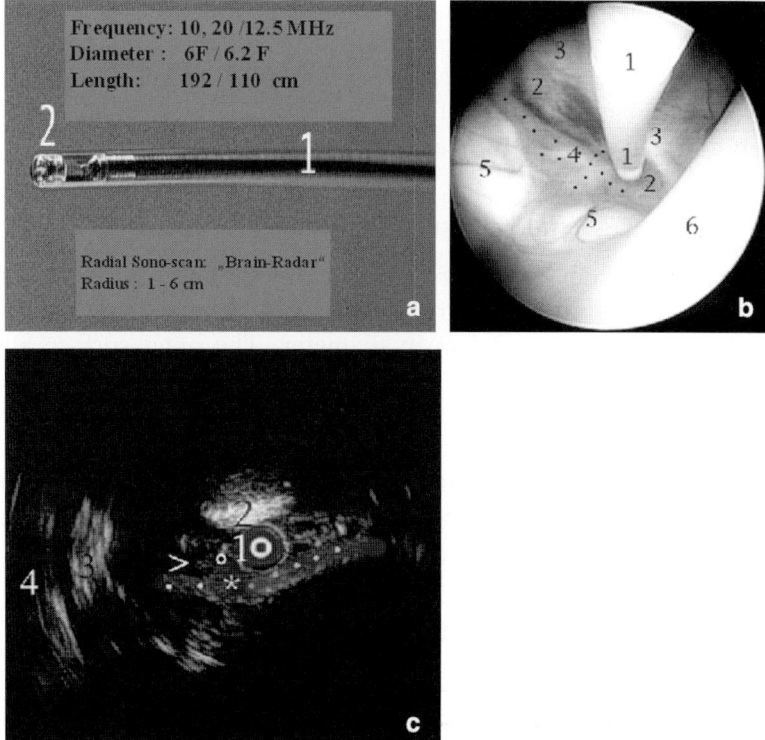

Fig. 18. (a) Probe for endosonography (*1* Sono-catheter, *2* Sono-Mini-Probe). (b) anatomic view of the third ventricle during neuroendosonography (*1* Sono-Mini-Probe, *2*
Cinereum (premammillar membrane), *3* Dorsum of Sellae, *4* Basilar Head (beneath
floor of 3. ventricle), *5* Mammillary Bodies, *6* Hypothalamus. (c) *1* Sono-Mini-Probe,
2 Dorsum of Sellae, *3* Oculomotor Nerve, *4* Tentorial Notch, * Basilar Artery, ..
Right and Left P1

and straight or curved microscissors. Their use is finalized chiefly for tissue
biopsy and fenestration of membranes. Flexible instruments for flexible-
steerable fiberscope may have the disadvantage that they do not open gradually. They snap, with a short delay between the action of the hand and the
effect at the tip of the instrument (Hellwig *et al.* 1998b).

Dilatation of the Stoma

After perforation of the floor, the hole obtained is usually no larger than
the outer diameter of the instrument used (<2 mm). If left like this, it is
bound to close rapidly because of the inflammatory reaction induced by
the thermal or mechanical injury inflicted for perforation and the consequent formation of glial scar tissue.

Grasping or Biopsy Forceps

Dilatation can be achieved by introducing a grasping or biopsy forceps into the hole closed, then carefully opening it. This technique is usually relatively safe, but has several drawbacks. The dorsal surface of the forceps is smooth, so that the edges of the hole slip on this surface during opening: this necessitates several manoeuvres of opening and closing the forceps to obtain a satisfactory result, especially in the case of a thick, nontranslucent floor. These repeated manoeuvres can result in accidental grasping of a perforating vessel in the interpeduncular cistern.

Fogarty Balloon

Dilatation with a Fogarty balloon (usually 3 or 4 French) is much safer, since both the tip of the catheter and the surface of the balloon are smooth. For the same reason, the edges of the stoma slip very easily on this surface, requiring repeated inflations before the stoma is dilated to the largest diameter allowed by the balloon (4–5 mm). Fogarty catheter is the cheapest and most common used device; the balloon must be inflated with saline. This allows the balloon to inflate very gradually, without the sudden inflation that is quite traumatic and is usually observed when only air is used for inflation or too much air remains trapped within the balloon when saline is used.

Double Balloon Catheter

A balloon specifically designed for third ventriculostomy (Lighttouch balloon, Integra Neuroscience, Biot, France) offers the ideal solution since the inflatable part is a dumbbell-shaped silicone membrane. The narrowest part is marked with a black dot when the balloon is deflated. The catheter is introduced deflated into the stoma and advanced until the black dot is at the level of the floor. Unlike Fogarthy balloon, the double balloon catheter must be inflated with air. The proximal part inflates first, then the distal one (Fig. 19). The floor membrane remains trapped between the two balloons and the stoma is gently dilated with further air inflation to the largest extent allowed with one single maneuver. This technique remains by far the simplest, safest, and fastest and should be recommended. However, the outer diameter of the available double balloon catheter is larger than 1 mm and cannot be used in small instrument channels.

The "Urological" Device

A variant of the balloon technique has been proposed using instruments designed for stone extraction in urological endoscopic surgery (Wong and

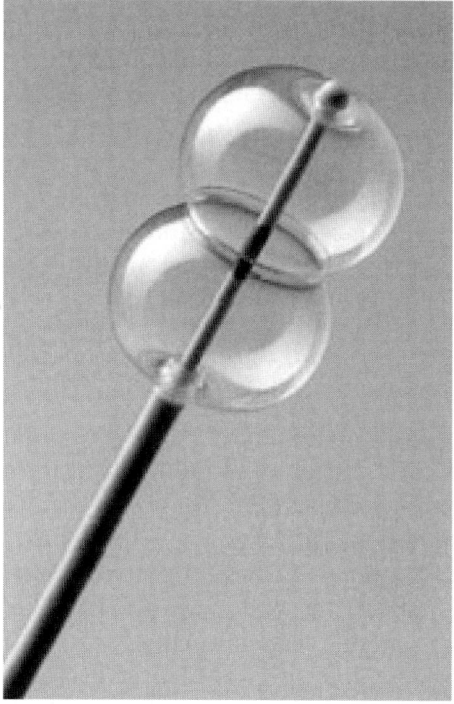

Fig. 19. The light-touch balloon is dumbbell shaped, in order to allow the floor membrane not to slide over the smooth surface of a normal Fogharty balloon. With this shape one single inflation is possible to obtain the largest dilatation (image courtesy of Integra Neuroscience, Plainsboro, NJ)

Lee 1996). The device is composed by a four-flat-wire basket tip that has a 1-mm outer diameter when closed. When opened it has a basket-like shape that enlarges the stoma by tearing the edges. The risk of catching a small vessel within the wires of the basket during closure reduces the advantages of this device compared to a Fogarty balloon.

Decq Forceps

As already discussed this is an useful tool to both perforate and dilate the floor. The inner surface of the forceps is smooth avoiding accidental catching of vessels during closure, while the outer surface presents indentations, avoiding the slipping of the edges of the stoma during opening, allowing dilatation with one single movement.

Opening of Liliequist's Membrane

The opening of Liliequist's membrane requires delicate surgical manipulation, since it can be more difficult to perforate than the floor of the third ventricle itself. Bipolar coagulation should be preferred; monopolar coagulation should be avoided because of the proximity of the basilar artery. In any case, only smooth instruments should be used at this level and the grasping forceps should also be avoided.

Disclosures: The authors have no financial interest in the material presented and discussed in this chapter.

Indications

A correct clinical indication is the most important factor influencing the success rate of ETV. Advances in MR imaging have improved the understanding of the specific physiopathologic mechanism at the base of the different forms of hydrocephalus consequently representing an essential adjunct for the correct patient selection. A more accurate preoperative diagnosis of mixed types of communicating hydrocephalus, where obstructive factors may still play a role, has lead to extend the indications for ETV to patients once excluded, namely subjects with posthemorrhagic hydrocephalus, postinfectious hydrocephalus and hydrocephalus in myelomeningocele. However, several points are still under discussion. The first is the role of age. In past years, many authors reported low success rates in infants independently from hydrocephalus etiology (Buxton et al. 1998a, Buxton et al. 1998b, Hopf et al. 1999b, Kim et al. 2000). Actually an increasing number of papers have challenged this assumption recently. In fact, when infants with pure obstructive hydrocephalus are considered, success rates up to 85% were reported (Beems and Grotenhuis, 2002, Javadpour et al. 2001). A second important point that has merged in recent years is that the primary hydrocephalus features may change in time in some patients; for example, "reopening" of subarachnoid spaces may occur in cases of posthemorrhagic or postinfectious hydrocephalus in children who initially presented with purely communicating forms of hydrocephalus; an acquired aqueductal stenosis is propounded as the long term main etiological factor accounting for the hydrocephalus persistence (Cinalli et al. 1998, Smyth et al. 2003).

Pure Obstructive Hydrocephalus

Aqueductal Stenosis

Aqueductal stenosis (AS) is responsible for 6–66% of cases of hydrocephalus in children and 5–49% in adults (Hirsch et al. 1986). Many classifications of this clinical condition were proposed; the first and most important distinction is between primary stenosis, due to intrinsic pathology of the aqueduct itself and secondary stenosis, which follows a compression of the aqueduct from mass lesions (i.e. tectal tumors, pineal region tumors,

Galen vein region vascular malformations) or in patients with communicating hydrocephalus from excessive dilatation of the occipital horns.

Primary Aqueductal Stenosis

Although pure primarily aqueductal stenosis is considered as a congenital condition, hydrocephalus secondary to AS is frequently diagnosed during adolescence or in adulthood. Different "evolution theories" have been proposed to explain this phenomenon. Head trauma, a small subarachnoid hemorrhage or a viral infection with benign meningitis have all been claimed as anatomic causes of aqueductal stenosis decompensation (Cinalli *et al.* 2004a, Jellinger, 1986, Lapras *et al.* 1986). A functional mechanism might also contribute: progressive enlargement of the lateral and third ventricles would lead to a distortion of the brainstem and kinking of the aqueduct (Nugent *et al.* 1979, Raimondi *et al.* 1976).

Secondary Aqueductal Stenosis

CSF flow obstruction in pure obstructive hydrocephalus may be the consequence of aqueductal compression by lesions arising within or around the Sylvian aqueduct. Tectal gliomas and hamartomas are actually considered one of the most frequent causes of this kind of condition. Unlike the majority of other diffuse brainstem tumors tectal tumors usually have an indolent clinical course, often remaining stable in size for several years (Chapman, 1990, Pollack *et al.* 1994).

Hydrocephalus in children with tectal plate tumors commonly becomes symptomatic when the tumor is small in size; indeed a circular restriction of the aqueduct occurs due to supependymal tumor growth. Brainstem clinical signs are often lacking; in most of the patients the management of hydrocephalus warrants a favorable long-term outcome (Cinalli *et al.* 2004a).

Differently from what occurs with tectal plate tumors, pineal region tumors are usually larger in size when hydrocephalus becomes clinically manifest. In these cases, Sylvian aqueduct obstruction is generally the result of a distortion of the whole mesencephalon from outside by part of the tumor (Rieger *et al.* 2000). Vascular malformations are a rare cause of secondary Sylvian aqueduct obstruction; abnormal draining veins of midbrain arteriovenous malformations may obstruct the CSF flow crossing the aqueduct so as large fusiform aneurysms of the basilar artery. Hydrocephalus in children with vein of Galen aneurysms has a multifactorial origin; beside venous hypertension a tectal plate compression and tonsillar herniation may contribute to the development of hydrocephalus (Russell, 1940). On these grounds, ETV may be considered in children with persis-

tent hydrocephalus after aneurysm/AVM endovascular treatment according to Cinalli *et al*. (Cinalli *et al*. 2004a).

Pathogenesis of Hydrocephalus in Patients With Aqueductal Stenosis

It is actually accepted that CSF flow through the aqueduct is not unidirec-
tional but has a systolic and a dyastolic cycle. According to recent MRI
studies, a downward displacement of CSF would occur during systole.
The decrease of blood volume and the recoil of CSF displaced in the lumbar sac during the dyastolic phase of the cardiac cycle would lead to a partial upward reversal of CSF flow. A net downward movement of CSF will
result, corrisponding to CSF production (0.0067 ml × s) (Enzmann and
Pelec, 1991, Quencer *et al*. 1990). Changes in aqueductal size and shape
alter this volume flow rate and pattern. In aqueductal stenosis a decrease
of CSF flow and an increase up to ten times of its normal velocity occurs,
with a consequent increase of wall shear stresses (Jacobson *et al*. 1999).
Increasing pressure within the aqueduct may be responsible for secondary
gliosis and for a progressive narrowing; in the "forking form", CSF flow
turbulence and stasis further contribute to CSF accumulation. The consequent increase of CSF pulsations during systole lead to periventricular
edema, partial displacement of blood from parenchymal vessels and further
accumulation of CSF in the lateral and third ventricles (Cinalli *et al*.,
2004a).

The role of CSF pulsations in the pathogenesis of ventricular dilatation
in this kind of hydrocephalus is confirmed by a previous experimental work
done at our institution. In this experience, the effects of increasing the amplitude of the intraventricular CSF pulse pressure in lambs was studied.
Intraventricular pulse pressure was artificially modified by inserting a pulsating balloon inside the lateral ventricles; modifications of ICP waves,
mean CSF pressure and ventricular anatomic structures were recorded.
No significant changes in mean CSF pressure were observed. On the contrary, ICP waves were significantly modified, with maximal changes when
the mechanical pulses were in phase with the physiological (arterial) ones.
Moreover, an increase of ventricular volume, with the pathological evidence of stretching of the ependymal layers and periventricular white
matter spongiosis was already obtained when the amplitude of the pulse
intraventricular pressure was increased three times the basal value. In the
animals, in which the intraventricular pulse pressure was mechanically
increased as much as six times the control value, a necrotic lesion of the
periventricular structures, associated with the ventricular dilatation, occurred (Pettorossi *et al*. 1978).

An alternative pathophysiological mechanism has been proposed in
infants in order to explain the relatively less satisfactory results of ETV

in this specific subset of patients. Oi *et al.* (Oi *et al.* 2004) prospectively analyzed 12 infantile cases of aqueductal stenosis with CT ventriculo-cisternography. Nine of them revealed to have a predominant transependy-mal intraparenchymal CSF outflow, with a minor involvement of the aque-duct pathway. On these grounds, the authors proposed that CSF dynamics undergo an evolutionary mechanism: a transependymal CSF absorption (also defined as minor pathway) would be favored in neonates and infants due to the incomplete development of Pacchionian bodies; once ended, the maturation process of subarachnoid spaces the CSF flow through the aque-duct (major pathway) and to the cranial vault would prevail.

Analysis of Results of ETV in Patients With Primary and Secondary Aqueductal Stenosis

Aqueductal stenosis is considered the ideal indication for third ventriculos-tomy. When patients with an exclusively obstructive triventricular hydroce-phalus are selected, the success rate is quite homogeneous and stable, being between 63% and 92% in most series (Böschert *et al.* 2003). The influence of age on the final result has been long discussed. Jones *et al.* (Jones *et al.* 1994) reported a 40% success rate of ventriculostomy in children with triv-entricular hydrocephalus operated on within the age of two years. For the same pathology with a clinical onset within the age of two years, but oper-ated on later in childhood and adolescence the success rate increased to 71%. Kim *et al.* (Kim *et al.* 2000) reported age as a significant predictor of outcome. Of the six patients younger than one year of age, only two (33.3%) had good outcomes, whereas 19 out of 23 patients older than one year of age (82.6%) were successfully treated with ETV. Similar results were reported by Grunert *et al.* (Grunert *et al.* 2003): in their personal ex-perience a statistically worse success rate was found in infants (22.3%) if compared with children aged over one year (71.4%) and adults (81.6%). In-complete development of the subarachnoid spaces and consequent impair-ment of CSF absorption are the main factors which are claimed to explain why ETV fails more frequently in infants (Buxton *et al.* 1998a, Buxton *et al.* 1998b, Hopf *et al.* 1999b, Kim *et al.* 2000). However, considering exclusively children with pure obstructive hydrocephalus, other factors might have influenced the reported ETV failure rate in infants. First of all, in most of these series a second endoscopic view was never or not always performed, in order to estabilish ventriculostomy patency and this factor might further have negatively influenced the overall outcome. In a multicenter study Siomin *et al.* (Siomin *et al.* 2001) reported repeat ETV as effective as primary procedures with an overall 65% success rate. Javad-pour *et al.* (Javadpour *et al.* 2001) analyzed the ETV success rate in a series of 21 infants with obstructive hydrocephalus. The success rate in 7 infants with congenital aqueductal stenosis was 71%; a redo third ventriculostomy was performed with success in one of the two initial failures with a final success rate of 86%. In a recent personal communication, Wagner *et al.* (Wagner and Koch, 2004) confirmed this finding, reporting the occlusion of the ventriculostomy site as the most frequent factor causing ETV failure in infants. The lower ICP with slowing of CSF flow through the stoma might explain ventriculostomy obstruction in this subset of patients (Böschert *et al.* 2003).

Another important point to consider is the time for defining ETV failure. According to Beems and Grotenhuis (Beems and Grotenhuis, 2002), infants have a longer adaptation time if compared with their older counterpart. Fourty-five percent of the 35 patients, aged less than two years, who underwent successful ETV at their institution required a relatively long adaptation time (mean adaptation time: 1 week). This is important to recognize, because it means that these young patients should not be treated too soon using a shunting device if the ETV does not lead to immediate relief of the symptoms.

Finally, concerning older series, a further factor which may have lead to a lower success of ETV in infants is the adopted surgical technique. In their earliest experience Cinalli *et al.* (Cinalli *et al.* 1999) reported a higher failure rate in infants (<6 months of age) operated on under ventriculographic guidance. In a subsequent series of 119 cases with triventricular hydrocephalus who underwent endoscopic third ventriculostomy, no significant difference was found between the two groups with an overall success rate of 79% (<6 months) and 71% (>6 months) respectively. Direct visualization of the ventricle floor and the anatomical landmarks of the interpeduncular cistern under magnified conditions allow a greater safety and efficacy of the procedure. Moreover, the larger size and more precise location of the stoma in the floor of the third ventricle might play a role in young patients, in whom the risk of secondary obstruction may be high.

A further discussed point is the influence of AS etiology on the success rate of ETV procedures. Most authors report ETV as effective in primary as in secondary forms of aqueductal stenosis (Boschert *et al.* 2003, Fukuhara *et al.* 2000, Hopf *et al.*, 1999b, Rieger *et al.* 2000). Pople *et al.* (Pople *et al.* 2001) reported ETV to be effective in the control of the hydrocephalus in 17 out of 18 cases with pineal region tumors. Macarthur *et al.* (Macarthur *et al.* 2001) referred a long-term (median follow-up: 12 months) control of the hydrocephalus in 39 out of 47 pediatric patients (82.9%) with secondary neoplastic AS. Similar results were reported by Rieger *et al.* (Rieger *et al.* 2000) in a series of 7 patients with pineal region tumors. A successful control of the hydrocephalus was obtained in all cases; however six of the seven children also underwent tumor removal few days after the neuroendoscopic procedure, this factor possibly contributing to the long term results. In contrast with what previously said, Goh and Abbott (Goh and Abbott, 2000) recently reported a 49% failure rate of ETV in 63 patients with tumoral AS (mean follow-up: 11.4 months). This result compared unfavorably with their 65% success rate obtained in children with primary aqueductal stenosis. An analysis of the different series does not seem to show differences in the prevalence of tumor location. Neverthless none of the reported authors details histological types nor the presence or absence of subarachnoid tumor seeding, a factor which might specifically influence the success of ventriculocisternostomy. This is particularly true for some pineal region tumors (i.e. germinal cell tumors) and it is confirmed by the fact that when exclusively benign lesions are selected (i.e. tectal plate gliomas) the success rates of ETV are almost homogeneous (ranging between 65% and 100%) and comparable with those obtained in patients with primary AS (Grunert et al. 2003, Macarthur et al. 2001). Moreover, as part of the therapeutic protocol, endoscopic tumor biopsy was combined with ETV in selected cases of some of these series; either minor intraoperative bleeding and/or increase in CSF proteins which can follow the bioptic step of endoscopic surgical procedures may condition the functional patency of the ventriculostomy site.

Another point that has been investigated in recent years is the role of ETV as secondary procedure in children with AS and previously implanted malfunctioning or infected shunt. Most series report results comparable to those obtained in primary procedures with overall success rates ranging from 62.9 to 82.3% (Böschert et al. 2003, Cinalli et al. 2004). The relative variability of these results can again be ascribed to different selection criteria and different methods in the evaluation of results. Böschert et al. (Böschert et al. 2003) found that two of the three failures in their seventeen patients series occurred in children with shunt malfunction and a previous history of shunt infection, suggesting a concurrent CSF absorption impairment in the pathogenesis of the hydrocephalus for these cases. However, other authors have pointed out that previous shunt infections or infected shunt malfunctions do not influence outcome. Jones et al. (Jones et al. 1990) reported on 4 patients having ETV for AS and infected shunt malfunction having a 75% success rate. Buxton et al. (Buxton et al. 2003) referred that 33% of their 88 patients had previously suffered one or more shunt infection, but they did not observe significant differences in their outcome if compared with those who did not have a history of an infected shunt.

Regarding evaluation of results Cinalli et al. (Cinalli et al. 1998) reported 30 patients, having a 76.7% ETV success rate, but many patients in their series were left with a potentially patent shunt in situ; only seven patients were left in the initial stages to rely on the NTV alone (6 with their shunts removed and one with it clipped). More stringent criteria were used by Buxton *et al.* (Buxton *et al.*, 2003) who retrospectively analyzed a personal series of 27 children with AS who underwent secondary third ventriculostomy at their institution. The shunt was removed or tied off in the neck in all cases; the final success rate was 62.9%.

Hydrocephalus in Posterior Cranial Fossa Tumors

The incidence of preoperative hydrocephalus in children with posterior fossa tumors is around 60–80%; however only 15–25% of these patients require a permanent "shunting" procedure. Younger age, the severity of hydrocephalus at diagnosis, midline tumor localization, degree of tumor removal and the use of substitute dural grafts are all factors considered to increase this risk (Sainte-Rose *et al.* 2001).

Pathogenesis

The development of hydrocephalus in children with posterior fossa tumors is directly related at diagnosis with the nature and localization of the tumor. Tumors arising or secondarily filling the IV ventricle primarily obstruct the ventricular cavity and its foraminal outlets. On the other hand, cerebellar hemisphere tumors induce an anatomical distortion of the fourth ventricle with a secondary occlusion of CSF pathways. In patients with malignant tumors, subarachnoid seeding may contribute, impairing CSF absorption. Though, logically, tumor removal should re-estabilish the communication between the fourth ventricle cavity and the subarachnoid spaces one-fourth to one third of these patients remain hydrocephalic (Sainte-Rose 2004). In the immediate postoperative period the surgical subarachnoid hemorrhage and the presence of cerebellar swelling may contribute to increase the resistance in CSF circulation. Subsequently the development of adhesions at the level of the fourth ventricle outlets and the adjacent cisterns may permanently alter CSF dynamics.

Management Strategies: The Role of ETV

Different protocols have been proposed for the management of hydrocephalus in children with posterior fossa tumors. Actually, most pediatric neurosurgeons use a combination of corticosteroids, early tumor surgery and external ventricular drainage where needed (Fritsch *et al.* 2004, Sainte-Rose 2004). Children with persistent postoperative hydrocephalus can be alternatively treated with ETV or shunt implantation. An important drawback of this protocol is the risk of CSF infections which is related with external ventricular drainage positioning (10% in the series of Rappaport and Shalit; 4.9% in the series of Schmid) (Rappaport and Shalit, 1989, Schmid and Seiler, 1986); upward brainstem herniation and intracranial hemorrhages have been also described.

Due to its relatively recent employment there is a noticeable lack of information about the role of ETV in the management of persistent hydrocephalus after posterior fossa tumor surgery. The reported successes vary from 50% (Jones *et al.* 1990, Ruggiero *et al.* 2004) to 100% (Sainte-Rose *et al.* 2001).

An alternative management strategy has been proposed by Sainte-Rose *et al.* (Sainte-Rose *et al.* 2001). Sixty-seven patients affected by posterior fossa tumors and hydrocephalus underwent endoscopic third ventriculostomy prior to tumor removal; postoperatively four of them (6%) required a second shunting procedure. A comparable group of 82 children underwent tumor removal as first surgical step; the rate of postoperative hydrocephalus in this subset of patients was 26.8%, significantly higher than in the first group ($p = 0.001$). However, the same authors acknowledge that the routine application of preoperative third ventriculostomy results in a proportion of patients undergoing an "unnecessary" procedure. The overall "shunting" rate in their preoperative ventriculostomy group was 106%. A further possible factor against preoperative ETV is the risk that surgical subarachnoid hemorrhage may induce the occlusion of the stoma within the floor of the third ventricle. Indeed, the role of preoperative ETV in "preventing" postoperative hydrocephalus has not been confirmed in more recent series. Ruggiero *et al.* (Ruggiero *et al.* 2004), performed ETV in 20 children with posterior fossa tumors preoperatively. One of the procedures was complicated by intraventricular bleeding requiring an external ventricular drainage and subsequent VP shunt implantation. Three of the remaining 19 patients developed postoperative hydrocephalus with a final 20% postoperative shunting rate. This percentage was comparable in their experience with the 15% rate of persistent hydrocephalus in children who underwent tumor removal as first surgical step.

Hydrocephalus With Possible Subarachnoid Spaces Impairment

Posthemorrhagic Hydrocephalus of Premature Infants

Between twenty and seventy-four percent of infants suffering an intraventricular haemorrhage (IVH) will go on to develop posthemorrhagic hydrocephalus (Boop 2004). It is important to remember that progressive ventricular dilatation in these patients does not represent always a pressure-related phenomenon. It may indeed represent an ex vacuo manifestation related to loss of brain substance as a result of venous infarction or periventricular white matter ischemia; this kind of evolution is more frequent when an hypoxic-ischemic encephalopathy is associated. In some other instances, a combination of volume loss and slowly progressive hydrocephalus occurs, making it difficult to determine whether to treat or follow the infant conservatively (Siomin *et al.* 2002). These factors may help to explain the variability of reported posthemorrhagic hydrocephalus rates and the differences in the surgical indications that may be found in the literature.

Pathogenesis

Posthemorrhagic hydrocephalus is presumed to develop as a consequence of the breakdown of blood products and cellular debris within the ventricular system. These blood products in turn cause chemical arachnoiditis and a fibrotic reaction within the ventricles and the arachnoid granulations, leading to granular ependymitis and adhesive arachnoiditis (Hill and Volpe, 1981). CSF flow studies seem to confirm this hypothesis, suggesting that it is an impairment of CSF circulation over the cerebellum and up to the cerebral vault subarachnoid spaces which is implicated in the pathogenesis of hydrocephalus in premature infants. This would explain why, in most cases, the increased intracranial pressure and full fontanel can be ameliorated with lumbar punctures temporarily (Roland and Hill, 1997). However, in some instances, infants with IVH develop a purely triventricular hydrocephalus. Kreusser *et al.* (Kreusser *et al.* 1985) attempted serial lumbar punctures as temporary management of hydrocephalus in 16 infants; this procedure was unsuccessful in four of them who all presented a triventricular hydrocephalus at neuroradiological investigations. On these grounds, the authors suggested that, in selected cases, a blood debris may alter CSF circulation by obstructing the posterior third ventricle and/or the Sylvian aqueduct selectively. This hypothesis seems to be confirmed by the recent description of infants with an apparently pure aqueductal stenosis that had suffered a fetal IVH, documented by prenatal MRI and/ or by neonatal endoscopic findings previously (Beni-Adani *et al.* 2004). An evolutionary mechanism has also been suggested for cases initially presenting with an apparently communicating hydrocephalus, and hence primarily shunted, who underwent a successful third ventriculostomy at the time of shunt malfunction. In these specific cases, a reopening of subarachnoid spaces may occur as a consequence of the reabsorption of subarachnoid blood debris (Siomin *et al.* 2002, Smyth *et al.* 2003); moreover, the CSF produced by the choroid plexus of the fourth ventricle, during the shunting period, may maintain the absorptive function of the arachnoid granulations, furtherly improved by the persisting reduction of intraventricular pressure. Consequently, at the time ETV is performed, the access to previously impaired CSF absorption spaces can be obtained, by-passing a persistent obstruction at the level of the Sylvian aqueduct and/or the IV ventricle outlets (Siomin *et al.* 2002).

Results of ETV in Infants and Children
With Posthemorrhagic Hydrocephalus

In the past, a history of hemorrhage was considered as a contraindication for ETV, because most patients were considered to suffer a communicating form of hydrocephalus. With the extension of the indications, an increasing number of patients with hydrocephalus and a history of ventricular hemorrhage has been included in ETV management protocols. Actually, the most relevant paper that can be found in the literature is the one of Siomin *et al.* (Siomin *et al.* 2002) reporting the results of a multicentric retrospective study on the role of ETV in children with posthemorrhagic and postinfectious hydrocephalus. Overall 36 children suffered a history of hemorrhage with an ETV success rate of 55.6%. A striking difference was documented between primary and secondary ETV procedures. At a mean follow-up of 1.87 ± 1.6 years, the 13 posthemorrhagic patients who had primarily received a VP shunt had a 100% ETV success rate. On the other hand, all the premature infants in whom ETV was performed as first line of treatment subsequently required a shunt. This result is confirmed by other authors. Buxton *et al.* (Buxton *et al.* 1998b) reported on 16 infants (mean corrected age: 8.9 months) with posthemorrhagic hydrocephalus who all underwent ETV as primary treatment for their hydrocephalus with a success rate of 30%. More recently Smyth *et al.* (Smyth *et al.* 2003) described a 71.4% success rate in 7 children (mean age: 9.2), all the successes being documented after secondary ETV procedures.

Javadpour *et al.* (Javadpour *et al.* 2001) suggested that a further mechanism which may negatively influence the results of ETV in infants with posthemorrhagic hydrocephalus is their low intracranial pressure; ICP compensating mechanisms due to expanding opened sutures may impair to build up a sufficient pressure gradient across the stoma and the arachnoid granulations, with secondary insufficient CSF reabsorption and ventriculostomy closure. In their experience only three out of ten infants were successfully treated with ETV, two of them requiring two endoscopic procedures, because of first closure of the ventriculostomy.

Two further factors that influenced the outcome of infants who had undergone ETV in the multicentric study of Siomin *et al.* (Siomin *et al.* 2002) were a history of associated CSF infection (the overall ETV success rate in this specific subset of patients was 10%) and the time interval between the onset of hydrocephalus and ETV (mean temporal gap of 6.34 ± 8.31 years successful results versus 3.15 ± 5.16 years for treatment failures). The latter factor probably influenced the results obtained by Beems and Grotenhuis too (Beems and Grotenhuis 2002). In contrast with previous reports, these authors did not find significant difference between the results of primary (44.4%) and secondary (50%) ETV procedures; however the time interval from the hydrocephalus clinical appearance and the surgical operation was less than two years in the children who required a second procedure subsequently.

Postinfectious Hydrocephalus

In the past, postinfectious hydrocephalus was commonly considered as a communicating form of hydrocephalus, the predominant pathogenetic factor being the inflammatory obstruction of the basal cisterns and cerebral vault subarachnoid spaces. The progressing knowledge of pathophysiology in different CNS infectious diseases has lead to a reevaluation of this clinical entity that should actually be regarded as a complex disease.

Pathogenesis

The consequences of a CNS infectious disease on CSF dynamics depend on the age of the child at the time of the primary infection (prenatal, neonatal, postnatal) as well as on the infectious agent (bacterial, viral, parasitic).

Prenatal Infections. Toxoplasmosis is one of the most frequent prenatal infections involving the CNS. In this condition, the parasites invade and destroy the ependymal lining of the ventricles. In acute cases extensive necrosis of the cerebral hemispheres can be observed with associated significant cerebral tissue loss, which contributes to the ventricular enlargement. Most authors consider post-toxoplasmosis hydrocephalus a consequence of severe leptomeningeal inflammation blocking the subarachnoid spaces (Kaiser 1985, Ciurea *et al.* 2004). However, a significant percentage of these children present with purely triventricular hydrocephalus. According to the experimental model of Stahl (Stahl *et al.* 1997), the inflammatory debris may selectively obstruct, in these cases, the Sylvian aqueduct. Moreover, ventricular enlargement would lead to a compression of the midbrain further limiting CSF flow to the IV ventricle.

Another prenatal infection which is commonly related to CNS diseases is Cytomegalovirus infection. However, the occurrence rate of hydrocephalus in these children is relatively low (10–15%) (Ciurea *et al.* 2004). In most cases an extensive involvement of the cerebral structures takes place leading to a diffuse encephalomalacia and consequent ex vacuo ventricular dilatation. Patients with an active hydrocephalus usually present with a diffuse involvement of the leptomeninges, revealed by an extensive contrast enhancement on CT scans. Obstruction of the Sylvian aqueduct with consequent triventricular hydrocephalus was only occasionally described (Perlman and Argyle, 1992).

Neonatal Infections. The majority of CSF infections in the neonatal period are due to bacterial agents. A recent study on very low birth weight neonates with meningitis identified coagulase-negative staphylococci as the most common cause (43% of episodes), followed by other gram-positive bacteria (19%) and gram-negative bacteria (17%) (Doctor *et al.* 2001). The incidence of postinfective hydrocephalus in this specific subset of patients ranges between 50% and 75% (Ciurea *et al.* 2004). It usually occurs within 2–3 weeks following the initial diagnosis, but it may develop months or years after the acute phase of the disease. If hydrocephalus develops soon after the resolution of the infection it is more frequently of the communicating type with tetraventricular dilatation; obstructive forms have more commonly been described in patients developing late hydrocephalus. As in the posthemorrhagic cases, the hypothesis is that the sequence of fibrosis, thickening of the leptomeninges and consequent obliteration of the subarachnoid spaces may be reversible, with, however, the possible persistence of an impaired CSF flow at the level of the Sylvian aqueduct or the basal cisterns (Siomin *et al.* 2002). In some cases the inflammatory reaction may lead to a loculation of the CSF spaces and the formation of multiple intraventricular septations. This process is more frequent after gram-negative and mycotic infections which are associated with a more severe subependymal inflammatory reaction.

Postneonatal Infections. About 60–75% of postneonatal bacterial meningitis in children are caused by Haemophilus influenzae type B. However, hydrocephalus in these patients is uncommon occurring in less than 10% of the cases (Daoud *et al.* 1998). One of the hypothesis for this low rate is that Haemophilus B ventriculitis usually causes choroid plexuses atrophy, with consequently reduced CSF production (Ciurea *et al.* 2004). In contrast with Gram negative CSF infections, viral ventriculitis are frequently followed by the development of hydrocephalus. Viruses inclusions cause granular ependymitis and ependimal cell loss or fusion; this process frequently involves the Sylvian aqueduct leading to secondary aqueductal stenosis; however, some authors have pointed out that the damaged ependyma itself may contribute to this process because of the secondarily induced reduction in CSF transportation and replacement (Ciurea *et al.* 2004). In non endemic countries hydrocephalus, secondary to tuberculous meningitis is a rare occurrence.

It is more frequently of the communicating type and secondary to extensive meningeal involvement of the basal cisterns; in rare cases the mass effect of an intracranial tuberculoma may lead to direct or secondary aqueductal stenosis (Shoeman *et al.* 2000).

Results of ETV in Infants and Children
With Postinfectious Hydrocephalus

A relatively low number of cases of infants and children with postinfectious hydrocephalus treated with ETV was reported in literature, mostly as a subgroup of combined series. In the multicentric study of Siomin *et al.* (Siomin *et al.* 2002), children with a history of meningitis, ventriculitis or shunt infection represented only 2.1% (27 cases) of all third ventriculostomies performed in seven international neurosurgical centers; at a mean follow-up of 1.87 ± 1.6 years the rate of success was 55.6% (15 cases). The severity of the infection, the number of episodes (less or more than 2), location of the inflammatory process (meninges, ventricular system, CSF shunt device) or the type of the agent (bacterial, yeast, unknown culture) did not appear to affect outcome with statistical significance. On the contrary, a history of associated hemorrhage had a negative predictive impact (ETV success rate: 10%); actually, at a review of endoscopic video recordings in this last subgroup of patients demonstrated a high incidence (53.8%) of interpeduncular fossa adhesions, a factor which might have accounted for the failure of the procedure at least in some patients.

Age was a significant factor with regards to the outcome, younger patients showing a greater rate of treatment failures. This observation was confirmed in other series and has a particular significance in infants; when patients under 2 years of age are selectively considered the reported ETV success rate ranges between 0% and 44.4% (Fukuhara *et al.*, 2000, Siomin *et al.* 2002, Smyth *et al.* 2003). The role of the age factor is further supported by the relatively high success rate reported, on the contrary, for secondary ETV procedures. For example, Smyth *et al.* (Smyth *et al.* 2003) obtained a 60% success rate in five cases of postinfective hydrocephalus, three of them primarily shunted and then treated by ETV at the time of shunt malfunction. Similarly to what propounded in cases of posthemorrhagic hydrocephalus, also in post-infectious hydrocephalus the resolution of the acute inflammatory reaction and the temporary control of CSF dynamics assured by the presence of an extrathecal CSF shunt may allow an improvement of CSF absorption mechanisms with time. ETV would then enhance the CSF access to the subarachnoid spaces by-passing a possible persistent obstruction at the level of the aqueduct or fourth ventricle outlets.

Hydrocephalus Associated with Dandy-Walker Syndrome

According to the original definition of Dandy and Blackfan, the term Dandy-Walker syndrome (DWS) indicate the association of: 1) cystic dilatation of the fourth ventricle; 2) partial or complete absence of the cerebellar vermis and 3) hydrocephalus. The remarkable variability of the pathological features of the disease has posed some problems in distinguishing this entity from other types of posterior fossa cysts, such as persistent Blake's pouch, retrocerebellar cyst, megacisterna magna and arachnoid cysts. Further difficulties were determined by the consistent controversy in the meaning of the terminology adopted by the various authors. Actually the differential diagnosis is done on radiological criteria; according to Barkovich (Barkovich et al. 1989), the main distinguishing factor between these different pathologies is the extension of the communication between the cystic cavity and the fourth ventricle: 1) a wide communication is documented in Dandy-Walker and Dandy Walker variant syndromes which are also characterized by a more or less extended dysgenesis of the cerebellar vermis; 2) a limited communication via the vallecula can be found in patients with mega cisterna magna; 3) a complete absence of communication occurs in posterior fossa arachnoid cysts. Hydrocephalus is typical only of children with DWS; however it is not always, in fact, associated with the disease. Hirsch (Hirsch et al. 1984) reported a 90% incidence of an associated hydrocephalus in their clinical series, but they argued that it could be an overestimation, since children are referred to neurosurgeons only when they develop active hydrocephalus and require surgical treatment.

Pathogenesis

The relevance of IV ventricle outlets atresia as a pathogenetic factor of hydrocephalus in children with DWS is debated; indeed the foramina of Lushcka and Magendie have occasionally been found to be patent in these patients. It has also to be considered that more than 80% of Dandy-Walker infants are not hydrocephalic at birth (Cinalli et al. 2004b). Shaw et al. (Shaw et al. 1995) described atresia of one or both foramina of Lushcka in 20% of autoptic normal brains. Barr (Barr, 1948) observed nonpatency of the foramina of Magendie in 1% of autopsies. On these grounds other factors have been claimed to contribute to the development of hydrocephalus in these patients. Aqueductal stenosis, due to a primary development defect or secondary to herniation of the vermis may be important in some specific cases (Cinalli et al. 2004b). Furthermore, the obstruction of CSF flow can also be distal to the outlets of the fourth ventricle. Glasauer (Glasauer, 1975) noted at isotope cisternography that occasionally the subarachnoid space anterior to the medulla and basal cisterns was not patent. Other authors have confirmed that the subarachnoid spaces can be abnormally developed in children with DWS (Hirsch et al. 1984). Finally, the severe malformation of the posterior fossa with elevation of the tentorium, the torcular Herophili and the transverse sinuses can lead to a lenghtening of the venous sinuses and to their direct compression from the posterior fossa cyst (Cinalli *et al.* 2004b). Subsequently, the possible role of venous hypertension should not be underestimated.

Results of ETV in Children With Hydrocephalus and Dandy-Walker Syndrome

Due to the relative rarity of the disease the papers that can be found in the literature on this subject are essentially limited to case reports or description of very limited series of patients. Hirsch *et al.* (Hirsch *et al.* 1984) and Hoffman *et al.* (Hoffman *et al.* 1980) reported a 50% success rate (2 out of four patients joining the two series). All the endoscopic procedures were performed under an exclusively radioscopic control, a factor which might have influenced the correct conclusion of the procedure. More recently, a higher success rate was reported by Cinalli (Cinalli, 1999) who described three long term outcomes out of four personal cases.

In children with secondary aqueductal stenosis a combined ETV and aqueductal stenting placement has been suggested. Mohanty (Mohanty, 2003) successfully performed this kind of procedure in two out of three patients (66.6%). Intraoperative endoscopic confirmation of aqueduct patency might be useful in order to select the appropriate surgical procedure (Jodicke *et al.* 2003).

One of the conditions that seems to negatively influence the results of ETV in patients with Dandy-Walker syndrome is the presence of associated CNS malformations, such as agenesis of the corpus callosum which might preoperatively allow the escape of CSF into the convexity subarachnoid spaces. Contrast dynamic CT scans, MRI with flow studies and flow-sensitive phase-contrast cine MRI images might help to make the correct patient's selection (Cinalli *et al.* 2004b). Anatomic modifications of the interpeduncular cistern, the orientation of the third ventricular floor, the higher position of the tip of the basilar artery and the displacement of the brainstem against the clivus are the conditions which usually make the endoscopic procedure difficult to perform. Endoneurosonography and transendoscopic Doppler ultrasound were proposed as technical adjuncts in order to identify intraoperatively parenchimal and vascular landmarks (Jodicke *et al.* 2003). Controlled CSF leak through the endoscope may allow a slow decompression of the cyst, which in turn may reduce the mass effect on the brainstem and recreate the space for a safe ETV.

Constrictive Hydrocephalus

Hydrocephalus Patients with Myelomeningocele

The exact incidence of hydrocephalus in myelomeningocele patients is not known. In most cases it is not present at birth but it develops in the first few weeks or months of life. Postnatal neuroimaging studies, obtained before closure of the spinal defect, have documented the presence of hydrocephalus in 15.25% of the cases; however, the proportion of patients who subsequently require shunting reaches up to 80–90% in most surgical series. No correlation has been shown between the level of the lesion and the presence of hydrocephalus (Teo and Jones, 1996).

Pathogenesis

A variety of factors are implicated in the pathogenesis of hydrocephalus in children with myelomeningocele which characteristically may combine the features of the communicating and obstructive forms of hydrocephalus. The Chiari type II malformation, aqueduct stenosis, anomalous venous drainage, and the closure of an open myelomeningocele all contribute to the development of hydrocephalus. In particular, the hypoplasic posterior cranial fossa and the hindbrain anomalies, typical of the Chiari II malformation, usually result in an overcrowing of the nervous and vascular structures and reduction of the CSF spaces, with secondary caudal dislocation of the IV ventricle and cerebellar tonsils with the upper cervical canal (Sgouros 2004). The consequent increased resistance to venous outflow and venous hypertension, may account for the development of a "communicating" form of hydrocephalus. On the other side, the mechanical distortion and deformation of the brain stem can secondarily lead to a functional aqueductal stenosis. Furthermore the tonsillar herniation may also contribute to the obstructive component of this type of hydrocephalus, impairing CSF flow at the level of the fourth ventricle outlets. The relevant role of CSF obstruction in the pathogenesis of hydrocephalus in myelomeningocele patients seems to be confirmed by recent series reporting on intrauterine repair of opened spina bifida and the possible correlation between the severity of Chiari type II malformation and the occurrence of hydrocephalus. Tulipan *et al.* (Tulipan *et al.* 1998) and Sutton *et al.* (Sutton *et al.* 1999) described a significant reduction of the hindbrain herniation occurrence in children who underwent intrauterine closure of myelomeningocele if compared with historical controls. The incidence of associated hydrocephalus was reduced from 91% to 59% in the series of Tulipan. Only one of the nine survivors in the series of Sutton required ventriculoperitoneal shunting. Hence these authors postulated that the lower incidence of hydrocephalus in their series had to be related to the absence of the obstructing effect of the hindbrain herniation at the level of the foramen magnum. However other authors have claimed this assumption to be oversimplistic as the intrauterine procedure had induced significant changes in volume of the posterior fossa, with consequent improved flow through the aqueduct, improved compliance of CSF flow around the brain stem and the tentorial hiatus, and lower venous outflow pressure. Some authors claimed that the predominant features of hydrocephalus in children with myelomeningocele may change in time. In infants, the subarachnoid spaces deformation and immaturity combined with the increased venous outflow resistance would prevail, consequently justifying the placement of an extrathecal CSF shunt device. However, the venous hypertension would not be corrected by the procedure, subsequently leading to an accumulation of interstitial fluid which in turn would worsen the aqueductal stenosis and change the hydrocephalus form, mainly communicating, in a mainly obstructive type (Sgouros 2004).

Results of ETV in Myelomeningocele Patients

Third ventriculostomy has been proposed in the management of hydroce-phalus in children with MMC since mid 90's but its role remains controver-sial. Actually the most relevant series is still the one of Teo and Jones (Teo and Jones, 1996) who reported on 69 patients with an overall success rate of 72%. At a mean follow-up of 28 months (min 12 months; max 17 years), a significant worse outcome was found in patients aged less than six month (12.5% success rate) if compared with their older counterpart (80% success rate). The authors also described significant differences in success rates if ETV was performed as first line of treatment or secondary procedure (29% vs. 84% success rate). Though not reaching statistical significance, other factors that predicted a favorable result were the presence of a triven-tricular hydrocephalus, a diameter of the third ventricle > than 4 mm, the evidence of normal or slightly reduced subarachnoid spaces. On the other hand, the previous cerebrospinal fluid infection and/or of an intraventricu-lar hemorrhage were negative predictive factors. Mori *et al.* (Mori *et al.* 2003) also observed differences in the outcome in patients aged less than one year as compared with their older counterpart (25% vs. 90% success rate). Other papers in the literature, though based on single case reports or series with limited numbers of patients, have challenged the just men-tioned results, in particular the role of age and the lower success rate of pri-mary versus secondary ETV. Fritsch *et al.* (Fritsch *et al.*, 2003) reported a 50% success rate which was independent from children's age at surgery. Similar results were reported in the multicenter study of Portillo *et al.* (Portillo *et al.* 2004) who referred an overall success rate of 21% in a series of 19 myelomeningocele patients, a rate which was not related with age or time when ETV was performed. In a recent series, Tamburrini *et al.* (Tam-burrini *et al.* 2004) were not able to found significant differences between infants (<six months) undergoing primary ETV (overall success rate: 70%) and older patients who underwent third ventriculostomy as second-ary procedure at the time of malfunction of a previously inserted CSF

shunt device (overall success rate: 60%). Such discordant results suggest that particular caution should be paid in these type of hydrocephalus, when planning ETV.

On the other hand, the slow and insidious progression which characterizes the evolution of this specific form of hydrocephalus in a large portion of subjects, might lead to a wrong claiming of a successful result after the surgical operation in patients who actually continue to have hydrocephalus. Subsequently, strict neuroradiological, ophtalmological and neuropsychological follow-up control is mandatory.

Analysis of Outcome

ETV, by forming a communication between the third ventricle and the subarachnoid spaces, restores an almost "physiological" CSF circulation in case of obstructive hydrocephalus. The recent views on the pathophysiology of hydrocephalus (Greitz 2004) enhance the importance of the intracranial compliance, that is widely dependent on the free movement of the CSF between the cranial and the spinal subarachnoid spaces. Despite of the accuracy of patients selection, several reports show that in a significant number of cases the symptoms and signs of increasing intracranial pressure related to hydrocephalus are however not controlled by the procedure (Cinalli *et al.* 1999, Hirsh *et al.* 1986). The phenomenon is more frequently observed in the immediate post-operative period (early failures), being only sporadically observed in a later phase, usually within 5–6 years from the operation (late failures). Recurrence of hydrocephalus beyond this period has been reported exceptionally (Beems and Grotenhuis 2004, Tuli *et al.* 1999). Pre-requisite of a successful ventriculostomy is the patency of the distal subarachnoid pathways: the difficulty to detect pre-operatively the presence of coexisting obstruction in the basilar cisterns or in the subarachnoid space of the surface may account for the 25–40% failure rate still reported in literature (Cinalli *et al.* 1999, Fritsch and Mehdorn 2002, Javadpour *et al.* 2001). A reliable non-invasive pre-operative test is not available yet; invasive tests have been proposed, but they are difficult to propose on a routine basis, especially in children (Bech *et al.* 1999, Czosnyka *et al.* 1996, Magnaes 1982, Magnaes 1989). Therefore, generally failures continue to be detected post-operatively on the basis of clinical, instrumental and radiological data.

The evaluation of the results immediately following the procedure may be sometimes difficult. After the opening of the third ventriculostomy, the increased amount of circulating CSF may require a so-called "adaptation period" (Bellotti *et al.* 2001, Cinalli 2004, Hopf *et al.* 1999b) to circulate or to be re-absorbed. This could explain the persistence of symptoms and increasing ventriculomegaly at CT scan examination, that can be erroneously interpreted as failure of the procedure and indication to perform a VP shunt.

Unfortunately, a widely accepted method to accurately detect early failures is not available and management of post-operative intracranial hypertension is controversial. Accurate pressure measurements associated, when necessary, to CSF withdrawal may allow the patients to reach an equilibrium in CSF dynamics in safe condition.

The persistence of dilatation of the ventricular system can be often observed following ETV also in the late post-operative phase in otherwise asymptomatic subjects. Such a ventricular dilation raises the question of its possible influence on intellectual outcome, especially when ETV is performed in very young children. Even though the findings seem not to be correlated with adverse effects, the absence of long-term follow-up observation in large series suggests the opportunity of prospective, multicentric, randomized studies for the direct comparison with he already available long-term outcomes of extrathecal CSF shunting procedures.

Early Results

Clinical Signs and Symptoms

Symptoms and signs of intracranial hypertension resolve immediately following a successful procedure in the great majority of the operated on patients. In particular, Parinaud's sign, if present preoperatively, is expected to disappear from the earliest post-operative moment; persistence of this sign should be taken as a potential predictor of failure (Cinalli 2004, Schijns and Beuls 2002). Papilledema disappears within 2–3 weeks. Feng et al. (Feng et al. 2004) found that an early satisfactory clinical response (within two weeks after the surgical operation) provided a high correlation with the overall ETV success.

The possible persistence of clinical symptoms using high ICP can be a major pitfall in the evaluation of the patient and can lead to the wrong, premature conclusion of a failure of the technique, with consequent decision to implant or re-implant the CSF shunt device. Symptoms and signs of persistently high ICP are relatively rare in patients operated on primarily for primitive aqueductal stenosis or obstructive triventricular hydrocephalus of any nature. In these patients, the most frequent sign of increased ICP are swelling at the level of the coronal incision due to subcutaneous accumulation of CSF or, in some cases, CSF leaks. In our experience of ICP monitoring post-operatively, patients operated on primarily can tolerate very high levels of ICP without complaining of any symptom. The situation is different when ETV is performed on patients operated on for shunt malfunction or for slit ventricle syndrome. These patients may have signifi-

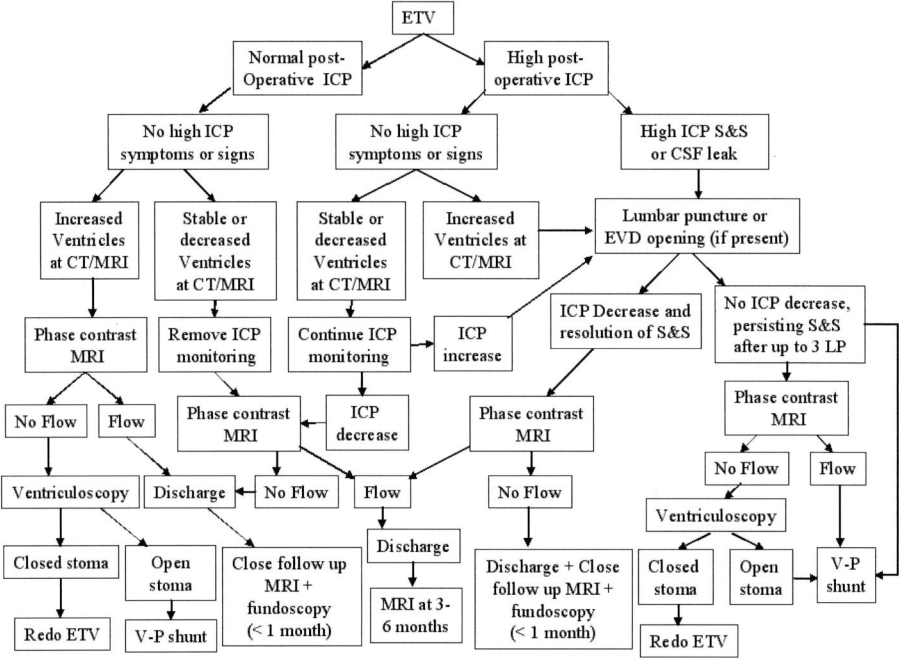

Fig. 20. Algorythm of evaluation of the early post-operative days following ETV and management of high ICP with or without symptoms and signs of increased intracranial pressure

cantly lowered intracranial compliance due to cephalocranial disproportion induced by the shunt, and may exhibit a progressive clinical deterioration as their ICP may tend to increase in the first 7–10 days after surgery. Consequently, ICP monitoring and safety external ventricular catheter (EVC) to be placed into the same corticotomy track produced by the endoscope should be considered as mandatory in these patients. In case of symptoms of high ICP seriated lumbar punctures should be performed according to the algorithm reported in Fig. 20. EVC should be kept closed and open only if lumbar punctures cannot manage the situation.

ICP Monitoring

The persistence of high ICP in the post-operative period following ventriculo-cisternostomy, is a well known phenomenon which was reported in the literature since the earliest descriptions of Matson's ventriculo-subarachnoid shunt (Matson 1969, Cinalli 2004). The explanations proposed so far are not conclusive, varying from the hypothesis of the progressive dilatation of the subarachnoid spaces not used to receive the circulation of the whole amount of CSF produced to that of the progressive adaptation of the compliance of the intracranial system, due to the "stiffness" of the brain. The last evenience would be frequent especially in patients shunted around birth and who had their shunt removed. Whatever the reason anyway, as a matter of fact ICP is usually higher than normal in the immediate post-operative period in several cases. In some patients, the average ICP levels can remain below 20 mmHg with, however, frequent spikes above this value. The values may then rapidly fall to even lower-than-normal level in very few days: these patients usually remain asymptomatic following ETV and lumbar punctures. EVC (if present) opening is rare if ever necessary. In other patients, mostly those who had their shunt removed and exhibit signs and symptoms of intracranial hypertension, the average ICP levels can be significantly higher. The transitory nature of the phenomenon in most instances may explain why some patients with EVC maintained at the same hydrostatic level show a high CSF output in the early post-operative days, followed by a gradual decreasing within the first two post-operatively weeks (Nishiyama et al. 2003).

Management of high ICP (see below) leads to resolution of symptoms in most cases, without the need of an extrathecal shunt. Thus, presence of symptoms of increased ICP in the post-operative period not always means that the procedure had failed (Bellotti et al. 2001, Cinalli 2004).

Neuroradiological Evaluation

Compared to the implant of an extrathecal CSF shunt device, the most important difference is the effect on the volume of the cerebral ventricles as demonstrated by the neuroimaging investigations. The ventricular size usually decreases rapidly and significantly following the shunt implantation. The volume decrease is much slower and smaller in scale after third ventriculostomy. A transient "adaptation period" (Bellotti et al. 2001, Hopf et al. 1999b) in which the CSF "finds its way" through the subarachnoid spaces could account for the persistence of the ventricular dilation in the early post-operative phase, whereas the lack of communication with a lower pressure cavity (e.g., right atrium, peritoneal cavity), and the nature of the ventricular dilation in case of hydrocephalus related to aqueductal stenosis, which often evolves over several years before becoming symptomatic (Fukuhara et al. 2000), might account for the persistent ventriculomegaly in later phases of the post-operative clinical course.

In spite of such a limitation, some radiological criteria that must be fulfilled for definition of success have been established (Cinalli 2004).
1. Reduction in ventricular size ranging from 10% to 50% must be observed from the first week (Schwartz et al. 1999), even if the ventricles remain large.

2. Periventricular lucency, if present before operation, must disappear.

3. CSF flow artifact must be visible on sagittal median T2-weighted fast spin-echo MRI sequences (Fukuhara *et al.* 2002, Kulkarni *et al.* 2000). CSF flow has been described in newborns with color Doppler ultrasonography (Wilcock *et al.* 1996).

4. The floor of the third ventricle, if bulging downward in the preoperative images, must be straight on postoperative images.

5. Atrial diverticula and pseudocystic dilatation of the suprapineal recess, if present preoperatively, must disappear or decrease significantly.

6. Pericerebral sulci, if not visible before operation, must reappear or increase in size.

When considering different neuroradiological indices, it has been shown that the width of the third ventricle is the quickest to decrease; such a volumetric modification remains stationary in time. A more than 15% decrease in size of the third ventricle within 1 month is considered the most reliable indicator of favorable outcome following third ventriculostomy (Buxton *et al.* 2002, Schwartz *et al.* 1996, Schwartz *et al.* 1999, Tisel *et al.* 2000).

The greater decrease of both the third ventricle (30%–40%) and the lateral ventricles (30%–32%) is already visible from the first postoperative week. The extent of ventricular volume decrease is in inverse correlation with the preoperative duration and magnitude of clinical symptoms (Figs. 21–23). A decrease in volume of less than 10% may be observed in

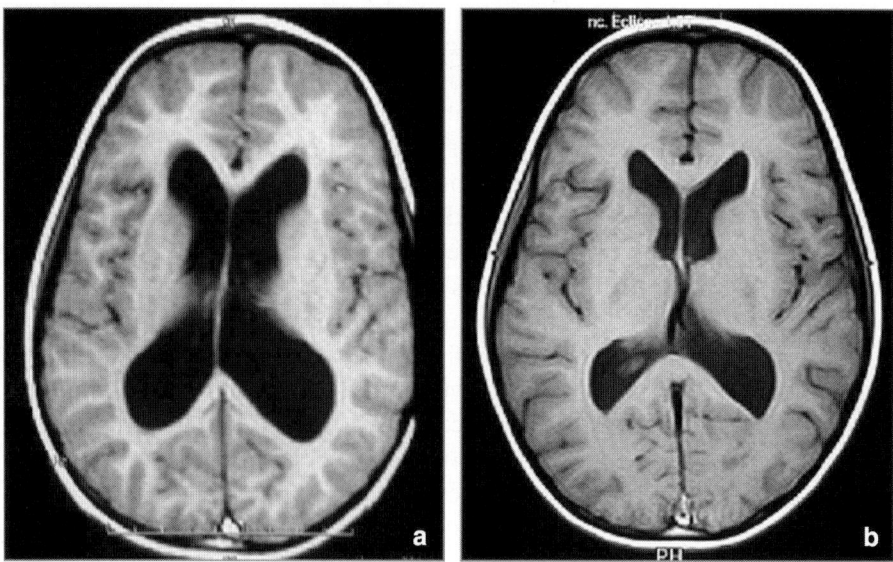

Fig. 21. Nine year old boy presenting with acute symptoms of intracranial hypertension (a) lasting for 1 week. After ETV, the ventricles return to near-normal size (b)

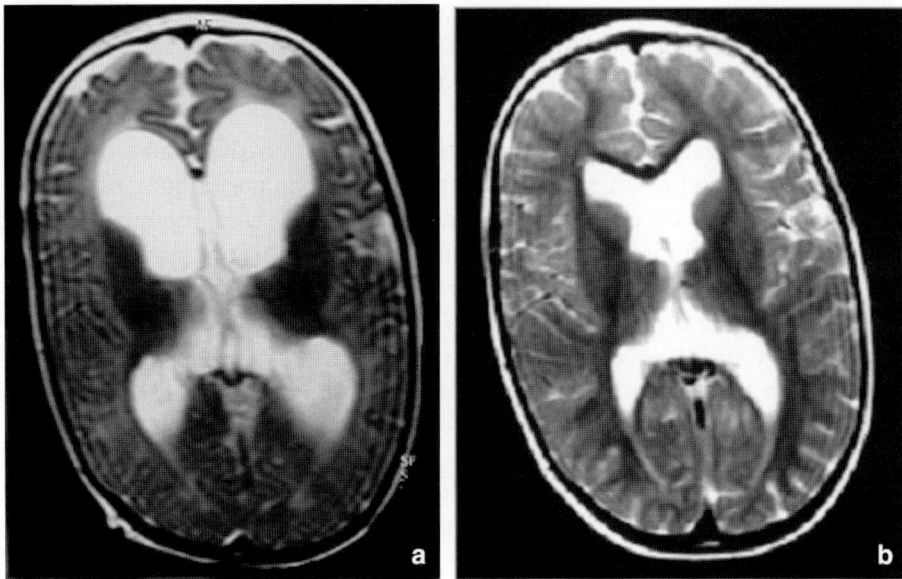

Fig. 22. Six month old boy presenting with a 4-month history of progressive macrocrania and vomiting. Pre-operative MRI shows significant ventricular dilatation (a), that regressed only partially after ETV (b)

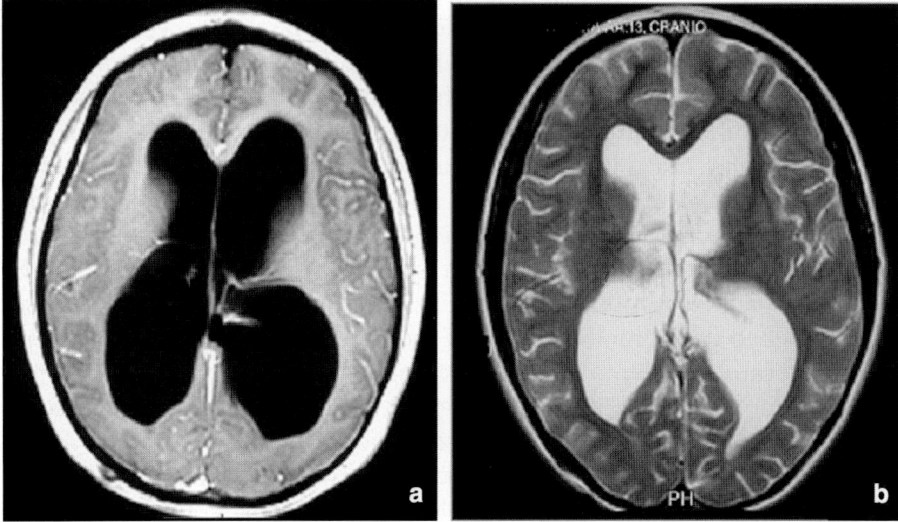

Fig. 23. Eleven year old girl, presenting with a 4-year history of intermittent headache, decreasing school performance and finally blurred vision and papilledema (a). After ETV there is almost no modification of the ventricular volume (b) in spite of an excellent clinical result

patients with long-standing chronic symptoms (Schwartz *et al.* 1999). The downward deviation and flattening of the brainstem reverts within 1 year, whereas the width and height of the lateral ventricles continue to decrease steadily for at least 2 years (Oka *et al.* 1993b).

Nevertheless radiographic evaluation in the immediate post-operative period is challenging. CT scan during the adaptation period may not reveal any modification or even serial scans may show re-enlargement of the ventricles following initial reduction in size, leading to the misdiagnosis of failure. Feng *et al.* (Feng *et al.* 2004) reported satisfactory results in 5 of 10 patients in which a persistent increase of the ventricular size was observed after surgery and they concluded that the size of the ventricles is not a good predictor in the evaluation of the outcome within 3 months following surgery. The same paradoxical re-enlargement of ventricular volume after an initial decrease has been described by other authors (Saint George *et al.* 2004).

Cine-phase contrast MRI, showing directly the patency of the stoma, could be useful as a prognostic factor for a successful ETV. However, the data of the literature are controversial: Kulkarni *et al.* (Kulkarni *et al.* 2000) demonstrated a statistically significant relation between evidence of postoperative aqueductal CSF on MRI and clinical success. Feng *et al.* (Feng *et al.* 2004) in a recent paper found a good correlation between the finding of a patent stoma and outcome. They documented that 12 of 16 patients (75%) in whom a positive flow void was observed on postoperative (within two weeks) cine MRI had a successful ETV, even if 4 patients, in whom a positive flow void was shown, experienced treatment failure. Other reports (Cinalli *et al.* 1999, Goumnerova and Frim 1997) had shown that cine MRI has not high sensibility and specificity in predicting outcome. The exam is more useful as an element of comparison to a previous similar exam in the same patient than evaluated independently. For this reason, we perform a cine PC MRI in the first post-operative week before discharge in order to have a baseline exam when the stoma is certainly patent to compare with similar studies carried out in the late post-operative period (Fig. 20). Early post-operative MRI is also useful in order to detect modifications of the profile of the third ventricle and of the lateral ventricle: in cases where cine PC is not conclusive and flow void artifact is not well visible on sagittal or coronal T2 turbo spin echo. The morphological analysis of the profile of the third ventricle is probably the most reliable evidence of occlusion of the stoma (Fig. 24a–g). The anatomical details can be visible well before symptoms and signs of increased intracranial pressure occur, and should prompt the surgeon to re-operate the patient even if asymptomatic in order to avoid situations of extreme emergency with possible fatal complications (Hader *et al.* 2002).

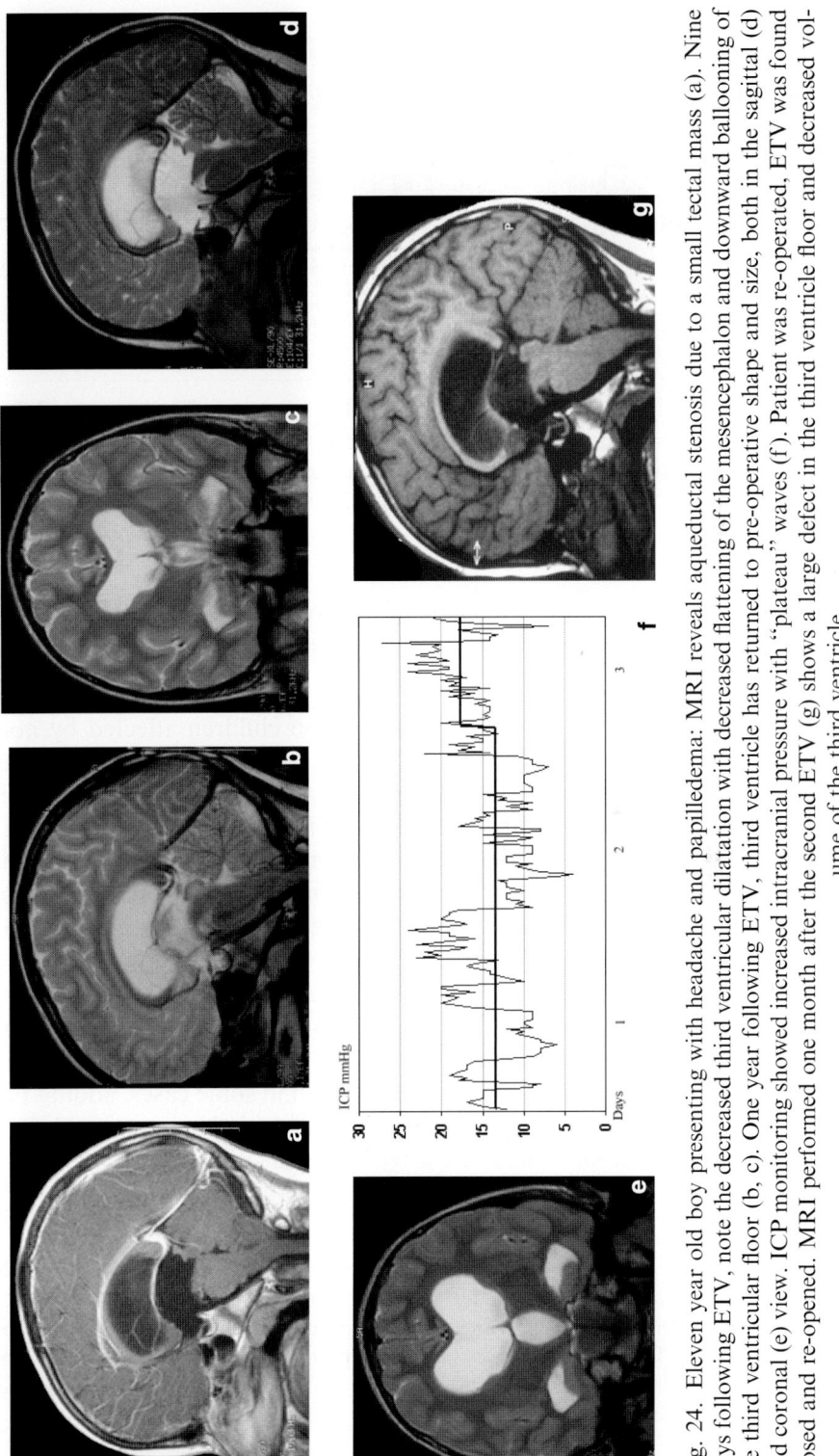

Fig. 24. Eleven year old boy presenting with headache and papilledema: MRI reveals aqueductal stenosis due to a small tectal mass (a). Nine days following ETV, note the decreased third ventricular dilatation with decreased flattening of the mesencephalon and downward ballooning of the third ventricular floor (b, c). One year following ETV, third ventricle has returned to pre-operative shape and size, both in the sagittal (d) and coronal (e) view. ICP monitoring showed increased intracranial pressure with "plateau" waves (f). Patient was re-operated, ETV was found closed and re-opened. MRI performed one month after the second ETV (g) shows a large defect in the third ventricle floor and decreased volume of the third ventricle

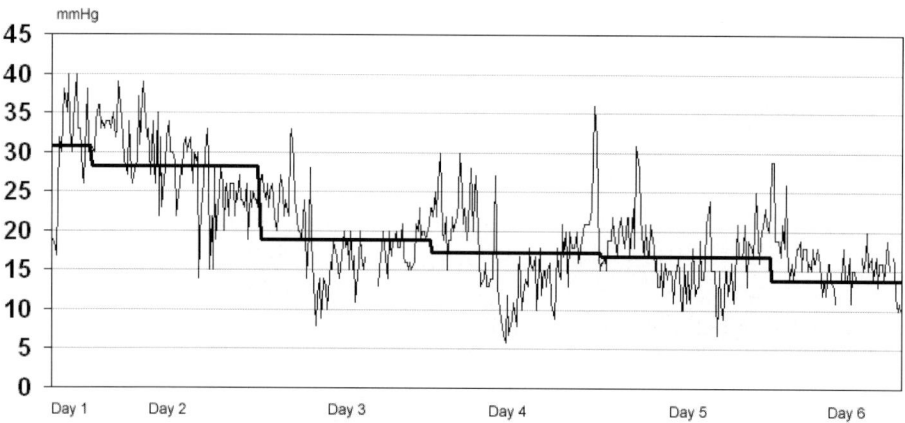

Fig. 25. Twelve year old girl affected by aqueductal stenosis. ICP monitoring following ETV shows very high values, asymptomatic, in the first 3–4 days with progressive,
spontaneous normalization

ICP Monitoring and Management of Increased ICP During the "Adaptation Period"

On the base of the continuous monitoring of ICP in the first two
weeks following ETV in a personal series of 56 children affected by non-
communicating hydrocephalus, we were able to distinguish two patterns
in the ICP behavior during the "adaptation period". In a first subgroup
of subjects, ICP returned immediately to the normal levels (<20 mmHg),
and remained normal for all the length of the monitoring: about half of all
our patients belonged to this group. In a second subgroup, ICP remained
or became high until the second post-operative day, to decrease in the subsequent days slowly (Fig. 25). In some cases, elevated values of ICP were
recorded until the 9th post-operative day, often associated with headache
and vomiting. In some children, the symptoms were so impressive to induce a suspect of failure. CT scan performed in this period demonstrated
an increasing dilatation of the ventricular system in some cases, adding further evidence to the suspicion of failure (Fig. 26a–e). However, such high
values of ICP in the first post-operative days, in our and in other series

Fig. 26. Eleven year old girl shunted at birth for neonatal hydrocephalus. Admitted
for shunt malfunction (a). After ETV and shunt removal, ventricles are smaller (b).
Four days after ETV the patient is severely symptomatic with high ICP values and
increased ventricular dilatation at CT scan (c). After lumbar puncture, ventricles
progressively decrease in size 6 days (d) and 9 days (e) after ETV

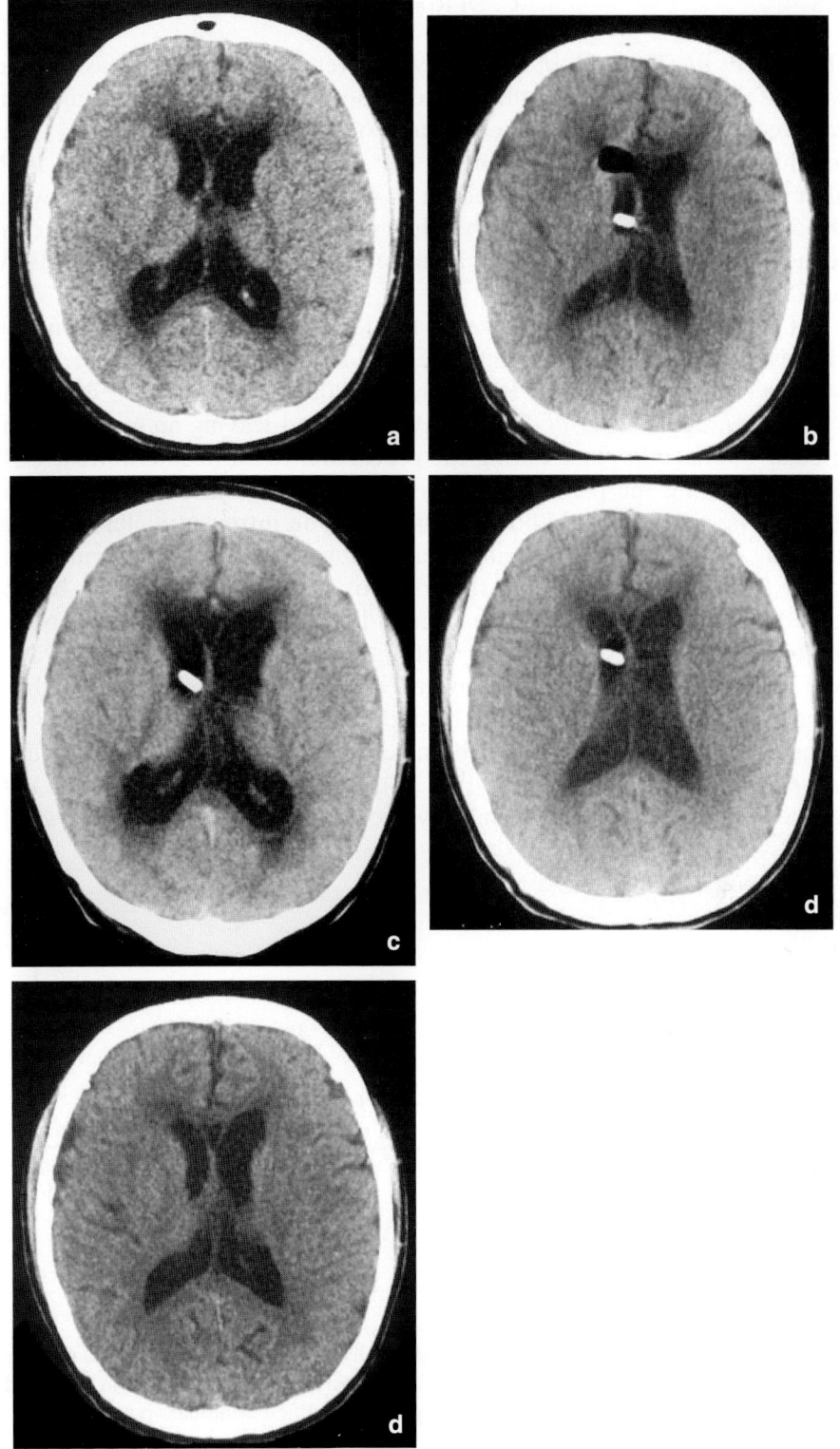

(Bellotti *et al.* 2001, Hopf *et al.* 1999b), were not necessarily associated with late failure. Actually, their incidence was not statistically significant when compared to that observed in cases of true failure. In the only 4 cases in which the procedure failed in the first 15 days (early failure), the ICP, after an initial decrease that lasted for 2–4 days, rapidly rose again between the 5th and 8th day to reach high values unaccettable for the patient, consequently suggesting the need for a further surgical procedure (second ETV or CSF shunt implant).

ICP monitoring is very useful to detect intracranial hypertension. However, how to adequately manage high ICP is still under debate. Some authors (Bellotti *et al.* 2001, Böschert *et al.* 2003, Hopf *et al.* 1999b, Nishiyama *et al.* 2003, Oi *et al.* 2000), have proposed leaving an external ventricular drainage in place during the procedure, to allow intermittent CSF drainage during the periods of pathological ICP elevation. According to these authors, this would allow transient drop in intraventricular pressure, allowing re-expansion of the intracranial subarachnoid spaces and facilitating CSF circulation toward the convexity. The same result can be obtained by performing CSF evacuation by lumbar taps under ICP monitoring (Cinalli 2004). Post-operative withdrawl CSF to restore a normal ICP was already suggested by Matson in 1969 (Matson 1969) and by others later (Hoffman *et al.* 1980, Jaksche and Loew 1986, Oka *et al.* 1995). In some cases lumbar tapping had been repeated 24–48 hours later; a third tap had been rarely necessary (Fig. 27). The positive effects on the ICP last longer than can be explained by the simple subtraction of the small

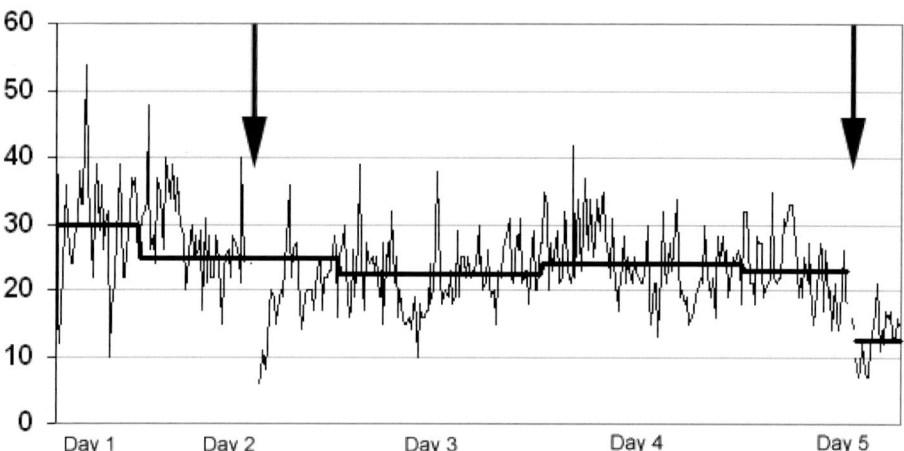

Fig. 27. Three year old boy shunted at birth for hydrocephalus and aqueductal stenosis. At the time of shunt malfunction, ETV was performed and shunt was removed. Post-operative ICP monitoring shows high ICP values. Lumbar punctures were performed (arrows) because of persisting symptoms and signs of increased ICP

amount of CSF, usually limited to 5–10 ml. One possible explanation is the increased compliance of the spinal subarachnoid spaces (SAS) and the increased pressure gradient between third ventricle and the SAS of the posterior cranial fossa induced by the CSF subtraction. This would facilitate CSF flow through the third ventriculostomy and, in turn, the reduction of the ventricular volume, so allowing the re-expansion of the intracranial subarachnoid spaces. In patients with intracranial mass lesions, in which ETV was performed before tumor removal, we refrained from using lumbar tapping, for the risk of tonsillar herniation and preferred to implant an external drainage in the ventricle to be utilized to remove CSF when necessary.

In conclusion ICP must return to normal values within 10 days after ETV. Should ICP remain elevated with persistent symptoms of intracranial hypertension, we suggest to perform at least two lumbar taps before concluding that the procedure failed (Fig. 27).

The algorythm of post-operative patient's evaluation we utilized in the first weeks after ETV is shown in Figure 4. A central role is attributed to ICP monitoring in subjects above one year of age. In newborns, ICP monitoring is very effectively replaced by fontanel examination.

Late Results

Review of the Literature

As shown on Table 5, different authors describe various definitions of failure by taking into account several factors, such as the need of an external CSF shunt, the clinical symptomatology, or the changes of the ventricular volume. According to this variability, outcome evaluation can differ as much as 30% in the same group of patients depending on the criteria used for definition of success. Moreover, in the same group of patients, different criteria of success have been reasonably used according to the different procedures. Most authors include neuroradiological criteria to confirm the clinical success (Beems and Grotenhuis 2004, Brockmeyer *et al.* 1998, Cinalli *et al.* 1999, Gorayeb *et al.* 2004, Teo 1998). According to these authors, the ventricular system must decrease in size or at least remain stable and radiographic signs of active hydrocephalus must disappear. In case of progressive dilatation of the ventricles (despite the absence of symptomatology) a further surgical procedure (ETV re-do or VP shunt insertion, according to the criteria discussed below) is warranted. Only Feng *et al.* (Feng *et al.* 2004) considered successful 5 cases in which clinical improvement occurred despite of the progressive dilatation of the ventricular system, and did not implant a shunt. Actually long-term effect of persistent ventriculomegaly (present in about 30% of patients) (Goumnerova and

Table 5. Third ventriculostomy studies

Author	N	Etiology	Patients <1 year	Previous shunt	History of hem./ infec.	Techn. failures	Definition of success/failure	Success, %	Follow up
Jones, 1990	24	AS 14 MMC 5 Tumor 5	5	NA	0	4	successful: not shunt insertion, normal head growth, no evidence of high ICP improved: no high ICP, but not normal head growth. failure: no improvement, shunt insertion	successful: 50% improved: 21% failed: 12% shunt free: 66.7%	13–121 mo (mean 68)
Kelly, 1991	16	AS 16	0	11	5	0	shunt freedom	94%	12–60 mo (mean 42)
Goumnerova, 1997	23	AS 12 tectal abn. 7 Other 4	NA	4	1	0	resolution of symptoms, no shunt	73%	7–44 mo (mean 17)
Brockmeyer, 1998	97	MMC 24 AS 19 tumors 10 tectal abn 9 post hem. 7 other/comm. hydroc. 28	NA	NA	7	26	complete shunt avoidance or removal. relief of symptoms. reduction or maintenance of ventricular size.	49% (34% including technical failures)	15–69 mo (mean 24)

							Criteria for success	Success rate	Follow-up
Hellwig, 1998	14	AS	NA	8	0	0	decline of clinical symptoms. No shunt	93%	1–43 mo
Teo, 1998	121 + 8[1]	MMC 55 AS 38 tumor 17 post hem. 5 post inf. 4 other 2	NA	NA	9	8[1]	1. resolution of clinical s/s and decrease of ventricular size on CT. or 2. resolution of clinical symptoms and signs and no change in ventricular size but improvement on psychometric testing.	73% (68% including technical failures)	6–72 mo (mean 22)
Cinalli, 1998	30[2]	AS 10 tumor 7 post men 5 post hem. 4 other 4	NA	30	13	3	shunt definitely removed. Resolution of signs and symptoms of high ICP without revision	76.7%	6–186 mo (mean 24)
Buxton, 1998	27	commun. hydroc. 21 non-com. hydroc. 9	27	0	10	0	absence of any subsequent procedure for the hydrocephalus	23%	6–42 mo

Table 5 (*Continued*)

Author	N	Etiology	Patients <1 year	Previous shunt	History of hem./infec.	Techn. failures	Definition of success/failure	Success, %	Follow up
Choi, 1999	71	AS tumor other	33 0 33 5	6	0	0	improvement in clinical status. No need of a shunt	91.5%	NA
Tuli, 1999	32	AS tumor	17 NA 15	0	0	0	failure defined according to the guidelines established in the shunt designed trial	56%	12–120 mo
Cinalli, 1999	119 + 94[3]	AS toxo pineal t. other	126 NA 23 15 49	NA	0	7 + 12[3]	resolution of clinical s/s. disappearance of all radiographic signs of active hydrocephalus	72% 65%[3]	0.2–108 mo (mean 25) 1–209 mo[3] (mean 76)[3]
Hopf, 1999	95	tumor AS hem/inf. other	42 4 40 9 4	25	9	2	improvement: partial or complete relief of symptoms	76%[4]	3–71 mo (mean 26)

Study	n	Etiology						Outcome definition	Success	Follow-up
Fukuhara, 2000	89	AS tumor hem. other	37 32 8 12	NA (<7)	32	22	0	failure: any condition requiring surgical intervention because residual or recurrent symptoms	67.4%	3–63 mo
Tisell, 2000	18	AS	16	0	3	0	0	unsatisfactory results: persistence or relapsing of symptoms compatible with hydrocephalus	50%	14–66 mo (mean 37)
Buxton, 2001	63	tumor AS other	22 18 23	0	24	6	0	absence of any subsequent procedure for the hydrocephalus	75%	0–84 mo (mean 37)
Helseth, 2002	58	prim AS seco AS tumors	22 22 14	NA	NA	0	NA	resolution of ICP s/s no shunt in the follow up	72%	30 mo
Siomin, 2002	101	post haem post inf. AS other	46 41 10 4	NA	56	101	0	resolution of ICP s/s no shunt in the follow up	57.4%	6–120 mo (mean 22)

Table 5 (*Continued*)

Author	N	Etiology	Patients <1 year	Previous shunt	History of hem./ infec.	Techn. failures	Definition of success/failure	Success, %	Follow up
Feng, 2004	58	tumor AS cyst hem/inf.	21 18 11 8	5	8	NA	partial or complete relief of symptoms. no shunt implantation	77.6%	3–41 mo (mean 24)
Gorayeb, 2004	36	AS chiari II other	11 36 11 14	NA	1	0	resolution of s/s of high ICP; normal head growth; no untoward finding on neuroimaging	64%	22–69 mo (mean 47)
Beems, 2004	380	NA	NA	NA	NA	4	definitive CSF shunt not implanted	77.1%	>6 mo

AS Aqueductal stenosis, *MMC* Myelomeningocele, *Hem* Hemorrhage, *Inf.* infection, *Comm. hydroc.* Communicating hydrocephalus, *S/s* Symptoms and signs.
[1] Technical failures.
[2] Seven of these procedures have been performed under ventriculographic guidance.
[3] Procedures performed under ventriculographic guidance.
[4] Including 6 patients requiring second ETV.

Frim 1997) on neuropsychological developmental is not fully known. Teo (Teo 1998) highlights the importance of maintaining a close vigilance on the patients whose ventricles remain large, by the means of neuropsychological testing. He recommended to shunt those individuals in which a neuropsychological deterioration is documented during the follow up, despite the absence of symptoms and signs of acute intracranial hypertension (see below).

Iantosca et al. (Iantosca et al. 2004), reviewing the influence of hydrocephalus etiology on outcome, found three groups with different rates of success: a group with high success rates (>75%), that included patients affected by acquired aqueductal stenosis, tumor, cyst or infectious lesions obstructing third ventricular outflow, tectal, pineal, thalamic or intraventricular tumor, shunt malfunction; a group of intermediate success rates (about 50%) that included patients affected by myelomeningocele (previously shunted), tumor obstructing fourth ventricle outflow, congenital aqueductal stenosis, Dandy-Walker malformation, slit ventricles syndrome, recurrent or intractable shunt infection/malfunction; and a group with a low success rate (<50%) that included patients affected by myelomeningocele (previously unshunted) and posthaemorrhagic/postmeningitc hydrocephalus. These authors, on the basis of their experience (Drake 1993, Iantosca et al. 2004) and that of others (Jones 1990), discourage the use of ETV in patients who have undergone prior radiation therapy, because poor success rates, altered anatomy (in particular thickness of the third ventricle floor) and increased risk of bleeding.

Thereafter, the best results are achieved in series in which only hydrocephalus secondary to aqueductal stenosis or benign mass lesions is considered (Choi et al. 1999, Cinalli et al. 1999, Hellwig et al. 1998a, Helseth et al. 2002, Kelly 1991). However, when only patients affected by obstructive triventricular hydrocephalus are selected the success rate is actually quite homogeneous and stable, being above 60–70% at any age (Fritsch and Mehdorn 2002, Gorayeb et al. 2004, Javadpour et al. 2001), even if a history of haemorrhage or infection is present (Siomin et al. 2002). All the series studied with actuarial method (Buxton et al. 2001, Cinalli et al. 1999, Elbabaa et al. 2001, Feng et al. 2004, Fukuhara et al. 2000, Helseth et al. 2002, Tuli et al. 1999) show a long term success rate above 70% with follow-up ranging from 4 years to 9 years (Fig. 28). Only Tuli et al. (Tuli et al. 1999) reported a 44% failure rate in a group of pediatric patients affected by obstructive tri-ventricular hydrocephalus. This may be secondary to the different definition of failure used in the different papers (Table 5). While most authors define failure as the need of extracranial shunt (table 1), Tuli et al. defined failure in a manner similar to the guidelines established in the shunt design trial (Drake et al. 1998), accounting for an unusually high rate of failure in selected patients. However, all long-term studies seem to

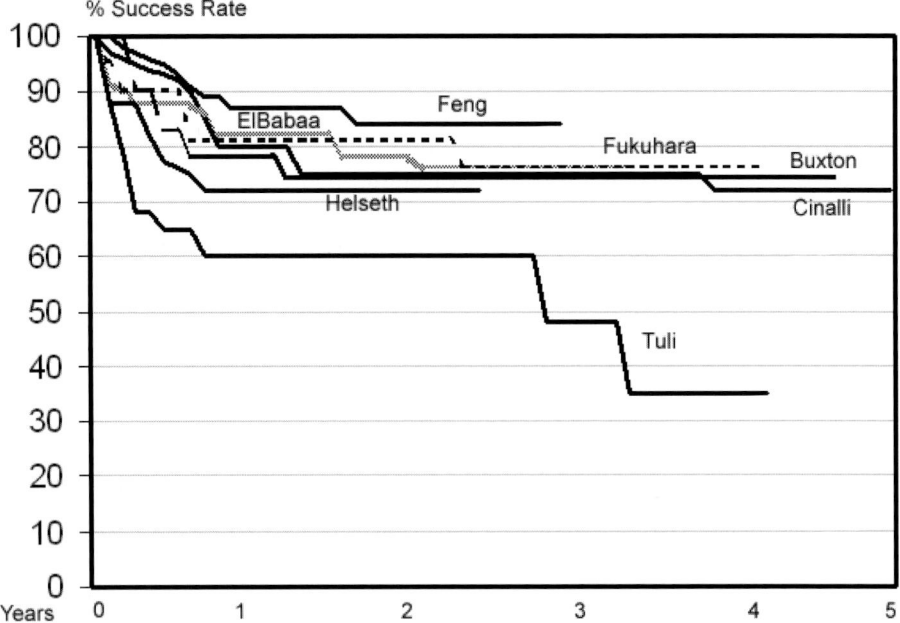

Fig. 28. Comparison of the seven actuarial studies available in the literature showing long term success rate studied with actuarial method following ETV in obstructive hydrocephalus. All the studies, except Tuli, remain between 70 and 85% at 5 years, no failures being observed beyond that date

show that most of the failures occur during the first year after surgery, and only sporadically after; basically no late failure has been described after the 5th year from ETV. Thus, close follow-up is mandatory in the first 5–6 years, and parents must be informed that immediate neurosurgical assessment is required if symptoms and signs of intracranial hypertension appear. Such failures can manifest with a very acute clinical onset, and if not recognized early may have dramatic consequences, leading to sudden death (Hader *et al.* 2002). As argued by Iantosca *et al.* (Iantosca *et al.* 2004), "education and regular follow up are important to counteract the false sense of security that may arise in the absence of a shunt".

Re-Obstruction of the Stoma

The lack of multicentric, prospective, randomized studies account for the current lack of reliable figures about the long term outcome of the children treated with ETV. However, Feng *et al.* (Feng *et al.* 2004) calculated a proportion of 75–80% functioning ETV at 1 year, Cinalli *et al.* (Cinalli *et al.* 1999) a functioning rate of 72% at 15 years, excluding technical failures.

Most failures of ETV occur in the first year following surgery; however, late failures have been reported, usually secondary to obstruction of the stoma by the growth of gliotic ependymal scarring tissue (Cinalli *et al.* 1999, Elbabaa *et al.* 2001, Hader *et al.* 2002, Hayashi *et al.* 2000, Koch *et al.* 2002, Mohanty *et al.* 2002, Siomin *et al.* 2001, Tuli *et al.* 1999).

If the patient is readmitted with recurrent signs of intracranial hypertension and increased ventricular dilatation on CT scan or MRI, the most likely diagnosis is obstruction or severe narrowing of the third ventriculostomy. This must be immediately confirmed by sagittal median T2-weighted fast spin-echo MRI sequences, which will show disappearance of the flow artifact and recurrence of the indirect signs of occlusion of the stoma (Elbabaa *et al.* 2001, Koch *et al.* 2002, Siomin *et al.* 2002). The treatment for obstruction can be reopening or enlargement of the stoma; this procedure carries the same success rate as the primary treatment (>65%) and should be preferred in a first instance to shunt implantation. If MRI at this stage shows good flow through the stoma, then a ventriculo-peritoneal shunt should be inserted.

The "delayed open-stoma" failure can rarely be observed also in children, especially in patients who have multiple potential etiological factors for hydrocephalus (e.g., hemorrhage, meningitis). This is the most insidious form of failure because of the difficulties in diagnosing it. Patients can present mild developmental delay that is sometimes difficult to diagnose without appropriate psychomotor testing. Radiology reveals large ventricles (stable at follow-up) and open stoma at MRI with clearly visible pericerebral CSF and no transependymal resorption. Clinically, papilledema is usually absent and only mild hyperreflexia can be observed, but progressive macrocrania in the first 2–3 years of life is almost the rule. ICP monitoring in these patients reveals abnormal baseline ICP values and B waves. After ventriculo-peritoneal shunting, patients usually show significant improvements in psychomotor testing.

Intellectual Outcome

Only a few studies focusing on neuropsychological outcome are available. Burtscher *et al.* (Burtscher *et al.* 2003) in a prospective study evaluated six patients affected by late onset aqueductal stenosis using standardized psychometric testing pre- and postoperatively. They found improvement of pre-operative deficits (usually combined deficits of memory and frontal-executive function) in all patients (full recovery in two, good recovery in three and moderate recovery in one). Ventricular size diminished in all these cases but never reached normal size. When comparable series of patients with aqueductal stenosis treated by insertion of VP shunt or ETV at the same Institution were analysed, no significant differences in the post-operative IQ or in the neurological, endocrinologic, social and behavioral status could be detected (Hirsch *et al.* 1986, Sainte-Rose 1992, Tuli *et al.* 1999). A controlled randomized study comparing neuroendoscopic versus non neuroendoscopic treatment of hydrocephalus in children seems to show that the outcome of the patients treated initially with a neuroendo-scopic procedure is significantly better than that of the patients initially treated with extrathecal CSF shunt implantation, but this observation requires further investigation in multicentric studies (Kamikawa *et al.* 2001b).

Complications

Although indications, surgical technique and technological advances re-garding ETV have been largely reported, the complications of the tech-nique have been the subject of a surprisingly scarce interest (Teo *et al.* 1996, Schroeder *et al.* 2002). Furthermore, only a few papers discuss the quality of the complications in term of cost-effectiveness among ETV and traditional treatments (Garton *et al.* 2002, Richard *et al.* 2000). Such a phe-nomenon depends on the low rate of complications itself but also reflects the bias of a recently "re-introduced" technique with the obvious limita-tions in the evaluation of the long-term outcomes Anyway, the data so far cumulated may allow a sufficiently reliable analysis and the comparison with those related to the traditional treatment of hydrocephalus, based on the placement of an extrathecal CSF shunt device.

The overall morbidity rate related to the shunting procedures is esti-mated to be about 40–50% in the patients operated on before the second year of age and near 30% in those ones treated afterwards (Drake *et al.* 2000). Mechanical malfunctions (60% of the cases), infections (1–40%, mean 8.5%), over/underdrainage (10%) and post-inflammation intraven-tricular septa (6%) are the most frequent causes of complication of the CSF shunt devices (Di Rocco *et al.* 1994, Drake *et al.* 2000). The reported overall complication rate associated with ETV is lower, ranging from 6 to 20% (Buxton and Jonathan 2000, Hopf *et al.* 1999b, Teo and Jones 1996). The experience of the neuroendoscopic team is one of the most influencing factor (Schroeder *et al.* 2002). Some authors report even a nil rate of com-plications (Boschert *et al.* 2003).

The complications of ETV may be divided into "significant" and "non-significant". Significant complications such as hemorrhages, post-operative neurological deficits, hypothalamic dysfunction, severe bradycardia, epi-lepsy, infections, CSF leak are more frequently observed after ETV than after shunt placement. On the other hand, infections are more commonly associated to the placement of CSF shunt devices.

In most cases, the significant complications of ETV are only transient. The rate of significant complications ranges around 7% of the cases with only one sixth of them leading to permanent neurological damages (Broggi et al. 2000, Cinalli et al. 1998, Fukuhara et al. 2000, Sainte-Rose and Chumas 1996, Schroeder et al. 2002, Teo et al. 1991, Teo et al. 1996). Many of these complications, often the most severe ones, occur during the opening of the floor of the third ventricle, a phenomenon which accounts for a large number of techniques and instruments proposed so far to carry out the ventriculocisternostomy (Decq et al. 2004, Guiot 1973, Kehler et al. 1998, Kunz et al. 1994, Lewis and Crone 1994, Oka et al. 1993b, Paladino et al. 2000, Rieger et al. 1996, Vandertop et al. 1998, Vries 1978, Wellons et al. 1999). On the other hand, non-significant complications are linked up to transient intra-operative accidents (small bleeding, least contusions of the ventricular wall or the brain, short episodes of bradycardia) that do not affect the success of the operation and do not cause any significant damages. Their rate ranges from 5 to 13% in the largest series (Schroeder et al. 2002, Teo et al. 1996) even if their exact occurrence is difficult to establish because they are not always reported or perhaps misunderstood. The loss of substance due to the introduction of the endoscopic instrumentation and its movements inside the brain parenchyma might be one of them (Aschoff 2002).

Data about the mortality rate for ETV show an incidence meanly lower than 1%, often equal to 0% in many series (Abtin et al. 1998, Hopf et al. 1999b, Javadpour et al. 2001, Schroeder et al. 2002, Siomin et al. 2001). This incidence is comparable to the very low one observed for extrathecal shunts (0.1%) (Di Rocco et al. 1994) and it confirms ETV is a safe surgical procedure. In fact, death usually occurs after catastrophic bleedings from large vessels or, less frequently, because of severe secondary complications (infections) in patients with poor clinical conditions (Schroeder et al. 2002, Singh et al. 2003). Nevertheless, deaths after late closure of the fenestrated floor of the third ventricle have been recently reported (Abtin et al. 1998, Hader et al. 2002, Javadpour et al. 2003, Jones et al. 1996). This is the most insidious complication of ETV, mainly because sudden, unexpected and potentially fatal. Nonetheless, a potential and simple solution, as placing a ventricular catheter and a subcutaneous reservoir, has been successfully used (Mobbs et al. 2003).

Hemorrhages

The control of the intraoperative bleeding during ETV may be challenging since the minimally invasive procedures often do not give the surgeon enough space to arrest it. Moreover, the chances to arrest a blood loss are reduced by a poor three-dimensional vision. Currently, however, the risk of severe hemorrhage during ETV is rather low (Abtin *et al.* 1998, Buxton and Jonathan 2000, Schroeder *et al.* 1999).

The damage to the vessels is brought through a mechanical (endoscope, balloon, forceps) or a thermal injury (electrocoagulation, lasers) (Abtin *et al.* 1998, Buxton and Jonathan 2000, McLaughlin *et al.* 1997, Schroeder *et al.* 1999). The risk of vascular damage is higher when monopolar/bipolar or laser probe are used as their thermal effect is not much predictable and it may add itself to the mechanical one (Vandertop *et al.* 1998). Correct instruments in expert hands and continuous irrigation are the only tools available to decrease this risk.

Intraventricular bleeding from the small subependymal vessels, due to the traumatic rhexis of these delicate vascular structures caused by the impact with the endoscopic instrumentation, is the most frequent hemorrhagic complication happening during ETV (Choi *et al.* 1999, Cinalli *et al.* 1999, Fukuhara *et al.* 2000, Gangemi *et al.* 1999, Sainte-Rose and Chumas 1996). The reported incidence ranges from 1 to 3% (Beems and Grotenhuis 2004, Fukuhara *et al.* 2000, Gangemi *et al.* 1999) but higher figures can be hypothesized (Koch and Wagner 2004, Schroeder *et al.* 2004). Usually, it is a non-significant complication that can be easily managed and stopped by (extensive) irrigation with lactate Ringer's solution or, less frequently, through a prudent coagulation of the bleeding vessel. On the contrary, mild hemorrhages (hypothalamic, thalamostriate, septal veins) sometimes make the irrigation ineffective, so affecting the operative visibility and the possibility to perform a safe electrocoagulation. In these cases, the procedure has to be abandoned and an extracranial CSF drainage is placed (Buxton *et al.* 1998b, Cinalli *et al.* 1999). This "technical" failure does not prevent the possibility to try again the ETV after the clearing of the CSF, with a risk of closure of the stoma following the aseptic ventriculitis just a little increased (Cinalli *et al.* 1999, Fukuhara *et al.* 2000, Hopf *et al.* 1999b). In some instances, the venous bleeding results in a subependymal clot that may clear up spontaneously without the need to abort the surgical procedure, as reported by Schönauer *et al.* (Schönauer *et al.* 2000). Otherwise, the management of the hematoma requires a microsurgical approach (Hopf *et al.* 1999b).

Cerebral hemorrhagic contusion, due to the damage toward the pial vessels at the introduction of the endoscope (Schroeder *et al.* 2002, Vandertop *et al.* 1998), and extradural hematoma (Choi *et al.* 1999, Cinalli *et al.* 1998) are uncommon.

The damage of large vascular structures is equally rare, occurring in about 1% of ETVs (Abtin *et al.* 1998, Cinalli *et al.* 1998, McLaughlin *et al.* 1997, Schroeder *et al.* 1999, Schroeder *et al.* 2002). The basilar artery and its branches (P1 segment of the posterior cerebral artery, perforating vessels) are usually involved but also the injury to the anterior cerebral artery or to the pericallosal branch has been described (Buxton and Jonathan 2000, Cohen 1993). The surgeon's prudence and her/his anatomical knowledge, besides her/his skill to succeed in "seeing" the arteries or their pulsations through the ventricular floor, can explain this low rate. It is recommended to perforate the floor of the third ventricle halfway between the infundibular recess and the mammillary bodies, in the midline, in order to minimize the risk of vascular as well as neurological and hypothalamic damage (Schroeder *et al.* 1999). The use of ultrasonic or neuronavigation guidance to reduce the incidence of vascular complications has been recommended (Broggi *et al.* 2000, Strowitzki *et al.* 2002). An unfavorable anatomy (small or too thick or excessively elastic floor of the third ventricle) and a poor intraoperative visibility are important risk factors. Basilar artery perforation appears as a intraventricular bloody flooding that suddenly obscures the operating field. The management consists of quick removal of the endoscope and placement of an external ventricular drain through which a copious irrigation must be performed till the bleeding is arrested. After that, the drainage can be use to infuse thrombin in order to achieve the lysis of the intraventricular clot (Abtin *et al.* 1998). Also blood replacement may be necessary and an adequate neuroanesthesiologic monitoring is mandatory. The preventive placement of a peelaway sheath in the third ventricle has been proved to be very helpful to avoid the blood accumulation, so preventing an herniation syndrome and saving the patient's life (Abtin *et al.* 1998). The radiological findings of post-operative vascular complications include the demonstration of more or less extensive subarachnoid hemorrhage, intraventricular clots, persistent hydrocephalus and possible pseudoaneurysm formation (McLaughlin *et al.* 1997, Schroeder *et al.* 1999). The hydrocephalus is usually treated by ventriculoperitoneal shunt but the traumatic aneurysm could need a craniotomy to be excluded (Abtin *et al.* 1998, McLaughlin *et al.* 1997). In spite of the reported successful management, the injury to the basilar artery remains a dramatic event, potentially fatal (Schroeder *et al.* 1999). Sometimes, fatal major intraventricular bleedings occur without the possibility to establish their source (Husain *et al.* 2003).

Neurological Disorders

Also the damage to the nervous structures resulting in both diffuse or focal nervous functional impairment are caused by mechanical and/or thermal injury, either directly or indirectly. The incidence of such a type of complication is, however, lower than that of vascular complications (Beems and Grotenhuis 2004, Schroeder *et al.* 2004). Indirect injuries derive from vascular ischemic impairment, sometimes as the consequence of one of the significant hemorrhagic complications seen before. Whatever the mechanism, the result is often limited to a transient morbidity (non-permanent significant complication). In most cases the damage involve the fornix, the thalami or the mammillary bodies (Schroeder *et al.* 2002). Actually, long term neurological complications are very rare as demonstrated by only anedoctical reports (two cases of hemiparesis and one case of midbrain damage (Jones *et al.* 1994), one case of confusion after herniation syndrome and one case of oculomotor palsy (Schroeder *et al.* 2002), one case of ankle clonus and speech delay (McLaughlin *et al.* 1997). Nevertheless, it is possible that some of the complications remain unreported (Schroeder *et al.* 2004).

Among the complications leading to diffuse impairment of cerebral function, the decrease in level of consciousness and confusion are the most common, mainly depending on focal diencephalic/brain stem lesions or extensive SAH (Brockmeyer *et al.* 1998, Buxton and Jonathan 2000, Fukuhara *et al.* 2000, Schroeder *et al.* 2002), and seldom resulting from postherniation syndrome (Beems and Grotenhuis 2004). Transient memory disfunctions seem to share a similar pathogenesis (Baskin *et al.* 1998, Choi *et al.* 1999, Ferrer *et al.* 1997, Handler *et al.* 1994). The fornix and the mammillary bodies, thanks to their neural connections to the hippocampus, exert an important role in the acquisition and consolidation of newly learned information. The contusions of these two endoscopic landmark during the surgical procedure may result in impairment of immediate memory, though rarely (Benabarre *et al.* 2001). Psychiatric complications of ETV (disinhibition, aggressive behavior) are exceptional and usually related to primary or secondary lesions of the frontal lobe (Benabarre *et al.* 2001, Buxton and Jonathan 2000). Also epilepsy is rare after ETV, in most cases resulting from an impaired general clinical condition (Siomin *et al.* 2001), from a previous extensive brain damage (Handler *et al.* 1994) or from associated metabolic disfunction (Vaicys and Fried 2000). Some authors even assert that ETV minimizes the risk of late seizures, compared to traditional shunts, thanks to the lack of a permanent irritating foreign body (ventricular drain) (Kramer *et al.* 2001).

Focal neurological deficits result from localized lesions, more often ischemic than mechanical, affecting the long neural pathways (internal capsule, brain stem), with consequent side motor impairment (hemiparesis) (Brockmeyer *et al.* 1998, Buxton and Jonathan 2000, Cinalli *et al.* 1999, Sainte-Rose and Chumas 1996, Teo *et al.* 1991), the III–VI cranial nerves and/or their nuclei, with consequent oculomotion palsy (Buxton and Jonathan 2000, Gangemi *et al.* 1999, Siomin *et al.* 2001), or the optic chiasm (hemianopia) (Beems and Grotenhuis 2004). The case of Horner's syndrome due to a probable hypothalamic suffering, reported by Fukuhara *et al.* (Fukuhara *et al.* 2000), and the case of peduncular hallucinosis (visual hallucinations resulting from midbrain, pontine or diencephalic lesions), described by Kumar *et al.* (Kumar *et al.* 1999), are still unique as complications of ETV.

Although uncommonly, focal and generalized neurological deficits, as consequence of both direct and indirect mechanism, can occur simultaneously. The case of a young patient, who underwent ETV because of a ventriculoperitoneal shunt twice complicated by subdural hematoma, is emblematic (Buxton and Jonathan 2000): the opening of the ventricular floor caused a bleeding from a perforating branch of the left anterior cerebral artery, with ESA and consequent vasospasm with bilateral cerebral infarction, and a lesion of the right III cranial nerve. The patient initially developed a right oculomotor palsy and drowsiness, both transient; afterwards, a left hemiparesis and behavior disfunction, both transient; finally, a late hyperphagia.

Brain herniation syndrome is unusual but potentially dangerous neurological complication of ETV, though only transient morbidity has been attributed to it so far. It can result from intraventricular overirrigation (Brockmeyer *et al.* 1998, Schroeder *et al.* 2002) or from sudden hydrocephalus resolution in case of posterior fossa tumor (upward herniation) (Sainte-Rose *et al.* 2001).

Hypothalamic and Neurovegetative Disfunction

As expected, the anatomy of the third ventricle makes the hypothalamus vulnerable during ETV. Besides all the other mechanisms seen before, the injuries to the hypothalamus can arise from the distortion of its nuclei caused by the strain of the ventricular floor or by overirrigation. This mechanism might be liable of one of the most frequent and less studied complications of ETV, the bradycardia. In fact, it has been found that the incidence of the phenomenon at the opening of the stoma may exceed a rate of 40%, because of the distortion of the hypothalamic autonomic nuclei (El-Dawlatly *et al.* 2000). However, this complication is clinically non-significant in most cases, rarely requiring a temporary interruption of the procedure. Seldom it leads to cardiac asystole and transient cardiac arrest (Fukuhara *et al.* 2000, Teo *et al.* 1996).

Ventricular tachycardia, followed by ventricular fibrillation and near-fatal cardiac arrest, reported in 14-years-old girl, was explained on the base of a similar mechanism (rapid and/or excessive distension of the third ventricle due to the irrigation) (Handler *et al.* 1994).

A few cases of transient respiratory arrest following ETV were laso described and attributed to the patient's general clinical condition after extubation rather than to the procedure itself (Enya *et al.* 1997, Siomin *et al.* 2001).

Diabetes insipidus is a common complication among the hypothalamic imbalances due to ETV but, since it is basically transient, its actual incidence is not well known (Schroeder *et al.* 2002, Teo and Jones 1996, Teo *et al.* 1996). The damage concerns the supraoptic and paraventricular nuclei or, more probably, their connections to the hypophyseal median eminence (Coulbois *et al.* 2001). It usually affects the patient for a few days after the operation and regresses spontaneously or after a short treatment by vasopressin (Coulbois *et al.* 2001, Teo and Jones 1996). To date, only three cases of persistent diabetes insipidus have been described (Di Roio *et al.* 1999, Schroeder *et al.* 2002, Beems and Grotenhuis 2004).

Other possible hypothalamic endocrinologic or metabolic disfunctions are uncommon. They include secondary amenorrhea (Teo *et al.* 1996), inappropriate secretion of antidiuretic hormone/hyponatriemia (Javadpour *et al.* 2001, Vaicys and Fried 2000), hyperphagia (Buxton and Jonathan 2000, Teo *et al.* 1996), and loss of thirst (Schroeder *et al.* 2002, Teo *et al.* 1996).

Transient hyperthermia is reported as possible result of hypothalamic manipulation, though it is often difficult to exclude a concomitant aseptic irritation of the ependyma (Sainte-Rose 1992, Sainte-Rose and Chumas 1996). In all the cases, however, it is mandatory to rule out a secondary infective complication when the fever persists after the first 24–48 postoperative hours. On the other hand, hypothermia, plausibly due to excessively cold irrigating fluid, is a not rare finding during the early recovery after ETV (Singh *et al.* 2003).

Other Complications

Infections account for the majority of non-specific complications of ETV. They are facilitated by the presence of a previous infected shunt or an external drainage more than depending on the endoscopic procedure itself (Fukuhara *et al.* 2000, Hopf *et al.* 1999b, Schroeder *et al.* 2002). According to the literature, the approximate overall rate ranges from 1 to 5% (Beems and Grotenhuis 2004, Schroeder *et al.* 2004, Teo *et al.* 1996) and includes wound infections (Jones *et al.* 1994, Siomin *et al.* 2001, Teo *et al.* 1991), ventriculitis (Abtin *et al.* 1998, Brockmeyer *et al.* 1998, Buxton *et al.* 1998b, Jones *et al.* 1994, Teo *et al.* 1991), and meningitis (Gangemi *et al.* 1999, Hopf *et al.* 1999b, Schroeder *et al.* 2002). This low incidence rate accounts for the current favor of ETV, especially when the data related to extrathecal CSF shunts (range 1–40%, mean 8.5%) (Whitehead and Kestle 2001) are taken into consideration. In spite of their generally benign course and the effectiveness of the antibiotic drugs (Schroeder *et al.* 2004), CSF infections following ETV are a very feared complication of the procedure because of the risk of arachnoiditis and consequent obliteration of the CSF pathways. Actually, postoperative meningitis resulted as an independent risk factor of failure of ETV from the multivariate analysis of Fukuhara *et al.* (Fukuhara *et al.* 2000).

CSF leak is a characteristic complication of ETV (Buxton *et al.* 1998b, Gangemi *et al.* 1999, Schroeder *et al.* 2002, Siomin *et al.* 2001, Tamburrini *et al.* 2004, Teo *et al.* 1996), especially in infants (thin skin, immature subarachnoid spaces) (Teo 2004). Though in many cases it can be successfully managed by intermittent tapping from lumbar or ventricular drainage (Cinalli 2004, Husain *et al.* 2003, Siomin *et al.* 2001), revealing the possibility of a gradual adaptation of the CSF hydrodynamics to the ETV (Nishiyama *et al.* 2003), this type of complication represents a risk factor for infections and, in particular, it may be the early sign of a failure of the procedure and the need of a shunt placement.

Subdural fluid collections (hematomas or hygromas) are reported as the second frequent complication involving the vascular structures after the intraventricular blood hemorrhages (Oka *et al.* 1993b, Fukuhara *et al.* 2000, Jones *et al.* 1994, Sainte-Rose and Chumas 1996, Teo *et al.* 1991). Their incidence, however, is relatively low, being less than 2% in large series (Genitori *et al.* 2004, Schroeder *et al.* 2002, Schroeder *et al.* 2004). These extracerebral collections are likely related to the quick drop of the intracranial pressure following the opening of the stoma (Sgaramella *et al.* 2004) and/or, more probably, to the possible acute CSF leak at the beginning or at the end of the procedure (Teo 2004). The phenomenon is especially common in children with a very thin brain cortex and massive hydrocephalus (Schönauer *et al.* 2000). They are quite often asymptomatic and do not require any further surgical procedure (Fukuhara *et al.* 2000, Schroeder *et al.* 2002). Only in a few cases, they need to be evacuated, producing a transient morbidity (Jones *et al.* 1990, Sgaramella *et al.* 2004). The case of a 3-months-old child, who developed a cardiorespiratory arrest because of a massive subdural collection after ETV and who was successfully treated by resuscitation and urgent aspiration of the fluid collection, is exceptional (Mohanty *et al.* 1997).

Finally, tension pneumocephalus is a rarely reported (Hamada *et al.* 2004) but dangerous complication of ETV, likely depending on the CSF leak during the procedure.

References

1. Abtin K, Thompson BG, Walker ML (1998) Basilar artery perforation as a complication of endoscopic third ventriculostomy. Pediatr Neurosurg 28: 35–41
2. Alberti O, Riegel T, Hellwig D, Bertalanffy H (2001) Frameless navigation and endoscopy. J Neurosurg 95: 541–543

3. Aschoff A (2002) III Ventriculostomy. EANS Winter Meeting, Rome, March 19th

4. Auer LA, Auer DP (1998) Virtual endoscopy for planning and simulation of minimally invasive neurosurgery. Neurosurgery 43: 529–548

5. Barkovich AJ, Kjos BO, Norman D, Edwards MS (1989) Revised classification of posterior fossa cysts and cyst-like malformations based on the results of multiplanar MR imaging. AJNR 10: 977–988

6. Barr (1948) Observations on the foramen of Magendie in a series of human brains. Brain 71: 281

7. Baskin JJ, Manwaring KH, Rekate HL (1998) Ventricular shunt removal: the ultimate treatment of the slit ventricle syndrome. J Neurosurg 88: 478–484

8. Bech RA, Bogeskov L, Borgesen SE, Juhler M (1999) Indications for shunt insertion or 3rd ventriculostomy in hydrocephalic children, guided by lumbar and intraventricular infusion tests. Child's Brain 15: 213–218

9. Beems T, Grotenhuis JA (2002) Is the success rate of endoscopic third ventriculostomy age dependent? An analysis of the results of third ventriculostomy in young children. Child's Nerv Sist 18: 605–608

10. Beems T, Grotenhuis JA (2004) Long term complications and definition of failures of neuroendoscopic procedures. Child's Nerv Syst 20: 868–877

11. Bellotti A, Rapana A, Iaccarino C, Schonauer (2001) Intracranial pressure monitoring after endoscopic third ventriculostomy: an effective method to manage the 'adaptation period'. Clin Neurol Neurosurg 103: 223–227

12. Benabarre A, Ibáñez J, Boget T, Obiols J, Martínez-Aran A, Vieta E (2001) Neuropsychological and psychiatric complications in endoscopic third ventriculostomy: a clinical case report. J Neurol Neurosurg Psychiatry 71: 268–271

13. Beni-Adani L, Bental Y, Ben-Sira L (2004) Does aqueductal stenosis really exist as an independent diagnosis or does it represent an old fetal hemorrhage misdiagnosed on real time fetal US? Child's Nerv Syst 20: 663

14. Boop FA (2004) Post-hemorrhagic hydrocephalus of prematurity. In: Cinalli G, Meixner W, Sainte-Rose C (eds) Pediatric hydrocephalus, Springer-Verlag, Milan, pp 121–132

15. Böschert J, Hellwig D, Krauss JK (2003) Endoscopic third ventriculostomy for shunt dysfunction in occlusive hydrocephalus: long-term follow up and review. J Neurosurg 98: 1032–1039

16. Bozzini P (1806) Lichtleiter; eine Erfindung zur Ashauung innerer Theile und Krankheiten nebst der Abbildung. In: Hufeland CW (ed) J der practischen Arzneykunde und Wundarrzneykunst, Berlin, 24: 107–124

17. Brockmeyer D, Abtin K, Carey L, Walker M (1998) Endoscopic third ventriculostomy: an outcome analysis. Pediatr Neurosurg 28: 236–240

18. Broggi G, Dones I, Ferroli P, Franzini A, Servello D, Duca S (2000) Image guided neuroendoscopy for third ventriculostomy. Acta Neurochir (Wien) 142: 893–899

19. Büki A, Doczi T, Veto F, Horvath Z, Gallyas F (1999) Initial clinical experience with a combined pulsed holmiumneodymium-YAG laser in minimally invasive neurosurgery. Minim Invasive Neurosurg 42: 35–40

20. Burtscher J, Bale R, Dessl A, Eisner W, Twerdy K, Sweeney RA, Felber S (2002) Virtual endoscopy for planning neuro-endoscopic intraventricular surgery. Minim Invasive Neurosurg 45: 24–31

21. Burtscher J, Bartha L, Twerdy K, Eisner W, Benke T (2003) Effect of endoscopic third ventriculostomy on neuropsychological outcome in late onset idiopathic aqueductal stenosis: a prospective study. J Neurol Neurosurg Psychiatry 74: 222–225

22. Burtscher J, Dessl A, Maurer H, Seiwald M, Felber S (1999) Virtual neuroendoscopy, a comparative magnetic resonance and anatomical study. Minim Invas Neurosurg 42: 113–117

23. Buxton N, Jonathan P (2000) Cerebral infarction after neuroendoscopic third ventriculostomy: case report. Neurosurgery 46: 999–1002

24. Buxton N, Macarthur D, Mallucci C, Punt J, Vloeberghs M (1998a) Neuroendoscopy in the premature population. Child's Nerv Syst 14: 649–652

25. Buxton N, Macarthur D, Mallucci C, Punt J, Vloeberghts M (1998b) Neuroendoscopic third ventriculostomy in patients less than 1 year old. Pediatr Neurosurg 29: 73–76

26. Buxton N, Macarthur D, Robertson I, Punt J (2003) Neuroendoscopic third ventriculostomy for failed shunts. Surg Neurol 60: 193–200

27. Buxton N, Turner B, Ramli N, Vloeberghs M (2002) Changes in third ventricular size with neuroendoscopic third ventriculostomy: a blinded study. J Neurol Neurosurg Psychiatry 72: 385–387

28. Chapman PH (1990) Indolent gliomas of the midbrain tectum. Concepts Pediatr Neurosurg 10: 97–107

29. Choi JU, Kim DS, Kim SH (1999) Endoscopic surgery for obstructive hydrocephalus. Yonsei Med J 40: 600–607

30. Cinalli G (1999) Alternatives to shunting. Child's Nerv Syst 15: 718–731

31. Cinalli G (2004) Endoscopic third ventriculostomy. In: Cinalli G, Maixner WJ, Sainte-Rose C (eds) Pediatric Hydrocephalus, Springer, Milan, pp 361–388

32. Cinalli G, Sainte-Rose C, Chumas P, Zerah M, Brunelle F, Lot G, Pierre-Khan A, Renier D (1999) Failure of third ventriculostomy in the treatment of aqueductal stenosis in children. J Neurosurg 90: 448–454

33. Cinalli G, Salazar C, Mallucci C, Yada JZ, Zerah M, Sainte-Rose C (1998) The role of endoscopic third ventriculostomy in the management of shunt malfunction. Neurosurgery 43: 1323–1329

34. Cinalli G, Spennato P, Cianciulli E, D'Armiento M (2004a) Hydrocephalus and aqueductal stenosis. In: Cinalli G, Meixner W, Sainte-Rose C (eds) Pediatric Hydrocephalus, Springer-Verlag, Milan, pp 279–294

35. Cinalli G, Spennato P, Del Basso De Caro ML, Buonocore MC (2004b) Hydrocephalus and the Dandy-Walker malformation. In: Cinalli G, Meixner W, Sainte-Rose C (eds) Pediatric Hydrocephalus, Springer-Verlag, Milan, pp 279–294

36. Ciurea AV, Coman TC, Mircea D (2004) Postinfectious hydrocephalus in children. In: Cinalli G, Meixner W, Sainte-Rose C (eds) Pediatric Hydrocephalus, Springer-Verlag, Milan, pp 201–218

37. Cohen AR (1993) Endoscopic ventricular surgery. Pediatr Neurosurg 19: 127–134

38. Coulbois S, Boch A-L, Philippon J (2001) Diabète insipide après ventriculo-cisternostomie par voie endoscopique. A propos d'un cas et revue de la littér-ature. Neurochirurgie 47: 435–438

39. Czosnyka M, Whitehouse M, Smielewski P, Simac S, Pickard JD (1996) Testing of cerebrospinal compensatory reserve in shunted and non-shunted patients: a guide to interpretation based on an observational study. J Neurol Neurosurg Psychiatry 60: 549–558

40. Dandy WE (1918) Extirpation of the choroid plexuses of the lateral and fourth ventricle in communicating hydrocephalus. Ann Surg 68: 569–579

41. Dandy WE (1922) An operative procedure for hydrocephalus. Bull Johns Hopkins Hosp 33: 189–190

42. Daoud AS, Omari H, al-Sheyyab M, Abuekteish F (1998) Indications and benefits of computer tomography in childhood bacterial meningitis. J Trop Pediatr 44: 167–169

43. Decq P (2004) Endoscopic anatomy of the ventricles. In: Cinalli G, Maixner WJ, Saint-Rose C (eds) Pediatric Hydrocephalus, Springer, Milan, pp 351–359

44. Decq P, Le Guerinel C, Palfi S, Djindjian M, Kéravel Y, Nguyen J (2000) A new device for endoscopic third ventriculostomy. J Neurosurg 93: 509–512

45. Di Rocco C, Marchese E, Velardi F (1994) A survey of the first complication of newly implanted CSF shunt devices for the treatment of nontumoral hydrocephalus. Cooperative survey of the 1991–1992 Education Committee of the ISPN. Child's Nerv Syst 10: 312–327

46. Di Roio C, Mottolese C, Cayrel V, Berlier P, Artru F (1999) Ventriculosto-mie du troisième ventricule et diabète insipide. Ann Fr Anesth Reanim 18: 776–778

47. Doctor BA, Newman N, Minich NM, Taylor HG, Fanaroff AA, Hack M (2001) Clinical outcomes of neonatal meningitis in very-low birth-weight infant. Clin Pediatr 40: 473–480

48. Drake JM (1993) Ventriculostomy for treatment of hydrocephalus. Neuro-surg Clin North Am 4: 657–666

49. Drake JM, Kestle JRW, Milner R, Cinalli G, Boop F, Piatt J Jr, Hainess S, Schiff SJ, Cochrane DD, Steinbok P, McNeil N (1998) Randomized trial of cerebrospinal fluid shunt valve design in pediatric hydrocephalus. Neuro-surgery 43: 294–303

50. Drake JM, Kestle JRW, Tuli S (2000) CSF shunts 50 years on – past, present and future. Child's Nerv Syst 16: 800–804

51. Duffner F, Shiffbauer H, Glenser D, Skalej M, Freudenstein D (2003) Anat-omy of the cerebral ventricular system for endoscopic neurosurgery: a mag-netic resonance study. Acta Neurochir 145: 359–368

52. Elbabaa SK, Steinmetz M, Ross J, Moon D, Luciano M (2001) Endoscopic third ventriculostomy for obstructive hydrocephalus in the pediatric popula-tion: evaluation of outcome. Eur J Pediatr Surg 11 (suppl. 1): 552–554

53. El-Dawlatly AA, Murshid WR, El-Khwsky F (1999) Endoscopic third ventriculostomy: a study of intracranial pressure vs heamodynamic changes. Minim Invas Neurosurg 42: 198–200
54. El-Dawlatly AA, Murshid WR, Elshimy A, Magboul MA, Samarkandi A, Takrouri MS (2000) The incidence of bradycardia during endoscopic third ventriculostomy. Anesth Analg 91: 1142–1144
55. Enya S, Masuda Y, Terui K (1997) Respiratory arrest after a ventriculoscopic surgery in infants: two case reports. Masui 46: 416–420 [Japanese]
56. Enzmann DR, Pelec NJ (1991) Normal flow pattern in intracranial and spinal cerebrospinal fluid defined with phase-contrast cine-MR imaging. Radiol 178: 467–474
57. Fay T, Grant FC (1923) Ventriculoscopy and intraventricular photography in internal hydrocephalus. JAMA 80: 461–463
58. Feng H, Huang G, Liao X, Fu K, Tab H, Pu H, Cheng Y, Liu W, Zhao D (2004) Endoscopic third ventriculostomy in the management of obstructive hydrocephalus: an outcome analysis. J Neurosurg 100: 626–633
59. Ferrer E, Santamarta D, Garcia-Fructuoso G, Caral L, Rumia J (1997) Neuroendoscopic management of pineal region tumors. Acta Neurochir 139: 12–21
60. Fischer RE, Biljana T (2000) Optical system design, McGraw Hill, New York
61. Foroutan M, Mafee MF, Dujovny M (1998) Third ventriculostomy, phase-contrast cine MRI and endoscopic techniques. Neurol Res 20: 443–448
62. Frazee JG, Shah AS (1998) Interactive surgery and telepresence. In: King W, Frazee J, De Salles A (eds) Endoscopy of the Central and Peripheral Nervous System, Thieme, New York, 1998, pp 243–251
63. Fritsch MJ, Mehdorn M (2002) Endoscopic intraventricular surgery for treatment of hydrocephalus and loculated CSF space in children less than one year of age. Pediatr Neurosurg 36: 183–188
64. Fritsch MJ, Mehdorn HM (2003) Indication and controversies for endoscopic third ventriculostomy in children. Child's Nerv Syst 19: 706–707
65. Fritsch MJ, Dorner L, Kienke S, Claviez A, Mehdorn HM (2004) Indication for endoscopic third ventriculostomy in children with posterior fossa tumors. Child's Nerv Syst 20: 668
66. Fukuhara T, Luciano MG, Kowalski RJ (2002) Clinical features of third ventriculostomy failures classified by fenestration patency. Surg Neurol 58: 102–110
67. Fukuhara T, Vorster SJ, Luciano MG (2000) Risk factors for failure of endoscopic third ventriculostomy for obstructive hydrocephalus. Neurosurgery 46: 1100–1111
68. Fukushima T, Ishijima B, Hirakaw K, Nakamura N, Samo K (1973) Ventriculofiberscope: a new technique for endoscopic diagnosis and operation. J Neurosurg 38: 251–256
69. Gangemi M, Donati P, Maiuri F, Longatti P, Godano U, Mascari C (1999) Endoscopic third ventriculostomy for hydrocephalus. Minim Invasive Neurosurg 42: 128–132

70. Garton HJL, Kestle JRW, Cochrane DD, Steinbok P (2002) A cost-effectiveness analysis of endoscopic third ventriculostomy. Neurosurgery 51: 69–78

71. Genitori L, Peretta P, Mussa F, Giordano F (2004) Endoscopic third ventriculostomy in children: are age and etiology of hydrocephalus predictive factors influencing the outcome in primary and secondary treated patients? A series of 328 patients and 353 procedures. II CURAC Congress, Munich, October 8th

72. Glasauer FE (1975) Isotope cisternography and ventriculography in congenital anomalies of the central nervous system. J Neurosurg 43: 18–26

73. Goh KYC, Abbott R (2000) Is endoscopic third ventriculostomy of benefit in tumour related aqueductal stenosis? Child's Nerv Syst 16: 127–128

74. Gorayeb RP, Cavalheiro S, Zymberg SY (2004) Endoscopic third ventriculostomy in children younger than 1 year of age. J Neurosurg (Pediatrics 5) 100: 427–429

75. Goumnerova LC, Frim D (1997) Treatment of hydrocephalus with third ventriculostomy: outcome and CSF flow patterns. Pediatr Neurosurg 27: 149–152

76. Grant JA (1998) An endoscopic view of the anterior commissure. Pediatr Neurosurg 28: 42

77. Greitz D (2004) Radiological assessment of hydrocephalus: new theories and implications for therapy. Neurosurg Rev 27: 145–165

78. Grunert P, Perneczki A, Resch K (1994) Endoscopic procedures through the foramen of Monro under stereotactic conditions. Mim Invasive Neurosurg 37: 2–8

79. Grunert P, Charalampaki P, Hopf N, Filippi R (2003) The role of third ventriculostomy in the management of obstructive hydrocephalus. Minim Invasive Neurosurg 46: 16–21

80. Guiot G (1973) Ventriculo-cisternostomy for stenosis of aqueduct of Silvius. Puncture of the floor of the third ventricle with a leucotome under television control. Acta Neurochir 28: 275–289

81. Hader WJ, Drake J, Cochrane D, Sparrow O, Johnson ES, Kestle J (2002) Death after late failure of third ventriculostomy in children. Report of three cases. J Neurosurg 97: 211–215

82. Hamada H, Hayashi N, Kurimoto M, Umemura K, Hirashima Y, Nogami K, Endo S (2004) Tension pneumocephalus after neuroendoscopic procedure – case report. Neurol Med Chir (Tokyo) 44: 205–208

83. Handler MH, Abbott R, Lee M (1994) A near-fatal complication of endoscopic third ventriculostomy: case report. Neurosurgery 35: 525–528

84. Hayashi N, Hamada H, Hirashima Y, Kurimoto M, Takaku A, Endo S (2000) Clinical features in patients requiring reoperation after failed endoscopic procedures for hydrocephalus. Minim Invasive Neurosurg 43: 181–186

85. Hecht J (1999) City of light: the story of fiber optics, Oxford University Press, New York

86. Heilman CB, Cohen AR (1991) Endoscopic ventricular fenestration using "saline torch". J Neurosurg 74: 224–229

87. Hellwig D, Haag R, Bartel V, Riegel T, Eggers F, Becker R, Bertalanffy H (1999) Application of new electrosurgical devices and probes in endoscopic neurosurgery. Neurol Res 21: 67–72

88. Hellwig D, Heinemann A, Riegel T (1998a) Endoscopic third ventriculostomy in treatment of obstructive hydrocephalus caused by primary aqueductal stenosis. In: Hellwig D, Bauer B (eds) Minimally Invasive Techniques for Neurosurgery, Springer, Berlin, pp 65–72

89. Hellwig D, Riegel T, Bertalanffy H (1998b) Neuroendoscopic techniques in treatment of intracranial lesions. Minim Invasive Ther Allied Technol 7: 123–135

90. Helseth E, Due-Tonnessen B, Egge A, Eide PK, Meling T, Lundar T, Froslie KF (2002) Behandling av Hydrocephalus med endoskopisk tredjeventrikkelstomi. Tidsskr Nor Loegeforen 122: 994–998

91. Hill A, Volpe J (1981) Seizures, hypoxic-ischemic brain injury, and intraventricular hemorrhage in the newborn. Ann Neurol 10: 109–121

92. Hirsch JF, Pierre-Kahn A, Renier D, Sainte-Rose C, Hoppe-Hirsch E (1984) The Dandy-Walker malformation. A review of forty cases. J Neurosurg 61: 512–522

93. Hirsch JF, Hirsch E, Sainte-Rose C, Renier D, Pierre-Kahn A (1986) Stenosis of the aqueduct of Sylvius. Etiology and treatment. J Neurosurg Sci 30: 29–36

94. Hoffman HJ, Harwood-Nash D, Gilday DL, Craven MA (1980) Percutaneous third ventriculostomy in the management of non-communicating hydrocephalus. Neurosurgery 7: 313–321

95. Hopf NJ, Grunert P, Darabi K, Busert C, Bettag (1999a) Frameless neuronavigation applied to endoscopic neurosurgery. Minim Invasive Neurosurg 42: 187–193

96. Hopf NJ, Grunert P, Fries G, Resch KDM, Perneczky A (1999b) Endoscopic third ventriculostomy: outcome analysis of 100 consecutive procedures. Neurosurgery 44: 795–806

97. Horowitz M, Ramzipoor K, Mair A, Miller S, Rappard G, Spiro R, Purdy P (2003) Experimental third ventriculostomy performed using endovascular surgical techniques and their adaptation to percutaneous intradural neuronavigation: proof of concept cadaver study. Neurosurgery 53: 387–392

98. Husain M, Jha D, Vatsal DK, Thaman D, Gupta A, Husain N, Gupta RK (2003) Neuroendoscopy surgery – experience and outcome analysis of 102 consecutive procedure in a busy neurosurgical center of India. Acta Neurochir 145: 369–376

99. Iantosca MR, Hader WJ, Drake JM (2004) Results of endoscopic third ventriculostomy. Neurosurg Clin N Am 15: 67–75

100. Jacobson EE, Fletcher DF, Morgan MK, Johnston IH (1999) Computer modelling of the CSF flow dynamics of aqueductal stenosis. Med Biol Eng Comput 37: 59–63

101. Jaksche H, Loew F: Burr hole third ventriculostomy (1986) An unpopular but effective procedure for treatment of certain forms of occlusive hydrocephalus. Acta Neurochir 79: 48–51

102. Javadpour M, Mallucci C, Brodbelt A, Golash A, May P (2001) The impact of endoscopic third ventriculostomy on the management of newly diagnosed hydrocephalus in infants. Pediatr Neurosurg 35: 131–135

103. Javadpour M, May P, Mallucci C (2003) Sudden death secondary to delayed closure of endoscopc third ventriculostomy. Br J Neurosurg 17: 266–269

104. Jellinger G (1986) Anatomomorphology of nontumoral aqueductal stenosis. J Neurosurg Sci 30: 1–16

105. Jodicke A, Berthold LD, Scharbrodt W, Schroth I, Reiss I, Neubauer BA, Boker DK (2003) Endoscopic surgical anatomy of the paediatric third ventricle studied using virtual neuroendoscopy based on 3-D ultrasonography. Child's Nerv Syst 19: 325–331

106. Jones FRC, Stening WA, Brydon M (1990) Endoscopic third ventriculostomy. Neurosurgery 26: 86–92

107. Jones RFC, Kwok BCT, Stening WA (1996) Endoscopic III ventriculostomy. How long does it last? Child's Nerv Syst 12: 364–365

108. Jones RFC, Kwok BCT, Stening WA, Vonau M (1994) The current status of endoscopic third ventriculostomy in the management of non-communicating hydrocephalus. Minim Invasive Neurosurg 37: 28–36

109. Kaiser G (1985) Hydrocephalus following toxoplasmosis. Z Kinderkir 40 (Suppl. 1): 10–11

110. Kamikawa S, Inui A, Kobayashi N, Kuwamura K, Kasuga M, Yamadori T, Tamaki N (2001b) Endoscopic treatment of hydrocephalus in children: a controlled study using newly developed Yamadori-type ventriculoscopes. Minim Invasive Neurosurg 44: 25–30

111. Kamikawa S, Inui A, Kobayashi N, Tamaki N, Yamadori T (2001c) Intraventricular hemorrhage in neonates: endoscopic findings and treatment by the use of our newly developed Yamadori-type 8 ventriculoscope. Min Invas Neurosurg 44: 74–78

112. Kamikawa S, Inui A, Tamaki N, Kobayashi N, Yamadori T (2001a) Application of flexible neuroendoscope to intracerebroventricular arachnoid cyst in children – Use of videoscope. Minim Invas Neurosurg 44: 186–189

113. Kehler U, Gliemroth J, Knopp U, Arnold H (1998) How to perforate safely a resistant floor of the third ventricle? Technical note. Minin Invasive Neurosurg 41: 198–199

114. Kelly PJ (1991) Stereotactic third ventriculostomy in patients with nontumoral adolescent/adult onset aqueductal stenosis and symptomatic hydrocephalus. J Neurosurg 75: 865–873

115. Kim SK, Wang KC, Cho BK (2000) Surgical outcome of pediatric hydrocephalus treated by endoscopic III ventriculostomy: prognostic factors and interpretation of postoperative neuroimaging. Child's Nerv Syst 16: 161–169

116. Koch D, Grunert P, Filippi R, Hopf N (2002) Re-ventriculostomy for treatment of obstructive hydrocephalus in cases of stoma dysfunction. Minim Invasive Neurosurg 45: 158–163

117. Koch D, Wagner W (2004) Endoscopic third ventriculostomy in infants of less than 1 year of age: which factors influence the outcome? Child's Nerv Syst 20: 405–411

118. Kramer U, Kanner AA, Siomin V, Harel S, Constantini S (2001) No evidence of epilepsy following endoscopic third ventriculostomy: a short-term follow up. Pediatr Neurosurg 34: 121–123

119. Kreusser K, Tarby T, Kovnar E, Taylor DA, Hill A, Volpe JJ (1985) Serial lumbar punctures for at least temporary amelioration of neonatal posthemorrhagic hydrocephalus. Pediatrics 75: 719–724

120. Kulkarni AV, Drake JM, Armstrong DC, Dirks PB (2000) Imaging correlates of successful endoscopic third ventriculostomy. J Neurosurg 92: 915–919

121. Kumar R, Behari S, Wahi J, Banerji D, Sharma K (1999) Peduncular hallucinosis: an unusual sequel to surgical intervention in the suprasellar region. Br J Neurosurg 13: 500–503

122. Kunz U, Goldman A, Bader C, Waldbaur H, Oldenkott P (1994) Endoscopic fenestration of the 3rd ventricular floor in aqueductal stenosis. Minim Invasive Neurosurg 37: 42–47

123. Laikin M (2001) Method of lens design, Dekker, New York

124. Lang J (1992) Topographic anatomy of preformed intracranial spaces. Acta Neurochir (Wien) (suppl.) 54: 1–10

125. Lapras C, Bret P, Patet JD, Huppert J (1986) Hydrocephalus and aqueductal stenosis. Direct surgical treatment by interventriculostomy (aqueduct cannulation). J Neurosurg Sci 30: 47–53

126. Lespinasse VL (1910) In: Davis L (ed) Neurological Surgery, Lea & Febinger, Philadelphia, 1936, p 405

127. Lewis AI, Crone KR (1994) Advances in neuroendoscopy. Contemp Neurosurg 16: 1–6

128. Liu CY, Wang MY, Apuzzo MLJ (2004) The physics of image formation in the neuroendoscope. Childs Nerv Syst 20: 777–782

129. Longatti P, Martinuzzi A, Fiorindi A, Malobabic S, Carteri A (2003) Endoscopic anatomic features of the triangular recess. Neurosurgery 52: 1491–1494

130. Macarthur DC, Buxton N, Vloeberghs M, Punt J (2001). The effectiveness of neuroendoscopic interventions in children with brain tumors. Child's Nerv Syst 17: 589–594

131. Magnaes B (1982) Cerebrospinal fluid hydrodynamics in adult patients with benign non-communicating hydrocephalus: one-hour test shunting and balanced cerebrospinal fluid infusion test to select patients for intracranial bypass operation. Neurosurgery 11: 769–775

132. Magnaes B (1989) Hydromechanical testing in non-communicating hydrocephalus to select patients for microsurgical third ventriculostomy. Br J Neurosurg 3: 443–450

133. Matson DD (1969) Neurosurgery of Infancy and Childhood, 2nd edition, Charles C Thomas, Springfield, Illinois

134. McLaughlin MR, Wahlig JB, Kaufmann AM, Albright AL (1997) Traumatic basilar aneurysm after endoscopic third ventriculostomy: case report. Neurosurgery 41: 1400–1404

135. Miller MN (1992) Organisation of the neuroendoscopy suite. In: Manwaring KH, Crone KR (eds) Neuroendoscopy, vol 1, Mary Ann Liebert, New York, pp 9–15

136. Mixter WJ (1923) Ventriculoscopy and puncture of the floor of the third floor. Preliminary report of a case. Boston Med Surg J 1: 277–278

137. Mobbs RJ, Vonau M, Davies MA (2003) Death after late failure of endoscopic third ventriculostomy: a potential solution. Neurosurgery 53: 384–386

138. Mohanty A, Anandt B, Reddy MS, Sastry KVR (1997) Contralateral massive acute subdural collection after endoscopic third ventriculostomy – a case report. Minim Invasive Neurosurg 37: 28–36

139. Mohanty A, Vasudev MK, Sampath S, Radhesh S, Sastry, Kolluri VR (2002) Failed endoscopic third ventriculostomy in children: management options. Pediatr Neurosurg 37: 304–309

140. Mohanty A (2003) Endoscopic third ventriculostomy with cystoventricular stent placement in the management of Dandy-Walker malformation: technical case report of three patients. Neurosurgery 53: 1223–1229

141. Mori H, Nishiyama K, Tanaka R (2003) Endoscopic third ventriculostomy in shunt malfunction and slit-ventricle syndrome. Child's Nerv Syst 19: 690

142. Murshid WR (2000) Endoscopic third ventriculostomy: towards more indications for the treatment of non-communicating hydrocephalus. Minim Invas Neurosurg 43: 75–82

143. Nishiyama K, Mori H, Tanaka R (2003) Changes in cerebrospinal fluid hydrodynamics following endoscopic third ventriculostomy for shunt-dependent noncommunicating hydrocephalus. J Neurosurg 98: 1027–1031

144. Nitze M (1879) Eine neue Beobachtungs und Untersuchungs Methode für Harnröhre, Harnblse und Rectum. Wien Med Wochenschr 24: 649–652

145. Nobles A (1998) The physics of neuroendoscopic systems and the instrumentation. In: Jimenez DF (ed) Intracranial Endoscopic Neurosurgery, Park Ridge, IL: American Association of Neurological Surgeons, pp 1–12

146. Nugent GR, Al-Mefty O, Chou S (1979) Communicating hydrocephalus as a cause of aqueductal stenosis. J Neurosurg 51: 812–818

147. Oi S, Shibata M, Tominaga J, Honda Y, Shinoda M, Takei F, Tsugane R, Matsuzawa K, Sato O (2000) Efficacy of neuroendoscopic procedures in minimally invasive preferential management of pineal region tumors: a prospective study. J Neurosurg 93: 245–253

148. Oi S, Hamada H, Nonaka Y, Kusaka Y, Samii A, Samii M (2004) Proposal of a "evolution theory in CSF dynamics": a significant factor affecting the failure of neuroendoscopic ventriculostomy in treatment of hydrocephalus. Child's Nerv Syst 20: 662

149. Oka K, Go Y, Kin Y, Tomonaga M (1993a) An observation of the third ventricle under flexible fiberoptic ventriculoscope: normal structure. Surg Neurol 40: 273–277

150. Oka K, Go Y, Kin Y, Utsunomiya H, Tomonaga M (1995) The radiographic restoration of the ventricular system after third ventriculostomy. Minim Invasive Neurosurg 38: 158–162

151. Oka K, Tomonaga M (1992) Instruments for flexible endoneurosurgery. In: Manwaring KH, Crone KR (eds) Neuroendoscopy, vol 1, Mary Ann Liebert, New York, pp 17–28

152. Oka K, Yamamoto M, Ikeda K, Tomonaga M (1993b) Flexible endo-neurosurgical therapy for aqueductal stenosis. Neurosurgery 33: 236–243

153. Paladino J, Rotim K, Stimac D, Pirker N, Stimac A (2000) Endoscopic third ventriculostomy with ultrasonic contact microprobe. Minim Invasive Neurosurg 43: 132–134

154. Pavez Salinas A (2004) Description of endoscopic ventricular anatomy in meningomyelocele. Child's Nerv Syst (in press)

155. Perlman JM, Argyl C (1992) Lethal cytomegalovirus infection in preterm infants: clinical, radiological, and neuropathological findings. Ann Neurol 31: 64–68

156. Pettorossi VE, Di Rocco C, Mancinelli R, Caldarelli M, Velardi F (1978) Communicating hydrocephalus induced by mechanically increased amplitude of the intraventricular cerebrospinal fluid pulse pressure: rationale and method. Exp Neurol 59: 30–39

157. Pollack JK, Pang D, Albright AL (1994) The long term outcome in children with late-onset aqueductal stenosis resulting from benign intrinsic tectal tumors. J Neurosurg 80: 681–688

158. Pople IK, Athanasiou TC, Sandeman DR, Coakham HB (2001) The role of endoscopic biopsy and third ventriculostomy in the management of pineal region tumours. Br J Neurosurg 15: 305–311

159. Portillo S, Zuccaro G, Fernandez-Molina A, Houssay A, Sosa F, Konsol O, Jaimovich R, Olivella E, Ledesma J, Guevara M, Ajler G, Picco P (2004) Endoscopic third ventriculostomy in the treatment of pediatric hydrocephalus. A multicentric study. Child's Nerv Syst 20: 666–667

160. Putnan TJ (1943) The surgical treatment of infantile hydrocephalus. Surg Gynecol Obst 76: 171–182

161. Quencher RM, Donovan Post MJ, Hinks RS (1990) Cine MR in the evaluation of normal and abnormal CSF flow: intracranial and intraspinal studies. Neuroradiology 32: 371–391

162. Raimondi AJ, Clark SJ, McLone DG (1976) Pathogenesis of aqueductal occlusion in congenital murine hydrocephalus. J Neurosurg 45: 66–77

163. Rappaport ZH, Shalit MN (1989) Perioperative external ventricular drainage in obstructive hydrocephalus secondary to infratentorial brain tumours. Acta Neurochir 96: 118–121

164. Resch KDM (2003) Endo-neuro-sonography: first clinical series (52 cases). Childs Nerv Syst 19: 137–144

165. Resch KDM, Perneczky A (1998) Endo-neuro-sonography: basics and current use. In: Hellwig D, Bauer BL (eds) Minimally Invasive Techniques for Neurosurgery, Springer, Berlin Heidelberg New York, pp 21–31

166. Resch KDM, Reisch R (1997) Endo-neuro-sonography: anatomical aspects of the ventricles. Minim Invasive Neurosurg 1: 2–7

167. Rhoton AL Jr (2002) The lateral and third ventricles. Neurosurgery 51 (suppl. 1): 207–270

168. Richard HK, Seeley HM, Okane C, Pickard JD (2000) ICP 11, Cambridge, July 22–26th

169. Riegel T, Alberti O, Hellwig D, Bertalanffy H (2001) Operative management of third ventriculostomy in cases of thickened, non-translucent third ventricular floor: technical note. Minim Invas Neurosurg 44: 65–69

170. Riegel T, Alberti O, Retsch R, Shiratori V, Hellwig D, Bertalanffy H (2000) Relationships of virtual reality neuroendoscopic simulations to actual imaging. Minim Invasive Neurosurg 43: 176–180

171. Riegel T, Freudenstein D, Alberti O, Duffner F, Hellwig D, Bartel V, Bertalanffy H (2002) Novel multipurpose bipolar instrument for endoscopic neurosurgery. Neurosurgery 51: 270–274

172. Rieger A, Rainov NG, Sanchin L, Schoop G, Burkert W (1996) Ultrasound-guided endoscopic fenestration of the third ventricular floor for non-communicating hydrocephalus. Minim Invasive Neurosurg 39: 17–20

173. Rieger A, Rainov NG, Brucke M, Marx T, Holz C (2000) Endoscopic third ventriculostomy is the treatment of choice for obstructive hydrocephalus due to pediatric pineal tumors. Minim Invas Neurosurg 43: 83–86

174. Rohde V, Gilsbach JM (2000) Anomalies and variants of the endoscopic anatomy for third ventriculostomy. Min Invas Neurosurg 43: 111–117

175. Rohde V, Krombach GA, Struffert T, Gilsbach JM (2001) Virtual MRI endoscopy: detection of anomalies of the ventricular anatomy and its possible role as a presurgical planning tool for endoscopic third ventriculostomy. Acta Neurochir 143: 1085–1091

176. Roland E, Hill A (1997) Intraventricular hemorrhage and posthemorrhagic hydrocephalus: current and potential future interventions. Clin Perinatol 1: 589–605

177. Ruggiero C, Cinalli G, Spennato P, Aliberti F, Cianciulli E, Trischitta V, Maggi G (2004) Endoscopic third ventriculostomy in the treatment of hydrocephalus in posterior fossa tumors in children. Child's Nerv Syst 20: 828–833

178. Russel DS, Nevin S (1940) Aneurysm of the great vein of Galen causing internal hydrocephalus: report of two cases. J Pathol Bacteriol 51: 447–448

179. Saint George E, Natarajan K, Sgouros S (2004) Changes in ventricular volume in hydrocephalic children following successful endoscopic third ventriculostomy. Childs Nerv Syst 20: 834–838

180. Sainte-Rose C (1992) Third ventriculostomy. In: Manwaring KH, Crone K (eds) Neuroendoscopy, Mary Ann Libert, New York, pp 47–62

181. Sainte-Rose C, Chumas P (1996) Endoscopic third ventriculostomy. Tech Neurosurg 1: 176–184

182. Sainte-Rose C, Cinalli G, Roux FE, Maixner V, Chumas PD, Mansour M, Carpentier A, Bourgeois M, Zerah M, Pierre-Kahn A, Renier D (2001) Management of hydrocephalus in pediatric patients with posterior fossa tumors: the role of endoscopic third ventriculostomy. J Neurosurg 95: 791–797

183. Sainte-Rose C (2004) Hydrocephalus in pediatric patients with posterior fossa tumours. In: Cinalli G, Meixner W, Sainte-Rose C (eds) Pediatric Hydrocephalus, Springer-Verlag, Milan, pp 155–162

184. Scavarda D, Bednarek N, Litre F, Koch C, Lena G, Morville P, Rousseaux P (2003) Acquired aqueductal stenosis in preterm infants: an indication for neuroendoscopic third ventriculostomy. Child's Nerv Syst 19: 756–759

185. Schijns OE, Beuls EA (2002) Parinaud's syndrome as a sign of acute obstructive hydrocephalus: recovery after acute ventriculostomy. Ned Tijdschr Geneeskd 46: 1136–1140

186. Schmid UD, Seiler RW (1986) Management of obstructive hydrocephalus secondary to posterior fossa tumours by steroids and subcutaneous ventricular catheter reservoir. J Neurosurg 65: 649–653

187. Schoeman JF, Honey EM, Loock DB (1996) Raised ICP in a child with cryptococal meningitis: CT evidence of a distal CSF block. Child's Nerv Syst 12: 568–571

188. Schönauer C, Bellotti A, Tessitore E, Parlato C, Moraci A (2000) Traumatic subependymal hematoma during endoscopic third ventriculostomy in a patient with a third ventricle tumor: case report. Minim Invasive Neurosurg 43: 135–137

189. Schroeder HWS, Gaab MR (1999) Intracranial neuroendoscopy. Neurosurg Focus 6 (4): Article 1

190. Schroeder HWS, Niendorf W-R, Gaab MR (2002) Complications of endoscopic third ventriculostomy. J Neurosurg 96: 1032–1040

191. Schroeder HWS, Oertel J, Gaab MR (2004) Incidence of complications in neuroendoscopic surgery. Child's Nerv Syst 20: 878–883

192. Schroeder HWS, Wagner W, Tschiltschke W, Gaab MR (2001) Frameless neuronavigation in intracranial endoscopic neurosurgery. J Neurosurg 94: 72–79

193. Schroeder HWS, Warzok RW, Assaf JA, Gaab MR (1999) Fatal subarachnoid hemorrhage after endoscopic third ventriculostomy. Case report. J Neurosurg 90: 153–155

194. Schwartz TH, Ho B, Prestigiacomo CJ, Bruce JN, Feldstein NA, Goodman RR (1999) Ventricular volume following third ventriculostomy. J Neurosurg 91: 20–25

195. Schwartz TH, Yoon SS, Cutruzzola FW, Goodman RR (1996) Third ventriculostomy: post-operative ventricular size and outcome. Minim Invasive Neurosurg 39: 122–129

196. Sgaramella E, Castelli G, Sotgiu S (2004) Chronic subdural collection after endoscopic third ventriculostomy. Acta Neurochir 146: 529–530

197. Sgouros S (2004) Hydrocephalus with myelomeningocele. In: Cinalli G, Meixner W, Sainte-Rose C (eds) Pediatric Hydrocephalus, Springer-Verlag, Milan, pp 133–144

198. Shaw CM, Alvord EC (1995) Hydrocephalus. In: Duckett S (ed) Pediatric Neuropathology, Williams and Wilkins, Baltimore pp 149–211

199. Shiau JSC, King WA (1998) Neuroendoscopes and instruments. In: Jimenez DF (ed) Intracranial Endoscopic Neurosurgery, Park Ridge, IL: American Association of Neurological Surgeons, pp 13–27

200. Singh D, Gupta V, Goyal A, Singh H, Sinha S, Singh AK, Kumar S (2003) Endoscopic third ventriculostomy in obstructed hydrocephalus. Neurology India 51: 39–42

201. Siomin V, Weiner H, Wisoff J, Cinalli G, Pierre-Khan A, Sainte-Rose C, Abbott R, Elran H, Beni-Adani L, Ouaknine G, Constantini S (2001) Re-

peat endoscopic third ventriculostomy: is it worth trying? Child's Nerv Syst 17: 551–555

202. Siomin V, Cinalli G, Grotenhuis A, Golash A, Oi S, Kothbauer K, Weiner H, Roth J, Beni-Adani L, Pierre-Kahn A, Takahashi M, Mallucci C, Abbott R, Wisoff J, Constantini S (2002) Endoscopic third ventriculostomy for patients with cerebrospinal fluid infections and/or hemorrhage. J Neurosurg 97: 519–524

203. Siomin V, Constantini S (2004) Basic principles and equipment in neuro-endoscopy. Neurosurg Clin N Am 15: 19–31

204. Smyth MD, Tubbs RS, Wellons JC 3rd, Oakes WJ, Blount JP, Grabb PA (2003) Endoscopic third ventriculostomy for hydrocephalus secondary to central nervous system infection or intraventricular hemorrhage in children. Pediatr Neurosurg 39: 258–263

205. Stahl W, Kaneda Y (1997) Pathogenesis of murine toxoplasmic hydrocephalus. Parasitology 114: 219–229

206. Stellato TA (1992) History of laparoscopic surgery. Surg Clin North Am 72: 997–1002

207. Strowitzki M, Kiefer M, Steudel W-I (2002) A new method of ultrasonic guidance of neuroendoscopic procedures. J Neurosurg 96: 628–632

208. Sutton LM, Adzick NS, Bilaniuk LT, Johnson MP, Crombleholme TM, Flake AW (1999) Improvement in hindbrain herniation demonstrated by serial fetal magnetic resonance imaging following fetal surgery for myelomeningocele. JAMA 282: 1826–1831

209. Tamburrini G, Caldarelli M, Massimi L, Ramirez-Reyes G, Di Rocco C (2004) Primary and secondary third ventriculostomy in children with hydrocephalus and myelomeningocele. Child's Nerv Syst 20: 666

210. Teo C, Jones RFC, Stening WA (1991) Neuroendoscopic third ventriculostomy. In: Matsumoto S (ed) Hydrocephalus: pathogenesis and treatment. Springer, Berlin Heidelberg New York Tokyo, pp 65–78

211. Teo C, Jones R (1996) Management of hydrocephalus by endoscopic third ventriculostomy in patients with myelomeningocele. Pediatr Neurosurg 25: 57–63

212. Teo C, Rahman S, Boop FA, Cherny B (1996) Complications of endoscopic neurosurgery. Child's Nerv Syst 12: 248–253

213. Teo C (1998) Third ventriculostomy in the treatment of hydrocephalus: Experience with more than 120 cases. In: Hellwig D, Bauer B (eds) Minimally Invasive Techniques for Neurosurgery. Springer, Berlin Heidelberg New York Tokyo, pp 73–76

214. Teo C (2004) Complications of endoscopic third ventriculostomy. In: Cinalli G, Maixner J, Saint-Rose C (eds) Pediatric Hydrocephalus, Springer, Milan, pp 411–420

215. Tirakotai W, Bozinov O, Sure U, Riegel T, Bertalanffy H, Hellwig D (2004) The evolution of stereotactic guidance in neuroendoscopy. Child's Nerv Syst 20: 790–795

216. Tisell M, Almstrom O, Stephensen H, Tullberg M, Wikkelso C (2000) How effective is endoscopic third ventriculostomy in treating adult hydrocephalus caused by primary aqueductal stenosis? Neurosurgery 46: 104–110

217. Tuli S, Alshail E, Drake JM (1999) Third ventriculostomy versus cerebrospinal fluid shunt as a first procedure in pediatric hydrocephalus. Pediatr Neurosurg 30: 11–15
218. Tulipan M, Hernanz-Schulman M, Bruner JP (1998) Reduced hindbrain herniation after intrauterine myelomeningocele repair: a report of four cases. Pediatr Neurosurg 29: 274–278
219. Vaicys C, Fried A (2000) Transient hyponatriemia complicated by seizures after endoscopic third ventriculostomy. Minim Invasive Neurosurg 43: 190–191
220. van Aalst J, Beuls EAM, van Nie FA, Vles JSH, Cornips EMJ (2002) Acute distortion of the anatomy of the third ventricle during third ventriculostomy. J Neurosurg 96: 597–599
221. Vandertop WP, Verdaasdonk RM, Van Swol CFP (1998) Laser assisted neuroendoscopy using a neodymium-yttrium aluminium garnet or diode contact laser with pretreated fiber tips. J Neurosurg 88: 82–92
222. Vinas FC, Dujovny N, Dujovny M (1996) Microanatomical basis for the third ventriculostomy. Minim Invas Neurosurg 39: 116–121
223. Vinas FC, Fandino R, Dujovny M, Chavez V (1994) Microsurgical anatomy of the supratentorial arachnoidal trabecular membranes and cisterns. Neurol Res 16: 417–424
224. Vries J (1978) An endoscopic technique for third ventriculostomy. Surg Neurol 9: 165–168
225. Wagner W, Koch D (2004) Why does third ventriculostomy fail more often in infants? Child's Nerv Syst 20: 666
226. Wellons JC III, Bagley CA, George TM (1999) A simple and safe techniques for endoscopic third ventriculocisternostomy. Pediatr Neurosurg 30: 219–223
227. Wharen REJ, Anderson RE, Scheithauer B, Sundt TM (1984) The Nd:YAG laser in neurosurgery: Part 1. Laboratory investigations: dose-related biological response of neural tissue. J Neurosurg 60: 531–539
228. Whitehead WE, Kestle JRW (2001) The treatment of cerebrospinal fluid shunt infections. Pediatr Neurosurg 35: 205–210
229. Wilcock DJ, Jaspan T, Punt J (1996) CSF flow through third ventriculostomy demonstrated with colour Doppler ultrasonography. Clin Radiol 51: 127–129
230. Willems PW, Vandertop WP, Verdaasdonk RM, van Swol CF, Jansen GH (2001) Contact laser-assisted neuroendoscopy can be performed safely by using pretreated 'black' fibre tips: experimental data. Lasers Surg Med 28: 324–329
231. Wong TT, Lee LS (1996) A method of enlarging the opening of the third ventricular floor for flexible endoscopic third ventriculostomy. Child's Nerv Syst 12: 396–398
232. Yamakawa K (1995) Instrumentation for neuroendoscopy. In: Cohen AR, Haines SJ (eds) Minimally invasive techniques in neurosurgery. Williams & Wilkins, Baltimore, pp 6–13

Minimally Invasive Procedures for the Treatment of Failed Back Surgery Syndrome

P. Mavrocordatos and A. Cahana

Department of Anesthesiology, Pharmacology and Intensive Care, Geneva, Switzerland

With 4 Figures

Contents

Abstract

Failed back surgery syndrome has become unfortunately a common clinical entity. FBSS does not have one specific treatment because it does not have one specific cause. Some features are shared with chronic low back pain (CLBP) and some pathological processes are specific. Both pathologies are leading causes of disability in the industrialized world and costly medical and surgical treatments are continuously used despite their limited efficacy. Nonetheless, evidence based practice guidelines are systematically developed.

In this chapter we cautiously review the vast, complex and at times contradictory literature regarding the treatment of FBSS. Interventional Pain literature suggests that there is moderate evidence (small randomized or non randomized or single group or matched case controlled studies) for medial branch neurotomy and limited evidence (non experimental one or more center studies) for intra-discal treatments in mechanical low back pain. There is moderate evidence for the use of transforaminal epidural steroid injections, lumbar percutaneous adhesiolysis and spinal endoscopy for painful lumbar radiculopathy and spinal cord stimulation and intrathecal pumps mostly after spinal surgery. In reality there is no gold standard for the treatment of FBSS but, these results seem promising.

Keywords: Failed back surgery; back pain; discography; nerve blocks; spinal cord stimulation; radio frequency lesions.

Introduction

Back pain is the most common cause of activity limitation in adults younger than 45 years, the second most frequent reason for visits to the physician, the fifth ranking cause of admission to hospital and the third most common cause of surgical procedures (Van Tulder *et al.*, 2002). A letter to the British Medical Journal in October 2003, Dr Lina Talbot reported: "Every general practitioner has one – a patient who has had back surgery but hasn't improved". Around 20000 cases of failed back syndrome are produced each year in United Kingdom (Talbot, 2003).

Failed Back Surgery Syndrome (FBSS) does not implicate only surgery but also the medical pathway that leads to it (Fritsch *et al.*, 1996). This syndrome constitutes a heterogeneous group of patients which have either their original cause of pain amenable to treatment or their original causes of pain non-amenable to surgery due to induced anatomical changes (Waguespack *et al.*, 2002). Possible causes include correct operation but wrong diagnosis; correct diagnosis but wrong operation and wrong diagnosis and wrong operation; but also, correct diagnosis and correct surgery. The track to this clinical disaster is often paved with approximate diagnos-

tics, precipitation and lack of strategy. Yet with appropriate care, its incidence could be reduced. This clinical entity is endemic and deserves the
highest priority (Porter *et al.*, 1997). This chapter focuses on minimally
invasive treatments but also on the prevention of the so-called Failed
Back Surgery Syndrome.

Back pain is essentially a benign self-limiting condition. "Red flag conditions" are rare and may essentially be ruled out by taking a detailed history. Once this critical issue has been carefully considered, spinal surgery is
never an emergency. This leaves us with the opportunity to review all the
possible alternatives to improve the natural history of the disease before
any non-reversible procedure is performed.

All the efforts must converge towards making a correct diagnosis. The
past 15 years have brought essential diagnostic tools and important many
studies show that the precision of diagnostic in chronic back pain can be performed (Schwarzer *et al.*, 1994; Schwarzer *et al.*, 1995a). These tests are not
meant to replace surgery if indicated or conservative medical treatment
when appropriate, their role is to assure a coherent working hypothesis.

In the early 80's, 80% of "low back pain" patients were classified as
"non-specific low back pain" (Kirwan, 1989). This was not a clear classification but probably the best at time. Today, thanks to a well defined
taxonomy, precision diagnostic blocks and technological progresses, only
20% of patients remain in this "non-diagnostic category" (Merskey *et al.*,
1994).

Another important step was to recognize that the classical "history –
physical examination – radiological imaging" triad is necessary but not sufficient to determine the origin and the mechanism of pain (Strendler *et al.*,
1997). History taking and physical examination did not improve during
the last decades and all the possible progresses had to come either from
advances in technology or from a different diagnostic approach. Although,
anamnesis and clinical examination remain essential to the patient and the
physician, they must be also oriented on the mechanism of pain and not
only on its putative anatomical cause. This is particularly true in sub-acute
and chronic situations for which, most of the time; a sole anatomical aeti-
ology cannot be identified. Finally, minimally invasive procedures, such
as spinal cord stimulation, may be superior to re-operation (North *et al.*,
1994).

Prevalence and Cost

The prevalence of FBSS should be placed in the context of low back pain
in general (Anderson *et al.*, 1999; Bressler *et al.*, 1999). The economic environment and local beliefs have an important influence on the type of treatment offered to low back pain patients. Comparing rates of back surgery in
eleven countries, Cherkin and al demonstrated an almost linear increase in

spinal surgery with the per capita number of orthopaedic and neurosurgeons in the country (Cherkin et al., 1994). However, the adequate ratio neurosurgeon-orthopaedic surgeons/population needed per capita has not been defined and most probably cannot be determined.

The United States National Council on Compensation Insurance in Healthcare estimates the costs of work-related low back pain 8.8 billion US$, not taking into account lost work, lost tax revenue, and indemnity (Williams, 1998). Most costly are diagnostic procedures (25%), surgery (21%), and physical therapy (20%). The past 20 years have witnessed significant changes in the indications for, and use of, instrumentation in lumbar spine surgery. Between 1979 and 1990 there has been an increase of over 55% in the incidence of spine surgery for chronic low back pain (Gibson et al., 1999).

Diagnostic Process in Chronic Low Back Pain

In failed back patients, looking back often reveals important lacks in the diagnostic process. Anatomical and radiological observations do not focus on the mechanism of pain and often conclusions are drawn only from history taking and physical examination.

Patient's History

Although no physician would deny patient's history is essential, there is no evidence to support that this helps in establishing a correct diagnosis, moreover, the best method of history taking in chronic low back pain has neither been defined nor validated. History must assess the patient in a bio-psycho-social context, particularly in FBSS (Guzman et al., 2001). Psycho-social "red flags", called yellow flags must be searched and a complete evaluation of the patient is mandatory if they are present (Deyo et al., 1992). Moreover, we recommend an interdisciplary approach for these patients.

After unsuccessful surgery, with or without added pain, all history must be reviewed even before the operation because unfortunately often FBSS means failed diagnostic. From the biological point of view, the localisation and quality of the pain must be established.

Localisation: The origin of the main pain should be clearly defined, is the pain coming truly from the back? Couldn't it be buttock pain or loin pain? If it is back pain, is lumbar spinal, sacral-spinal or lumbo-sacral spinal pain (Merskey et al., 1994). This precision is important since each condition suggests different diagnostics. If more than one pain is present, a link between them should not be presumed before a clear history has been drawn for each of them. If pain is clearly in the leg, could it be somatic referred pain? In order this question the quality of the pain will help in this regard.

Quality: somatic pain is characterized by deep, dull pressure-like pain and it must be differentiated from radicular neurogenic shooting or lancinating pain. Neurogenic pain will lead to a different diagnostic strategy and probably to another treatment (Fukui *et al.*, 1997).

The other elements of history are all indicative without being essential.

The mode of onset is not diagnostic. Spontaneous or explosive start is more alarming and serious conditions as infection, fracture and tumour must be ruled out, but in chronic low back pain these pathologies have usually already been eliminated, especially if the patient had spinal surgery.

The initial clinical presentation of the pain helps dividing patients with predominant *low back* versus *leg pain* and this may influence the diagnostic strategy which differs between the two groups.

Intensity of the pain is not a good indicator of the severity of the disease, but a comparison to baseline Visual Analog Scores (VAS) is useful to follow the patient along the treatment course.

Duration of pain is a more complex issue. For patients with chronic pain unlike acute pain, a multidisciplinary approach is essential.

An exhaustive list of therapeutic and diagnostic procedures that have been performed is mandatory. Not only must the individual procedures be listed but also the order in which they were performed.

Precipitating, aggravating and relieving factors have not been shown to have an important diagnostic value. *Difficult social or psychological conditions* must be evaluated and in the case of failed-back, interdisciplinary evaluation may raise crucial pitfalls. Most chronic pain patients have to some extent psycho-social distress. This may be only an aggravating factor or a more causal disorder. If not all pain patients need a psychosocial evaluation, failed back patients are probably good candidates for such an approach. In these patients, suffering and distress may be severe, and social context is most of the time disturbed as a consequence of the disease and the loss of self-esteem (Guzman *et al.*, 2001).

Physical Examination

The reliability of a clinical sign is usually evaluated using a K score. K score measures the agreement between two individual observers and is always less than or equal to 1.0 (Cohen, 1960). In rare situations, K can be negative and this is the sign that two observers agree less than it would be expected just by chance. K scores inferior to 0.2 signs a poor agreement; between 0.2 and 0.4 slight agreement, 0.4 to 0.6 moderate agreements, 0.6 to 0.8 good agreement and 0.8 to 1.0 very good agreement. In low back pain evaluation, K scores range from 0.1 to 0.6. Compared to the neurological exam ranging between 0.6 and 0.9, it seems almost unreliable (Bogduk *et al.*, 2002 a).

These considerations demonstrate the limits of the clinical observation and shows how overconfident experienced clinicians may feel about there daily practice (Dreyfuss *et al.*, 1996).

The patient however expects the physical exam, it means attention and care. Moreover, it is essential to orientate the physician towards more precise procedures but should never be considered independently.

Radiological Findings

Before surgery, it has been clearly demonstrated that plain x-ray of the lumbar spine, with or without the associated clinical examination, is not a valuable tool (Simmons *et al.*, 2003). The lumbar spinal x-ray is not only of little value in the diagnostic of degenerative changes, it is also an insensitive method for diagnosing serious conditions in a general population of patients with low back pain (Van Den Bosch *et al.*, 2004). Despite guidelines recommending its limited use, it is still often requested by general practitioners and even by specialists. This habit is not only expensive but may give a false impression of security to the patient and to his physician.

CT-scan imaging does not appear to affect treatment modalities in chronic back pain (Gilbert *et al.*, 2004). Moreover, many asymptomatic patients have a pathological CT-scan (Wiesel, 1986). After surgery, the value of plain x-rays is not expected to be higher than before the operation. The value of MRI and CT-scan depends on numerous factors. For early post-operative complications, the validity of such exams is unquestionable, early complications like haematoma and infections are the perfect target for investigation. For recurrent chronic pain few months after spinal surgery, the evidence is not there. Two situations must be differentiated. First, is when surgery did not improve the pre-operative pain. In this case, no new radiological approach will do better than the preoperative one. Second, surgery has worsened the situation, then a new mechanical component must be looked for. In this situation, imaging the spine may offer new information. This new input should however be considered very cautiously for it might not be the cause of the worsening of pain.

MRI plays a key role in the investigation of chronic back pain and even more in FBSS. It reveals and excludes more lesions than either plain films or CT scan (Gilbert *et al.*, 2004). Although most of the MRI will reveal mainly degenerative changes, some features may help to determine the "pain generator". High density zone (HIZ) is a feature that can occur in the posterior annulus of the lumbar intervertebral disc. It is seen in T2-weighted images. It constitutes the appearance in sagittal section of the circumferential portion of a grade IV annular fissures (Ricketson *et al.*,

1996). Although HIZ will not detect all cases if internal disc disruption, when present, it is unlikely to be false-positive. This sign does not discriminate between patients with or without back pain, but applies to patients who do have back pain. This sign strongly suggests discogenic pain when present (Ito *et al.*, 1998). Endplate changes also provide diagnostic indices for internal disc disruption (Braithwaite *et al.*, 1998).

Normal MRI may also be useful as long as the discs are considered. Normal discs are unlikely to be the source of pain and the diagnostic investigations should first focus on other aetiologies. This may reduce the use of more invasive diagnostic procedures such as provocative discography.

Minimally Invasive Approaches Diagnostic Procedures for Low Back Pain

The precision diagnostic approach was developed to determine in conjunction with other diagnostic tools the cause or the causes of pain in low back pain. By stimulating or anesthetising specific structures, needle procedures can determine precisely the source of the patient's pain (Steindler, 1938). These procedures can target the source of pain and unlike imaging studies determine whether the structure is generating pain or not. This approach is subject to control in order to ensure the validity of the test in each and every patient. The procedure requires fluoroscopy and special skills such as the ability to deliver a needle accurately and safely to the targeted structure.

Epidemiologically, three causes of back pain are predominant with or without surgery. Discogenic pain, Facet joint pain and SI joint pain (Manchikanti *et al.*, 1999 a). For these three aetiologies, three test procedures are available in the investigation of chronic low back and FBBS pain: Discography, Medial Branch blocks and Sacro-iliac (SI) joint blocks.

Many FBSS patients present with a mixed clinical picture and multiple tests may be needed to determine the "pain generator".

Provocative Discography

Discography involves the injection of radiographic contrast into the nucleus of an intervertebral disc. This invasive procedure is justified only if it provides new information that cannot be obtained by less invasive options. Discography does not compete with CT or MRI in the diagnosis of disc herniation. It is not only an anatomical diagnostic but mainly a functional test.

After a classical evaluation of the patient, including radiological exams, even when diseased structures have been identified with MRI for instance, most of the time, we still don't know which structure causes pain. We still need a way to *reproduce* the pain as we try to do during the physical exam;

we still want a symptom related response. We then need to target the cause of the pain and its specific origin.

Provocation discography is achieved by distending the disc from the inside using medium contrast. Diseased discs are painful (Walsh *et al.*, 1990). Although originally believed to be due to increased pressure on nerve roots in patients with herniations, pain occurs in patients with no evidence of herniation or disc-bulge and so must arise from the disc itself. Moreover, the reproduction of pain cannot be ascribed to a chemical effect of contrast medium or spillage of contrast medium into the epidural space, for it occurs without spillage, or if normal saline is used instead of contrast medium (Coppes, 1997).

Discography is performed under local anaesthesia; no or minimal sedation is required or desired. Heavily sedated patients may give partial to inadequate answers to the test. The patient, under sterile conditions lies prone. A posterolateral approach is used to enter the disc at the desired level. A well trained operator is necessary to perform a discography; a painful procedure due to inexperience will preclude a good and valid evaluation. A 22 G to 25 G needle 13 to 17 cm is used to enter the disc. Under the C-arm, lateral and a-p views are used to check the exact place of the tip of the needle. The contrast medium is injected into the disc and intensity and quality of pain are recorded as well as the pressure needed to induce pain (McNally *et al.*, 1996). Discography findings are classified in two groups: symptomatic and radiological findings.

Symptomatic findings: An intact disc without any degenerative abnormalities will support pressures as high as 100 pounds per square inch; the injection is not very uncomfortable. In pathological conditions, the pressure needed to induce pain may vary a lot between subjects but should be below 50 PSI. Pain is recorded on a visual analog scale ranging from 0 being no pain to 10 rating unbearable pain. The patient should be blinded to the level of injection, not knowing if the control disc or the suspected disc is injected first. Evaluation must include quality of the pain, it should be similar to the usual patient's complain.

Radiological findings: The procedure must be completed by a post-discography CT-scan. This exam determines the grade of fissure of a disc. A non injected image does not give this information. CT-scan evaluation of a discogram is looking at the repartition and shape of the injected dye. It must be planned immediately after the discography (Bernard *et al.*, 1990).

The disc can be either intact or ruptured or may present with internal disc disruption (IDD). IDD presents in four different stages. Grade I, II, III extend to the inner, middle and outer third of the annulus fibrosus, grade IV also extends circumferentially around the annulus assuming the shape of a ship's anchor (Aprill *et al.*, 1992). These fissures have no relation with degeneration, are not age related. A strong correlation as been demonstrated between painful discography and IDD (Moneta *et al.*, 1994). The reason why IDD is painful is not clearly established. Probably, in grade II, III and IV, the degradated matrix of the nucleus may chemically irritate the nerve endings of the outer third of the annulus (Heggeness *et al.*, 1993). The second hypothesis is increased mechanical nociception due to mis-distribution of the charges on the diseased disc more sensitive to stress. These theories need further studies to be clearly demonstrated.

Provocative discography like any other diagnostic procedure in low back pain evaluation must not stand alone. It must be interpreted in the light of all the other information about the bio-psycho-social context of the patient. Viewed as an individual exam, this test has its limitations. The important issue is to be able to draw conclusions after a negative or a positive test.

In healthy young subjects with no pre-existing chronic painful illness, the false positive rate is extremely low. Walsh *et al.* in 1990 reported in a study on 10 volunteers. 16.7% of them had minimal pain on injection, 6.7% moderate pain and 3.3% "bad" pain (Walsh, 1990).

Further studies on older subjects suffering from chronic pain and on patients with significant psychometric features showed, as one would expect, higher false-positive rates. Carragee and al in 2000, conducted a prospective study including 30 patients. Little pain was elicited by low pressure injection of any anatomically normal disc. However, when discs although asymptomatic had fissuring of the annulus, the injection was painful. The main predictors of pain intensity were presence of chronic pain and abnormal psychometric scores (Carragee, 2000). As compared to the Walsh study, 40% of chronic pain group and 80% of the somatization group had at least one positive disc (Carragee, 2000).

In FBSS patients, discography is often used to evaluate recurrent or persistent back pain. Heggeness *et al.* reported in a retrospective study 83 patients who had undergone discography. 72% of them had a positive concordant pain response on injection of the previously operated disc. This may give a clue about the importance of the discogenic pain in FBSS patients (Heggeness, 1997).

Another study examined post-discectomy patients with or without persisting pain. 40% of the asymptomatic patients had positive injections on the previously operated level as compared to 63% in the symptomatic group. Moreover, considering the psychometric data, the rate of positive injections were the same in the two groups. Operated discs are painful in symptomatic as well in asymptomatic patients. Yet it remains true that concordant pain is reproduced in symptomatic patients. A damaged disc, symptomatic or not is usually painful when injected and according to the presence of associated aggravating factor, however the pain may be more intense (Carragee, 1999).

Future studies that focus on provocative discography should include a control discography on the adjacent level as proposed by the International Spinal Injection Society (ISIS). Since using a control discography with the provocative one, the false positive rate in normal discs is low even in the chronic pain population. Furthermore adding psychometric screening may help in reducing false-positive rate. As false-negative tests do not occur, we may conclude that all pathological discs are sensitive to provocative discography and with a good patient selection, reasonable diagnostic accuracy can be achieved. If the adjacent disc is used as a control, the specificity of this test will increase.

Medial Branch Blocks

Among the workers population, the prevalence of facet joint pain is 15% (Schwartzer, 1995 b). In an older group population, it increases up to 40% (Manchikanti, 1999 a). Not testing patients with back pain for zygoapophysal joint pain precludes the diagnostics in this proportion of patients and leads to further and perhaps futile investigations. After surgery, this remains valid and although the incidence is lower for other causes overpass this one, a significant proportion of patients will benefit from investigating the Z joint.

The zygoapophyseal joint is innervated by the medial branch of the dorsal rami (Fig. 1).

Provocative saline Z-joint injection has been shown to induce pain in the back, the buttock and even down the leg in healthy volunteers. Anaesthetizing medial branches prevented the induction of pain in similar conditions (Kaplan, 1998).

Medial branch blocks are achieved under fluoroscopy guidance by specifically placing a needle onto the nerve and inject 0,5 ml of local anaesthetic (Bogduk, 1997). Each z-joint is innervated by two nerves blocked separately to anesthetize the joint. Single diagnostic blocks have a 47% false-positive rate. To achieve validity, controlled blocks must be performed with two local anaesthesia agents, a short and a long acting one (15% false positive blocks). If anaesthesia of the joint lasts longer with the second agent, the test is valid (Schwarzer, 1994).

Sacro-Iliac Joint Blocks

The sacro-iliac joint is responsible for 15% of low back pain (Schwartzer et al., 1995 c). S-I joint block is performed under fluoroscopy and a needle is introduced in the joint cavity. A contrast medium is used to insure correct placement of the needle tip (Slipman et al., 2002). A control block is mandatory to reach validity (Maigne, 1996).

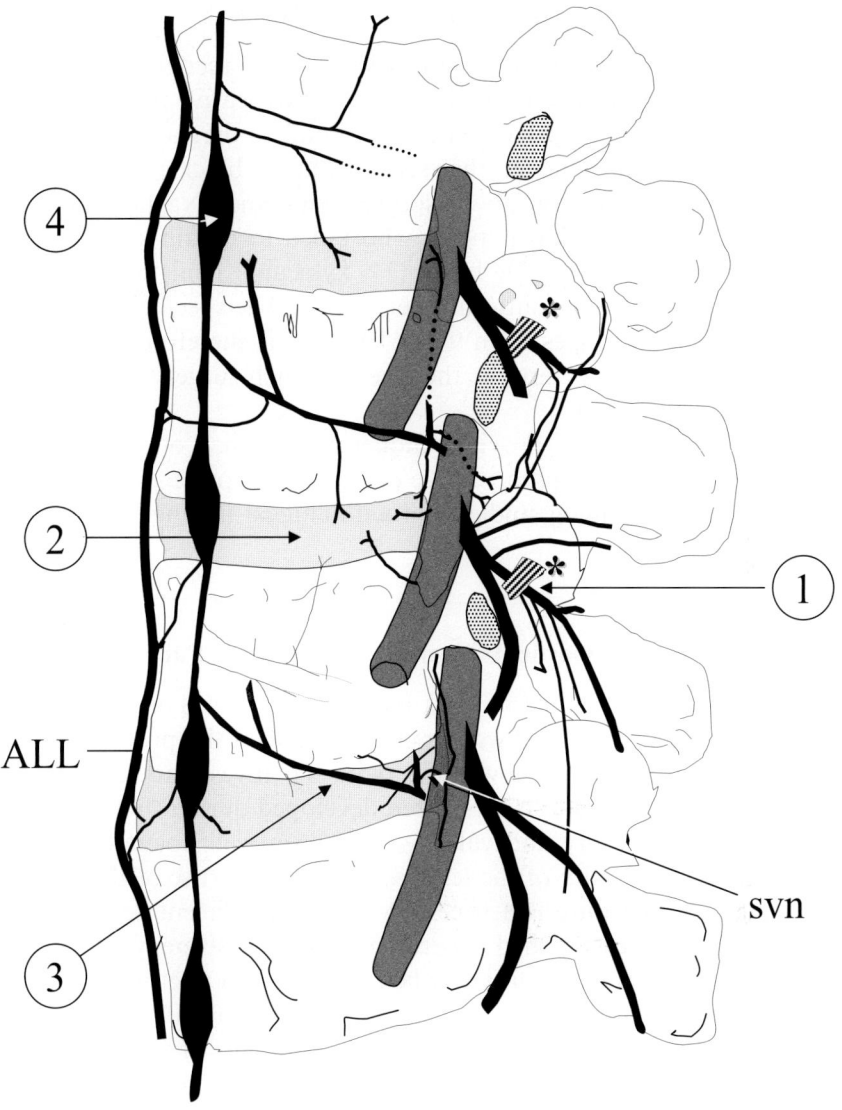

Fig. 1. Schematic drawing of lumbar spine nerve supply. *1* Medial branch dorsal ramus; *2* intervertebral disc; *3* communicating ramus; *4* sympathetic trunk; * = mamillo-accessory ligament

Minimally Invasive Approaches Diagnostic Procedures for Leg Pain

To investigate leg pain or low back and leg pain associated with or without FBSS, trans-foraminal root sleeve injections, lumbar sympathetic blocks and spinal cord stimulation testing may be essential diagnostic tools and frequently determine the treatment.

Trans-Foraminal Diagnostic Injections

The clinical usefulness of nerve root blocks has been recognized as early as 1938 by Steindler and Luck (Steindler, 1938). Provocative response and anaesthesia of the nerve root have both been used as diagnostic procedures. In 1992, Nachemson indicated that selective nerve block provided important prognostic information about surgical outcome (Nachemson, 1992). It has also been shown that patients who failed to obtain sustained relief of radicular pain following the block (2% xylocaïne 1 ml) were less likely to benefit from subsequent surgical intervention. The specificity of selective nerve root injection ranges from 94 to 100%. It is therefore a good prognostic factor, useful to determine the level in which surgery should be performed (Doodley, 1988).

The current literature provides moderate evidence of transforaminal epidural injections in the preoperative evaluation of patients with negative or non conclusive imaging studies, but with clinical findings of root irritation (Pang, 1998).

Algorithm for Diagnostic Assessment of Low Back Pain and FBSS

An algorithm to investigate low back pain must be based on the likelihood of the diagnosis. In 1995, Schwarzer et al. described the prevalence of the predominant aetiologies in low back pain. To investigate chronic back pain, minimally invasive tests have been developed during the last 15 years and there reproducibility and validity have been well documented (Bogduk, 2002 b). The quality of the test itself or the expertise of the physician performing the procedure is a necessary but not sufficient condition. The diagnostic must be established according to a clear strategy. Back versus leg pain must first be distinguished when possible and nociceptive differentiated from neuropathic pain. Physical examination will stress signs of radiculopathy versus pseudoradiculopathy. Although differentiating back from leg or radicular pain is particularly difficult to achieve in FBSS, the predominant features will determine the diagnosis process and later the treatment.

In each group, the next step consists in identifying the structure(s) responsible for the pain. The pain generator should be identified. In most non operated patients, a single cause of pain can be identified and treated. In operated patients, we face the possible overlap of multiple sources of pain. When surgery obviously has worsened the situation the question of a second and new pathological condition must be evaluated separately as a separate entity. Pain may be persist despite correct surgery for a correct diagnostic and further surgical treatment is impossible due to postoperative anatomical changes. The worst situation is when new symptoms follow sur-

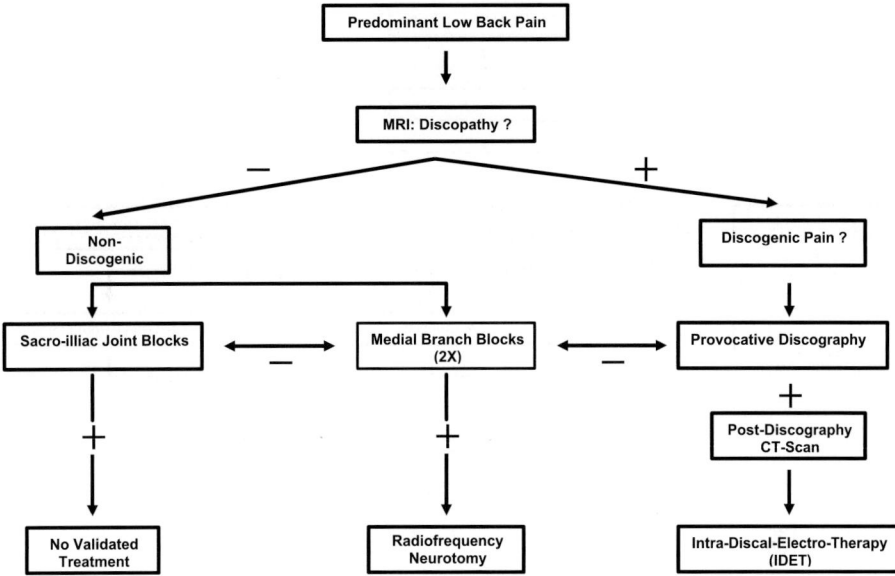

Fig. 2. Diagnostic and treatment algorithm for predominant or exclusive low back pain

gery and are added to the pre-existing, unrelieved pain. In FBSS particularly, low back and leg pain can be present simultaneously. Investigations must be conducted according to the predominant feature (leg or back pain) bearing in mind that leg pain may be triggered by low back structures and even neuropathic-like pain may be due to disc or zygoapophysal joint pathologies.

Predominant back pain (Fig. 2): In chronic low back pain, the first exam is MRI. This exam will give the likelihood to orientate towards a discogenic pain or not. When the MRI shows abnormal discs, the level must be determined with multiple provocative discographies if needed. As discussed above, discography will help symptomatically and radiologically to determine the painful level. Operated discs, symptomatic or not do respond to provocative discography. In symptomatic operated patients the value of this test is not reduced. Therefore, when an operated patient is symptomatic it is of interest to know if injecting dye in the operated disc reproduces patient's pain. It may still be the source of the pain. What we fail to have with this test is that asymptomatic disc may still generates pain as well.

In most cases, normal disc on MRI are not likely to cause pain. If they do, it is rare but when no other source of pain is found, it should not be forgotten that this hypothesis has not been tested.

When normal discs are demonstrated with MRI, the diagnostic strategy must be oriented towards other sources of pain, such as z-joints and SI

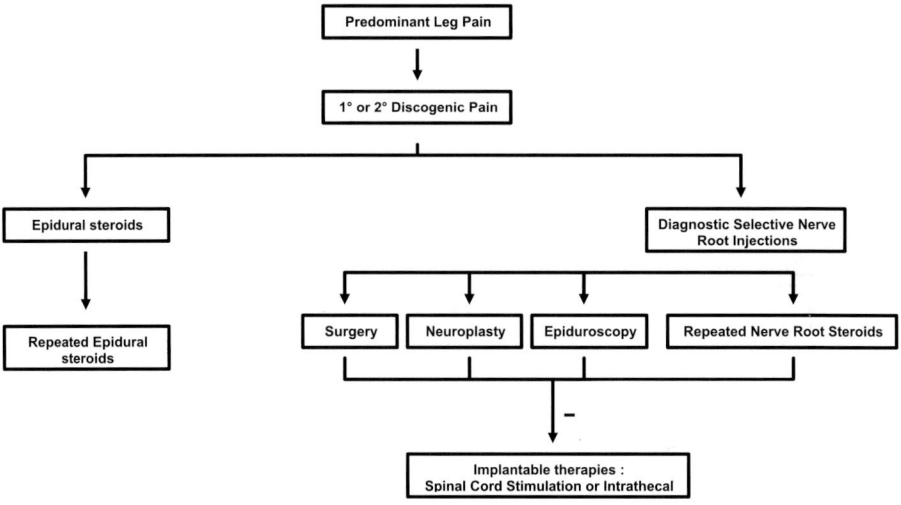

Fig. 3. Diagnostic and treatment algorithm for predominant or exclusive leg pain

joints. The choice between these two is made according to their prevalence, older patients are more prone to suffer from diseased z-joint than young workers.

Unfortunately other sources of pain particularly after surgery, arachnoiditis, adhesions, fusion masses, damaged small nerves and neuromas do not have specific diagnostic tests. Moreover, no specific treatment follows these diagnoses.

Predominant leg pain (Fig. 3): When pain is in the leg, the quality of pain is helpful. History and clinical exam will help differentiate radiculopathy versus pseudoradiculopathy and by localising the level of radiculopathy. MRI findings will improve the accuracy of the diagnosis. Nociceptive and neuropathic pain should be differentiated. Transforaminal local anaesthetics injections will confirm and define the precise level of the radiculopathy. When an autonomic feature is associated to the painful leg, lumbar sympathetic blocks will also be diagnostic before a adequate therapy is chosen.

Minimally Invasive Treatments for Low Back Pain

A diagnostic strategy properly conducted leads to specific treatment. For the predominant causes of back pain, treatment is available. When no diagnosis can be established, the treatment will be symptomatic and when all medical and conservative treatments have failed, minimally invasive approach including spinal cord stimulation and intrathecal drug delivery systems will be used.

Intra-Discal-Electro-Therapy (IDET)

It is estimated that in a substantial percentage of patients with chronic back pain the lumbar disk is the pain generator. IDET was developed as an alternative for selected patients with chronic discogenic pain who have failed all conservative treatments and to whom the next step offered was arthrodesis. Intra-discal electrotherapy was developed because this last treatment was not the perfect response to discogenic pain and because no specific treatment was available for internal disc disruption.

The indication for IDET is demonstrated discogenic pain. As explained in the "provocative discography" section of this chapter, painful, low pressure, dye injection into the disc followed by CT-scan imaging constitute the selection criteria for intra-discal therapy.

Under local anaesthesia and under fluoroscopy guidance, a 17 gauge needle is placed into the centre of the disc to be treated. A navigable intra-discal catheter with a temperature-controlled thermal resistive coil is then deployed through the needle and navigated intradiscally under continuous two-plane fluoroscopic control. The catheter is navigated as far as possible adjacent to the inner posterior annulus. Once placement is optimal the catheter temperature is gradually raised according to a uniform protocol to 90°C over a period of 13 minutes and maintained at 90°C for 4 minutes. The 90°C catheter temperature creates annular temperatures of 60–65°C. Some authors advocate the use of prophylactic intradiscal antibiotics but no evidence shows it utility (Fig. 4).

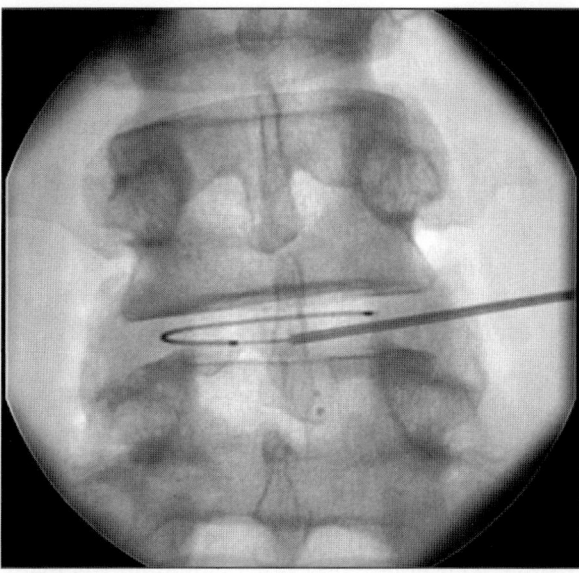

Fig. 4. A-P view of a placement of an Intra-discal elecrotherapy (*IDET*) catheter

The analgesic mechanism of IDET is thought to be related to the sealing of the radial fissures and to the destruction of the nerve endings in the annulus.

Observational studies have been conducted on IDET. Unfortunately, inclusion criteria did not include post discography CT-scan and control discography. However, significant improvement of pain scores and even improvement on disability questionnaire were observed. In 2002, Saal and al reported an outcome study on 58 patients with a two years follow-up. VAS, tolerance to the sitting position and SF-36 were reported at 6, 12, 24 months. At 24 months, 50% of patients reported a 4 point reduction on the VAS scale and similarly 78% of patients showed a 7 points improvement on the bodily pain scale of the SF-36 and 59% showed at least 14 points improvement. Moreover, these significant improvements were associated with no complication or adverse event (Saal et al., 2000; Saal et al., 2002).

A controlled study by Karazek and al on 36 patients with diagnosed internal disc disruption and a control group of 17 patients was conducted over 12 months. The control group was denied treatment by insurers and was treated with standard rehabilitation program. The control group did not improve except one patient. In the IDET group, after 12 months, 13 patients had no benefit and of the 23 that remained, 40% achieved 70% pain relief, 60% obtained at least 50% relief and 23% were completely relieved (Karazek et al., 2000). These results are encouraging for a still new and minimally invasive technique (Pauza et al., 2002).

Medial Branch Radio-Frequency Lesionning

A common technique to treat z-joint pain was the intra-articular steroids injection. This method did not pass the exam of controlled trial. Carette et al. demonstrated no more benefits with steroids than with intra-articular normal saline (Carette et al., 1991). Thus treatment of back pain by intra-articular steroids injection cannot be recommended.

The other method to prevent z-joint pain is denervation of the medial branches innervating the desired level. Two consecutive positive medial branch blocks with different local anaesthetics half-lives at the same levels predict an 80% pain reduction in 60% of patients 12 months after radio-frequency neurotomy (Dreyfuss et al., 1999). The treatment is achieved by lesionning the medial branches with a radio-frequency generator connected to a coated needle. The needle propagates a radio-frequency wave heating a small area around the non-coated to a preset temperature, generally 90°C for the defined time. This small area must surround the medial branch. To achieve perfect lesioning, the needle must be placed parallel to the nerve, since the heated lesion is an ovoid shape that develops around but not extends the needle-tip.

Minimally Invasive Treatments for Leg Pain

Therapeutic approaches to leg pain are closely related to their underlying mechanism. Leg pain arising for low back pathology can be either inflammatory, or neuropathic.

Therapeutic Epidural Injections

Epidural steroid injection is probably the most frequent procedure performed to treat radicular pain. Technique is simple, and safe. Complications occur and may be related to the needle placement or to the drug administered. They include infections, dural tap and very rarely neurological damage (Nelson *et al.*, 2001). Manchikanti *et al.* evaluated the effects of neuraxial steroids and found no significant effect of epidural steroids on weight and bone mass density (Manchikanti *et al.*, 2000). Moreover, the commonly available steroid preparation can be safely used in the epidural space (Dunbar *et al.*, 2002).

Three approaches to the epidural space are possible: Transforaminal, Inter-laminar or Caudal. The efficacy of epidural steroids injections has been questioned in many studies, most of them supporting the use of the technique (McQuay *et al.*, 1998; Devulder, 1999). However, many studies either prospective or retrospective mixed the results of these three different techniques and did not consider the possible differences in the spread of the medication in the epidural space. Since, this problem has been addressed and in each of the three approaches differences have been shown (Price *et al.*, 2000).

The most effective technique is probably the transforaminal approach (Karppinen *et al.*, 2001). It is however associated with the highest complication rate and has been recently questioned. The most worrying complication is related to inadvertent injection of steroid solution into the Adamkiewicz artery (Houten *et al.*, 2002). The entry of the artery into the foramen is subject to a high anatomical variability and enters between L2 to T9 in 85% of patients but may arise from the lower lumbar spine and even from as low as S1. To reduce the incidence of such complications, it is advisable to not only aspirate on the syringe but also to inject dye before injecting a solution with potential aggregates prone to induce small vessels occlusions.

The combined evidence of caudal epidural steroid injections with randomised trials and prospective and retrospective trails is strong for short-term relief and moderate for long-term relief. It is a safe technique and should always be performed under fluoroscopy. Two studies have specifically addressed the problem of FBSS. Revel in a study including 60 patients showed significant improvement of symptoms in 49% of patients against 19% in the control group (Revel *et al.*, 1996). However, another multicenter randomized study including 47 patients reported no short or long term benefit in this group of patients (Meadeb *et al.*, 2001).

Trans-laminar epidural injections show moderate evidence for short term relief and no evidence for long-term relief. This may be due to the repartition of the solution in the epidural space, probably remaining in the posterior epidural compartment. It may also be related to the fact that most inter-laminar procedure are performed without fluoroscopic guidance.

Percutaneous Epidural Neuroplasty (Racz Procedure) and Epiduroscopy

If the effect of epidural steroid injections is local, i.e. a direct effect on the injured nerve root or on the "leaky disc", it is essential that the steroid reach the site of injury. Historically, epidural steroid injections have been performed "blindly", without any radiological guidance, however many factors may prohibit steroids from reaching the intended nerve root, such as scarring, adhesions, adipose tissue and septa, which may be present in the operated and non-operated backs. Thus theoretically drugs injected into a scarred epidural space will follow the path of least resistance, away from the painful site.

Percutaneous epidural neuroplasty (Racz procedure): It seems rational to assume that mobilization or dissolution of fibrosis may remove barriers that prevent application of drugs. Epidural neuroplasty (also known as Racz procedure) consists of accessing the epidural space in a caudal or transforaminal approach, injecting non-ionic contrast material (thus performing an epidurogram) in order to detect "filling defects" in the epidural space. This is followed by gentle manipulation of a metal reinforced catheter in order to liberate adhesions ("filling the defects"), and then injecting the targeted medication (Heavner *et al.*, 1999). This procedure, which allows prolonged pain relief in refractory cases, has the advantage of targeted drug delivery, but has the disadvantage of an indirect, two dimensional vision of the presumed pathology.

The epidurographic diagnosis of spine pathology may be followed by neurolysis with the injection of corticosteroids, hypertonic saline and/or hyaluronidase. Two RCT and 3 retrospective evaluations showed pain relief up to a year, with cost effectiveness gains of up to 8,127 US$ per year per patient. When performed by appropriately skilled personnel this procedure has a low complication rate, however dural puncture, spinal cord compression, catheter shearing, hypertonic saline toxicity, infection and bleeding remain worrisome (Manchikanti *et al.*, 1999 b).

Spinal Endoscopy: Even when injection is done under fluoroscopy, the image obtained is two-dimensional and can be misleading. Thus, epidural endoscopy provides us a three-dimensional, real-time, color view of anatomy-pathology in the epidural space.

Access of the epidural space with a flexible fibre-optic catheter via the sacral hiatus appears to be safe and efficient (Geurts *et al.*, 2002). The procedure is done under local anaesthesia while continuously monitoring intra-epidural pressures, and patient's response. Normal nerve roots when touched cause paraesthesia, diseased ones pain, so patient report is essential while gently performing adhesiolysis. The technique allows examination of the epidural space and its contents, targeted injection of medication, lysis of scar tissue (adhesiolysis) and (potentially) retrieval of foreign bodies (Kitahata, 2002). As technology grows new possibilities such as minimally invasive surgery, intraoperative nerve stimulation and immuno-biological interference evolve, promising an important role of spinal endoscopy in the treatment of spinal pain.

In a prospective case series all patients undergoing epiduroscopy suffered from adhesions between nerve roots, dura and ligamentum flavum, 41% very dense, associated with previous surgery. If fibrosis is a result of chronic radiculitis, neurogenic inflammation and impaired fibrinolysis, repeat surgery will probably aggravate the situation and is thus ill advised. The authors hypothesize that adhesions obstruct radicular veins and interfere with the nervi vasorum, creating intra-neural edema and abnormal pain transmission. Dilution or "washing out" phospholipase A2 and synovial cytokines may also contribute to symptom improvement (Richardson *et al.*, 2001).

In another recent study Igarashi *et al.* showed that epiduroscopy reduces back and leg pain among 58 elderly patients suffering spinal stenosis. Pain relief lasted more than a year after the procedure without any neurological complications, especially in patients suffering from abundant adhesions (Igarashi *et al.*, 2004). This is of importance since persistent pain among patients suffering from FBSS is thought to be due to epidural scar. Furthermore, reservations about using this technique in patients suffering from a "restrictive" epidural space and thus fear from elevated intra-epidural pressures during the procedure, have been founded to be clinically debatable.

Spinal Cord Stimulation

Electrical stimulation has been used in the treatment of a variety of disease since the ancient Greeks. From the torpedo fish or "narke" inducing *narcosis* to the Faradization in the 18th century, electricity has been regarded as a therapeutic tool. In the end of the 19th and the early 20th century electricity has been disregarded in favor to emergent new pharmaceutical agents and it is not until the early 60's that electrotherapy reappeared. The publication by Wall and Merzack in 1965 of the "Gate control theory" of pain gave birth to the contemporary spinal cord stimulation (Melzack and Wall, 1965). The argument that electrical stimulation of large fibers would close the gate to input from the smaller diameter and unmyelinated A-delta and C fibers mediating pain was determinant to the success of SCS. Since, this hypothesis has been subject to criticism and we know now that it is not the only mechanism involve in pain control (Linderoth et al., 1999).

Spinal cord stimulation is achieved using a voltage-controlled pulse generator. It creates a potential difference between two outputs. The injected current is distributed in a 3-dimentional space made up of electrically conducting anatomical structures. The resulting 3-dimentional electric field can be represented by its potential distribution and by its current density distribution. These distributions can be visualized by isopotential line and isocurrent lines, respectively, as shown in the transverse section of spinal cord stimulation model. The stimulation induces mainly a depolarization of the nerve large myelinated fibers, both orthodromically and antidromically (Oakley et al., 2002).

The principle is to stimulate the dorsal column and interfere with the sensory information coming from the painful area. The analgesic mechanisms of SCS are however not clear. It is universally accepted that paresthesia coverage of the painful area, indicating the activation of the dorsal column, is necessary to obtain pain relief, it may however not be the mechanism of SCS. A possible stimulation target may be the dorsolateral funiculus which is known to contain descending pain controlling pathways. There is no convincing evidence for the involvement of opioid mechanism in the effect of SCS. Endorphins levels are not influenced by SCS and naloxone does not reverse pain relief induced by SCS (Meyerson et al., 1977). A possible role of GABA and adenosine in the analgesic action of SCS is suggested by animal and human studies indicating that GABA antagonists reverse partially the effect of SCS (Cui et al., 1996) and that a synergic effect of adenosine with SCS was observed (Cui et al., 1997).

The first spinal cord stimulator was placed in 1967 by Shealy by a D2-D3 laminectomy (Shealy et al., 1967). The first indication was cancer pain. Rapidly, it became clear that not all "pains" were sensible to SCS. Mainly, neuropathic pain was, nociceptive pain was not. Thanks to numerous publications on SCS, we now know that intermediate clinical states and other sympathically maintained pain may be responsive to SCS which has progressively gained acceptance in a number of clinical pain syndromes including FBSS (Krames, 1999).

Implantable devices have a place in treatment of FBSS patients when all other conservative and minimally invasive tests and therapies have failed to diagnose or treat a particular condition. This includes an important number of patients (North *et al.*, 2002).

A proper patient selection is essential to achieve adequate pain relief with SCS. History is essential for the appropriate selection of the candidates to SCS. Pain characteristics and neuropathic features, must be searched, psychological screening may be useful. The evaluation of these crucial elements may lead to a shorter trial period, resulting in less infection rate and therapeutic failures.

In FBSS patients, leg pain responds better than axial back pain to SCS and neuropathic better than nociceptive, mechanical pain, the later almost non responsive to SCS.

With SCS, the active electrode, the cathode or negative electrode must be located near the level of the spinal cord dorsal columns that anatomically represents the level to be stimulated. The electrode is therefore placed in the epidural space under fluoroscopy guidance and with a patient awake and anesthetized locally at the needle entry point. The Tuohy needle is inserted into the epidural space using the loss of resistance technique and advanced rostrally up to the desired level. At this point, the external stimulator is connected and the patient is asked whether the stimulation, the paresthesia felt is covering the painful area or not. When the pain is unilateral, the electrode is placed on the side of the patient's pain lateral to the midline on the homolateral dorsal column. If the pain is bilateral or axial, single or multiple electrodes must be placed on the midline or close to it.

SCS includes 3 components: The epidural electrode the connection between the epidural electrode and the battery and the Implantable-pulse-generator (IPG). A wide range of electrodes may be used. Two main categories are percutaneous leads and surgical leads, the later requiring laminotomy.

In failed-back patients, the implantation of the SCS is divided in three steps. The test electrode implantation performed under local anesthesia, the trial period ranging from one to four weeks and the IPG implantation commonly achieved under general anesthesia.

We think that placing a test lead without patient's collaboration leads to a higher failure rate. The only indications for a direct implant of a surgical electrode are recurrent displacement of percutaneous leads or of if a predicted target is in the area of prior surgery.

Trial period duration is debatable. Most authors recognize a one week test is minimal to obtain reasonable information to proceed to a definitive implantation. According to local practice the period extends from one to four weeks test. Criteria for a positive test are listed (Table 1).

The definitive implant requires connecting the implanted epidural lead to the connection, tunneled under the skin to the hypochondria where the IPG is placed.

Table 1. *SCS Screening Trial Criteria*

---

1. Minimum of 50% pain reduction in VAS score with test-lead implant
2. The area of induced paraesthesia must cover the area of pain
3. Paraesthesia well tolerated
4. Mood, sleep, activity improvement

Once the patient is implanted, treatment really begins. The surgical and trial periods are the easy part of the work. The follow-up of these patients is a dynamic process and may require long hours and programming is not always easy. Numerous consults may be needed and the willingness and patience of the physician and his team are essential.

Complications may be divided in 3 groups: surgical complications, device related and stimulation related complications.

Potential surgical complications include infection, spinal fluid leakage, hemorrhage and neurological injury. In 1995, Turner reviewed 31 studies referring between 0 to 12% infection rates, mean 5% (Turner *et al.*, 1995).

In over 20 years, North's group reported no major morbidity defined as neurological injury, meningitis or life-threatening infection (North *et al.*, 1993). Electrode migration is the most common complication occurring 24% of the time (Turner *et al.*, 1995). For this reason multichannel devices have been shown to be more reliable in this regard. It has also been advocated that paddle electrodes are more stable (North *et al.*, 1997). Although no randomized studies have been published, it seems that paddle electrodes are associated with improved long term effectiveness, particularly for low back pain. This region needs high voltage stimulation and the design of the paddle leads with the stimulating electrode directed towards the dura unlike the percutaneous electrodes which directs all the usable current towards the medulla. This problem is of utmost importance for the development of new technology: What we really needed is a percutaneous paddle-like electrode.

Other problems like discomfort due to inadequate IPG position in the abdomen needing repositioning are uncommon.

Stimulation related discomfort is rare as it usually precludes definitive implant. If stimulation is painful or bothers the patient during the trial period, it is usually not a successful test and the electrode is removed. Patients usual complaint is related to posture induced changes in the intensity of stimulation. Important reprogramming sessions are mostly related to electrode displacement.

Most studies on SCS for FBSS are retrospective. Turner *et al.* reviewed 41 articles reporting approximately 50–60% of patients with FBSS describing a >50% pain reduction from the use of SCS (Turner *et al.*, 1995). Hieu *et al.*, showed a long term efficacy in 63% of patients and fair in 22% after 42 months follow-up (Hieu *et al.*, 1994).

Although no controlled studies have been conducted on SCS, recent prospective series reinforced the role of SCS in FBSS. North conducted a randomized comparison of SCS with re-operation with a 6 months cross-over arm in the study. 51 patients with FBSS consented to randomization. This study demonstrated a significant difference between patients who opted for cross over from SCS to re-operation but not visa versa and concluded that SCS is a viable alternative to re-operation (North *et al.*, 1995).

Cost effectiveness can be evaluated comparing the estimated cost of therapy per year in groups treated by SCS versus alternative treatment. Bell *et al.* compared SCS versus surgeries and other alternative treatment over 5 years. The reduced demand for medical care of successfully SCS treated patients leads to the observation that SCS pays for itself in an average of 2.1 years (Bell *et al.*, 1997).

Considering that SCS is an end stage technique used in patients in whom everything has failed, SCS is an effective treatment, particularly considering the low complication rate. However, new technology developments are needed to allow percutaneous placement of more efficient electrodes in terms of energy sparing and precision of current distribution (Deer *et al.*, 2001).

Intrathecal Medications

The nature of back pain and the conjunction of nociceptive and neuropathic symptoms frequently reduce therapeutic margin of single or even complex medication, therefore, many FBSS patients fail to respond to oral or transcutaneous drug administration.

Nerve blocks have also limited efficacy, for these precise diagnostic tools do not always have a corresponding treatment. More invasive therapies must be cautiously examined for, as previously discussed in this review, the failure rate increases with the number of spinal re-operations and unless a specific target has really been identified recurrent surgery is not an option.

Intrathecal drug infusion is now well accepted as a treatment option when all conservative and etiologic treatment failed. These therapies have failed either because pain relief is inadequate or due to intolerable side effects.

When it comes to neuromodulation therapies, the choice between intrathecal medication and spinal cord stimulation is an important issue. SCS and Intrathecal drug infusion share common indications, but while SCS applies mainly to neuropathic symptoms, Intrathecal drug infusion also covers important nociceptives aspects of pain.

Once all other treatments have failed, a careful screening process of the candidates to an implantable therapy is needed. This screening can be divided in three steps.

The characteristics and localization of the pain must first be established. Low back versus leg pain and nociceptive versus neuropathic pain help in choosing the most appropriate approach between SCS and Intrathecal drug infusion. Hassenbusch *et al.* in a retrospective study in 1995 estimated that intrathecal infusion may be best for bilateral leg and back pain as compared to spinal cord stimulation (Hassenbusch *et al.*, 1995). No evidence has yet determined the adequacy of a particular treatment modality to select between spinal infusion and SCS, however, clinical practice is helpful in this regard. Although Intrathecal drug infusion may be efficient in a wide range of pain patterns and share common indications with SCS, the latter is easier for the patient and the physician. With SCS, no refills are needed, the patient may manage some stimulation parameters and there are no side-effects. Intrathecal drug delivery pumps need refilling and side-effects may be important. For these reasons, in common indications, it is only when SCS has failed that Intrathecal drug delivery should be used. For other indications like mixed pain patterns, Intrathecal drug infusion comes first.

Once the indication to Intrathecal drug delivery is determined, in a second step, patients must follow a medication trial and the most appropriate drugs must be tested.

The main principle is to first choose the most appropriate agent to the characteristics and localization of the pain. If not sufficient, it should be associated with a second medication. This second drug should be from another class of drugs. It should enhance the effect or complete the effect of the first drug by acting on other pain mechanisms like, for example, a local anesthetic if the first drug is an opioid.

Association of drugs may be required to achieve adequate analgesia but it will also complicate adaptations and changes of the medication as each drug concentration depends on the other. For example, to increase the delivery of one of three drugs mixed in the reservoir, the concentration of the others will need to be modified to keep their delivery flow constant. These sometimes complex therapies are needed and may be extremely efficient.

In most patients, *morphine* comes first. In a review of current practices Hassenbusch *et al.* determined that 98% of pain physicians who answered the questionnaire recalled using intrathecal morphine (Hassenbusch *et al.*, 2000). The national outcomes registry for low back pain collected prospective data on 136 patients with chronic low back pain treated using intraspinal infusion via implanted devices, 81% of whom received morphine. Oswestry Low Back pain disability scale ratings after 12 months improved by 47% in patients with back pain and in 31% in patients with leg pain (Deer T et al., 2004).

In Intrathecal drug delivery, besides side-effects of the infused drugs one may face other associated complications. Recent studies have confirmed the clinical observation that intrathecal morphine infusion was responsible for catheter-tip inflammatory masses. Coffey has recommended positioning the catheter tip in the lumbar thecal sac to minimize opioid dosage and concentration to the extent possible. It was also proposed to provide an attentive follow-up of patients to encourage early diagnosis and to reduce the risk of neurological injury in these patients (Coffey R J et al., 2002).

Bupivacaine used mostly in association with opioids is a local anesthetic agent. Its use and safety in neuropathic pain syndromes has been widely recognized.

Up to maximum doses of 30–35 mg/day side effects are rare. Beyond 30 mg/day, and according to the place of the catheter tip, hypotension and motor weakness may be severe.

Less frequently used than morphine are mixtures: morphine+bupivacaïne (68% of pain physicians), hydromorphone (58% of pain physicians), morphine-clonidine, morphine-bupivacaïne-clonidine. Fentanyl and sulfentanyl are also used alone or in mixed solutions. Combining drugs maximizes the effects and reduces the side-effects.

Although the above medications are used in a majority of patients new agents are in the pipeline and will soon be applied in clinical practice.

No definitive strategy has been established and the choice of the drug or the choice of the combination of drugs is specific to each and every patient. However, general principles are shared by pain specialists and guidelines have been proposed after reviewing current literature and practices by an expert panel in a polyanalgesic consensus conference in 2000, updated in 2003 (Bennett et al., 2000 a and Hassenbusch et al., 2004).

Although the acute cost of these implantable devices is high, the long term therapy is not more expensive than the conventional approaches (De Lissevoy et al., 1997).

New intra-thecal agents currently studied include midazolam, ketamine, neostigmine, gabapentine, ziconotide among others (Hassenbusch et al., 2004). These agents may be particularly helpful in the treatment of difficult neuropathic pain syndromes.

Conclusions and Future

The low back pain population includes a wide variety of patients (Walker, 2000). Not all patients should go through such diagnostic processes and treatments. 90% of acute back pain patients will resolve spontaneously in the first three months and among the reminders not all will suffer enough to necessitate such approaches. For the small portion of the patients needing invasive therapies, non reversible procedures should take place only when a valid diagnostic strategy has been undertaken. In chronic back pain patients, surgery is never an emergency.

The principal problems leading to FBSS can be classified in 4 categories.

Knowledge update: All physicians taking care of low back pain patients should be aware of the leading epidemiological causes of acute and chronic back pain, of the headlines of the diagnostic algorithm in chronic back pain and detect the biological and psychological red flags.

Common sense evidence: Relying on history, physical examination and non MRI radiological findings may lead to wrong diagnostic, false security and sometimes to the wrong operation. Common sense is needed to treat low back pain but some historical evidences should be reconsidered.

Diagnosis process: Shortcuts from radiological findings to spinal surgery are not acceptable for chronic low back pain patients. Unless the source of pain can be determined precisely and that source possesses at least a mechanical component, surgery has no role.

Surgery is not the ultimate solution: The surgical approach must be confronted to a recent RCT comparing lumbar instrumented fusion with cognitive intervention and exercise in patients with chronic low back pain due to disk degeneration. This study was unable to detect any difference after one year in pain, analgesic consumption, satisfaction and return to work rate (Brox *et al.*, 2003). Moreover, when evaluating surgical results, it is important to consider radiographic fusion and functional outcome separately, thus improvement rate following surgery remains non conclusive. A comprehensive review suggests that 68% of patients have a satisfactory outcome following lumbar fusion; however, long term follow-up of decompressive laminectomy for lumbar spinal stenosis has shown no difference in outcome between surgical and non-surgical treatments (Turner *et al.*, 1992) (Iguchi *et al.*, 2000).

An 18 year follow-up in patients with spondylolisthesis showed that surgical interventions are indicated only for radiculopathies (Matsunaga *et al.*, 2000).

Collaboration related: Interdisciplinary approach is essential to investigate patients before surgery and to insure an adequate follow-up after. On the biological point of view, if surgery is performed only after a proper algorithm is followed, the target related procedure has the place it deserves; the adequate treatment.

We do not think the strategy described above will reduce the number of surgical procedures, but hopefully it may lead to more precise diagnosis and this will allow a better patient selection.

The trend is now for less invasive techniques and the industry have redirected their efforts towards the development of minimally invasive approaches. This economical and technological input will give birth to new high tech instruments. New ideas arise from our daily practice and a critical and constructive spirit will contribute to reduce the morbidity linked to our still incomplete understanding of pain and disability.

References

1. Andersson GBL (1999) Epidemiological features of chronic low back pain. Lancet 354: 581–585
2. Aprill C, Bogduk N (1992) High-intensity zone: a diagnostic sign of painful lumbar disc on magnetic resonance imaging. Br J Radiol 65(773):361–369
3. Bell GKK, Kidd DH, North RB (1997) Cost-effectiveness analysis of spinal cord stimulation in treatment of Failed back surgery syndrome. J Pain symptom Manage 13(5): 286–295
4. Bennett G, Burchiel K, Buchser E, Classen A, Deer T, Du Pen S, Ferrante FM, Hassenbusch SJ, Lou L, Maeyaert J, Penn R, Portenoy RK, Rauck R, Serafini M, Willis KD, Yaksh T (2000) Clinical guidelines for intraspinal infusion: report of an expert panel. PolyAnalgesic Consensus Conference 2000. J Pain Symptom Manage 20(2): 37–43
5. Bernard TN Jr (1990) Lumbar discography and post-discography computerized tomography: refining the diagnosis of low-back pain. Spine 15: 690–707
6. Bogduk N (1997) International Spinal Injection Society guidelines for the performance of spinal injection procedures. Part 1: zygoapophyseal joint blocks. Clin J Pain 13: 285–302
7. Bogduk N (2002) Medical management of acute and chronic low back pain. An evidence based approach. Pain research and clinical management, vol 13. Elsevier Science BV
8. Bogduk N (2002) Diagnostic nerve blocks in chronic pain. Best Pract Res Clin Anaesthesiol 16(4):565–578
9. Braithwaite I, White J, Saifuddin A, Renton P, Taylor BA (1998) Vertebral end-plate (Modic) changes on lumbar spine MRI: correlation with pain reproduction at lumbar discography. Eur Spine J 7(5):363–368
10. Brox JI, Sorensen R, Friis A et al (2003) RCT comparing lumbar instrumented fusion with cognitive intervention and exercises in patients with CLBP due to disc degeneration. Spine 28(17):1913–1921
11. Carette S, Marcoux S, Truchon R et al (1991) A controlled trial of corticosteroid injections into facet joints for chronic low back pain. N Engl J Med 325: 1002–1007
12. Carragee E, Chen Y, Tanner C et al (1999) Provocative discography in patients after limited lumbar discectomy. A controlled, randomized study of pain response in symptomatic and asymptomatic subjects. Proceedings of the North American Spine Society, Chicago, IL, p 95–96
13. Carragee EJ, Chen Y, Tanner CM et al (2000) Provocative discography in patients after limited lumbar discectomy. Spine 25: 3065–3071

14. Cherkin DC, Deyo RA, Loeser JD, Bush T, Waddell G (1994) An international comparison of back surgery rates. Spine 19(11):1201–1206
15. Coffey RJ, Burchiel K (2002) Inflammatory mass lesions associated with intrathecal drug infusion catheters: report and observations on 41 patients. Neurosurgery 50: 78–87
16. Cohen J (1960) A coefficient of agreement for nominal scales. Educ Psychol Meas 20: 37–46
17. Coppes MH, Marani E, Thomeer RT et al (1997) Innervation of «painful» lumbar discs. Spine 22: 2342–2350
18. Cui JG, Linderoth B, Meyerson BA (1996) Effects of spinal cord stimulation on touched evoked allodynia involve GABAergic mechanisms. An experimental study in the mononeuropathic rat. Pain 66: 287–295
19. Cui JG, O'Connor WT, Ungersteldt U, Linderoth B, Meyerson BA (1997) Spinal cord stimulation attenuates augmented dorsal horn release of excitatory amino acids in mononeuropathy via a GABAergic mechanism. Pain 73:87–95
20. Deer TR (2001) Current and future trends in spinal cord stimulation for chronic pain. Curr Pain Headache Rep 5: 503–509
21. Deer T, Chapple I, Classen A et al (2004) Intrathecal drug delivery for the treatment of chronic low back pain: Report from the National Outcomes Registry for Low Back Pain. Pain Med 5: 6–13
22. De Lissovoy G, Brown RE, Halpern M et al (1997) Cost-effectiveness of long-term intrathecal morphine therapy for pain associated with failed back surgery syndrome. Clin Ther 19: 96–112
23. Devulder J (1998) Transforaminal nerve root sleeve injection with corticosteroids, hyaluronidase, and local anesthetic in the failed back surgery syndrome. J Spinal Disord 11: 151–154
24. Deyo RA, Rainville J, Kent DL (1992) What can the history and physical examination tell us about low back pain? JAMA 268: 760–765
25. Dooley JF, McBroom RJ, Taguchi T et al (1988) Nerve root infiltration in the diagnosis of radicular pain. Spine 13: 79–83
26. Dreyfuss P, Michaelson M, Pauza K et al (1996) The value of medical history and physical examination in diagnosing sacroiliac joint pain. Spine 21: 2594–2602
27. Dreyfuss P, Halbrook B, Pauza K, Joshi A, McLarty J, Bogduk N (1999) Efficacy and validity of radiofrequency neurotomy for chronic lumbar zygoapophyseal joint pain. Spine 24: 1937–1942
28. Dunbar SA, Manikantan P, Philip J (2002) Epidural infusion pressure in degenerative spinal disease before and after epidural steroid therapy. Anesth Analg 94: 417–420
29. Fritsch EW, Heisel J, Rupp S (1996) The failed back surgery syndrome. Reasons, intraoperative findings, and long-term results: A report of 182 operative treatments. Spine 21: 626–633
30. Fukui S, Ohseto K, Shiotani M, Ohno K, Karasawa H, Naganuma Y (1997) Distribution of referred pain from the lumbar zygapophyseal joints and dorsal rami. Clin J Pain 13(4):303–307

31. Geurts JW, Kallewaard JW, Richardson J, Groen GJ (2002) Targeted methylprednisolone acetate hyaluronidase/clonidine injection after diagnostic epiduroscopy for chronic sciatica: a prospective, 1-year follow-up study. Reg Anesth Pain Med 27(4): 343–352

32. Gibson JNA, Grant IC, Waddell G (1999) The Cochrane review of surgery for lumbar disc prolapse and degenerative lumbar spondylosis. Spine 24: 1820–1832

33. Gilbert FJ, Grant AM, Gillan MGC, Vale LD, Campbell MK, Scott NW, Knight DJ, Wardlaw (2004) Low back pain: influence of early MR imaging or CT on treatment and outcome-multicenter randomized trial. Radiology 231: 343–351

34. Guzman J, Esmail R, Karjalainen K, Malmivaara A, Irwin E, Bombardier C (2001) Multidisciplinary rehabilitation for chronic back pain: A systematic review. Br Med J 322: 1511–1516

35. Hassenbusch SJ, Stanton-Hicks M, Covington EC (1995) Spinal cord stimulation versus spinal infusion for low back and leg pain. Acta Neurochir [Suppl] 64:109–115

36. Hassenbusch SJ, Portenoy RK (2000) Current practices in intraspinal therapy – a survey of clinical trends and decision making. J Pain Symptom Manage 20(2):4–11

37. Hassenbusch SJ, Portenoy RK, Cousins M, Buchser E, Deer TR, Du Pen SL, Eisenach J, Follett KA, Hildebrand KR, Krames ES, Levy RM, Palmer PP, Rathmell JP, Rauck RL, Staats PS, Stearns L, Willis KD (2004) Polyanalgesic Consensus Conference 2003: an update on the management of pain by intraspinal drug delivery – report of an expert panel. J Pain Symptom Manage 27(6):540–563

38. Heggeness MH, Watters WC, Gray PM (1997) Discography of lumbar discs after surgical treatment for disc herniation. Spine 22: 1606–1609

39. Heavner JE, Racz GB, Raj PP (1999) Percutaneous epidural neuroplasty: Prospective evaluation of 0.9% NaCl vs 10% NaCl with or without hyaluronidase. Reg Anesth Pain Med 24: 202–207

40. Hieu PD, Person H, Houidi K, Rodrigez V, Vallee B, Besson G (1994) Treatment of chronic lumbago and radicular pain by spinal cord stimulation. Rev Rhum Ed Fr 61(4): 271–277

41. Houten JK, Errico TJ (2002) Paraplegia after lumbo-sacral nerve root block: report of three cases. The Spine J 2: 70–75

42. Igarashi T, Hirabayashi Y, Seo N, Fukuda H, Suzuki H (2004) Lysis of adhesions and epidural injection of steroid/local anesthetic during epiduroscopy potentially alleviate low back and leg pain in elderly patients with lumbar spinal stenosis. Br J Anaesth 93:181–187

43. Iguchi T, Kurihara A, Nakamaya J et al (2000) Minimum 10 year outcome of decompressive laminectomy for degenerative lumbar spinal stenosis. Spine 25: 1754–1759

44. Ito M, Incorvaia KM, Yu SF, Fredrickson BE, Yuan HA, Rosenbaum AE (1998) Predictive signs of discogenic lumbar pain on magnetic resonance imaging with discography correlation. Spine 23(11): 1252–1258; discussion 1259–1260

45. Kaplan M, Dreyfuss P, Halbrook B, Bogduk N (1998) The ability of lumbar medial branch blocks to anesthetize the zygoapophyseal joint. Spine 23: 1847–1852

46. Karasek M, Bogduk N (2000) Twelve-month follow-up of a controlled trial of intradiscal thermal annuloplasty for back pain due to internal disc disruption. Spine 25: 2601–2607

47. Karppinen J, Malmivaara A, Kurunlahti M et al (2001) Periradicular infiltration for sciatica. Spine 26:1059–1067

48. Kirwan EO (1989) Back pain. In: Wall PD, Melzack R (eds) Text book of pain, 2nd edn. Churchill Livingstone, Edinburgh, pp 335–340

49. Kitahata LM (2002) Recent advances in epiduroscopy. J Anesth 16: 222–228

50. Krames E (1999) Spinal cord stimulation: Indications, mechanism of action, and efficacy. Cur Rev Pain 3: 419–426

51. Linderoth B, Foreman R (1999) Physiology of spinal cord stimulation: review and update. Neuromodulation 3: 150–164

52. Maigne JY, Aivaliklis A, Pfefer F (1996) Results of sacroiliac joint double block and value of sacroiliac pain provocation tests in 54 patients with low-back pain. Spine 21: 1889–1892

53. Manchikanti L, Pampati V, Fellows B, Bakhit CE (1999 a) Prevalence of lumbar facet joint pain in chronic low back pain. Pain Physician 2: 59–64

54. Manchikanti L, Pampati V, Bakhit CE, Pakanati RR (1999 b) Non endoscopic and endoscopic adhesiolysis in post laminectomy syndrome: a one year outcome study and cost effectiveness analysis. Pain Physician 2: 52–58

55. Manchikanti L, Pampati V, Beyer C et al (2000) The effect of neuraxial steroids on weight and bone mass density: A prospective evaluation. Pain Physician 3:357–366

56. Matsunaga S, Iriji K, Hayashi K (2000) Neurosurgically managed patients with degenerative spondylolisthesis: a 10 to 18 year follow-up study. J Neurosurg 93: 194–198

57. McNally DS, Shackleford IM, Goodship AE, Mulholland RC (1996) In vivo stress measurement can predict pain on discography. Spine 15;21(22):2580–2587

58. McQuay HJ, Moore RA (1998) Epidural corticosteroids for sciatica. An evidence-based resource for pain relief. Oxford University Press, Oxford, New York, pp 216–218

59. Meadeb J, Rozenberg S, Duquesnoy B et al (2001) Forceful sacroccocygeal injections in the treatment of postdiscectomy sciatica. A controlled study versus glucocorticoid injections. Joint Bone Spine 68: 43–49

60. Melzack R, Wall PD (1965) Pain mechanisms: a new theory. Science 150: 971–978

61. Merskey H, Bogduk N (1994) (eds) Classification of chronic pain. Descriptions of chronic pain syndromes and definitions of pain terms, 2nd edn. IASP Press, Seattle

62. Meyerson BA, Boethius J, Terenius L, Wahlström A (1977) "Endorphine mechanisms in pain relief with intra-cerebral and dorsal column stimulation". In 3rd Meeting of the European society for stereotactic and functional Neurosurgery, Freiburg

63. Moneta GB, Videman T, Kaivanto K *et al* (1994) Reported pain during lumbar discography as a function of annular ruptures and disc degeneration. A re-analysis of 833 discograms. Spine 17: 1968–1974
64. Nachemson A (1992) Newest knowledge of low back pain: A critical look. Clin Orthop Rel Res 279: 8–20
65. Nelson DA, Landau WM (2001) Intraspinal steroids: history, efficacy, accidentality and controversy with review of United States food and drug administration reports. J Neurol Neurosurg Psychiatry 70: 433–443
66. North R, Kidd D, Zahurak M *et al* (1993) Spinal cord stimulation for chronic intractable pain: Experience over two decades. Neurosurg 32: 384–394
67. North RB, Kidd DH, Lee MS, Piantodosi S (1994) Spinal cord stimulation versus reoperation for failed back surgery syndrome: a prospective randomized study design. Acta Neurochir [Suppl] 64:106–108
68. North RB, Lanning A, Hessels R, Cutchis PN (1997) Spinal cord stimulation with percutaneous and plate electrodes: side effects and quantitative comparisons. Neurosurg Focus 2(1): Article 3
69. North RB, Wetzel FT (2002) Spinal cord stimulation for chronic pain of spinal origin. Spine 27: 2584–2591
70. Oakley J, Prager J (2002) Spinal cord stimulation: Mechanism of action. Spine 22: 2574–2583
71. Pang WW, Mok MS, Lin ML *et al* (1998) Application of spinal pain mapping in the diagnosis of low back pain – analysis of 104 cases. Acta Anaesthesiol Sin 36: 71–74
72. Pauza K, Howell S, Dreyfuss P *et al* (2002) A randomized, double-blind, placebo controlled trial evaluating the efficacy of intradiscal electrothermal annuloplasty (IDET) for the treatment of chronic discogenic low back pain: 6-month outcomes. In Proceedings of the International Spinal Injection Society Austin, Tx
73. Porter RW (1997) Spinal surgery and alleged medical negligence. JR Coll Surg Edinb 42: 376–380
74. Price CM, Rogers PD, Prosser AS *et al* (2000) Comparison of the caudal and lumbar approaches to the epidural space. Ann Rheum Dis 59: 879–882
75. Revel M, Auleley GR, Alaoui S *et al* (1996) Forceful epidural injections for the treatment of lumbosciatic pain with post-operative lumbar spinal fibrosis. Rev Rhum Engl Ed 63: 270–277
76. Richardson J, McGurgan P, Cheema S, Prasad R, Gupta S (2001) Spinal endoscopy in chronic low back pain with radiculopathy: a prospective case series. Anaesthesia 56: 447–484
77. Ricketson R, Simmons JW, Hauser BO (1996) The prolapsed intervertebral disc. The high-intensity zone with discography correlation. Spine 1;21(23):2758–2762
78. Saal JA, Saal JS (2000) Intradiscal electrothermal treatment for chronic discogenic low back pain. Spine 25: 2622–2627
79. Saal JA, Saal JS (2002) Intradiscal electrothermal treatment for chronic discogenic low back pain: prospective outcome study with a minimum 2-year follow-up. Spine 27(9):966–973

80. Schwarzer AC, April CN, Derby R, Fortin J, Kine G, Bogduk N (1994) The relative contributions of the disc and zygoapophyseal joint in chronic low back pain. Spine 19: 801–806
81. Schwartzer AC, Aprill CN, Derby R, Fortin J, Kine G, Bogduk N (1995 a) The prevalence and clinical features of internal disc disruption in patients with chronic low back pain. Spine 20: 1878–1883
82. Schwartzer AC, Wang S, Bogduk N, McNaught PJ, Laurent R (1995 b) Prevalence and clinical features of lumbar zygoapophyseal joint pain: a study in an Australian population with chronic low back pain. Ann Rheum Dis 54: 100–106
83. Schwarzer AC, Aprill CN, Bogduk N (1995 c) The sacroiliac joint in chronic low back pain. Spine 20: 31–37
84. Shealy CN, Mortimer JT, Reswick JB (1967) Electrical inhibition of pain by stimulation of the dorsal columns: preliminary clinical report. Anesth Analg 46(4):489–491
85. Simmons ED, Guyer RD, Graham-Smith A, Herzog R (2003) Radiograph assessment for patients with low back pain. The Spine J 3: 3–5
86. Slipman CW, Huston CW (2002) Diagnostic sacroiliac joint injections. In: Manchikanti L, Slipman CW, Fellows B (eds) Interventional pain management. Low back pain – diagnosis and treatment. ASIPP Publishing, Paducah, KY, pp 269–274
87. Steindler A, Luck JV (1938) Differential diagnosis of pain in the low back: Allocation of the source of the pain by the procaine hydrochloride method. JAMA 110: 106–113
88. Strendler LE, Sjoblom A, Sundell K, Ludwig R, Taube A (1997) Inter-examiner reliability in physical examination of patients with low back pain. Spine 22: 814–820
89. Talbot L (2003) Failed back surgery syndrome. BMJ 327: 985–986
90. Turner JA, Ersek M, Herron LD (1992) Patient outcomes after lumbar spinal fusions: a comprehensive literature synthesis. JAMA 268: 907–911
91. Turner JA, Loeser JD, Bell KG (1995) Spinal cord stimulation for chronic low back pain: a systematic literature synthesis. Neurosurgery 37(6): 1088–1095
92. Van den Bosch MAAJ, Hollingsworth W, Kinmonth AL, Dixon AK (2004) Evidence against the use of lumbar spine radiography for low back pain. Clinical Radiology 59: 69–76
93. Van Tulder M (2002) Low Back Pain. Best Prac Res Clin Rheumatol 16(5): 761–775
94. Schofferman J, Slosar P et al (2002) Etiology of long-term failures of lumbar spine surgery. Pain Med 3: 18–22
95. Walsh TR, Weinstein JN, Spratt KP et al (1990) Lumbar discography in normal subjects. J Bone Point Surg 72-A: 1081–1088
96. Wiesel SW (1986) A study of computer-assisted tomography. 1. the incidence of positive CATscans in asymptomatic group of patients. Spine 9:549–551
97. Williams DA, Feuerstein M, Durbin D et al (1998) Healthcare and indemnity costs across the natural history of disability in occupational low back pain. Spine 23: 2329–2336

Surgical Anatomy of Calvarial Skin and Bones— With Particular Reference to Neurosurgical Approaches

H. D. Fournier, V. Dellière, J. B. Gourraud, and Ph. Mercier

Laboratory of Anatomy, Faculty of Medicine, University of Angers, Angers, France

With 16 Figures

Contents

Abstract

This chapter on surgical anatomy is addressed to young neurosurgeons and could be used as an introduction to basic neurosurgical technique. It aims

to cover the basic anatomy relevant to making incisions in the scalp and
creating bone flaps, an essential preliminary to any form of intracranial
surgery. We will examine the anatomy of the scalp, its arterial and venous
supply and its nervous system, as well as providing some technical points
related to the cranial vault and the base of the skull. It will be explained
how a well-grounded knowledge of the anatomical details makes it possible
to execute correctly two of the most common approaches in neurosurgical practice, namely the pterional approach and an approach around the
sinuso-jugular axis.

Keywords: Anatomy; operative technique; pterional craniotomy; scalp; skull base;
vascularisation; venous sinuses.

Introduction

It has long been recognized that a key capacity for any operator who
wishes to gain access to a lesion within the skull is the ability to install
and center a bone flap as effectively and accurately as possible. Nothing
is as detrimental as a poorly oriented head or a badly centered flap. The
position of the incision into the scalp should therefore be first decided
according to the location of the lesion to be accessed as well as on the basis
of criteria pertaining to the creation of a suitable flap. Without losing sight
of this key point, two other principles need to be borne in mind when making incisions: firstly the need to preserve the vasculature and nervous supply of the scalp; and secondly, aesthetic considerations and the possibility
of hiding the scar. Our aim here is not to give technical details about
surgical procedures but rather to point out certain key anatomical features
which dictate basic neurosurgical technique.

Descriptive Anatomy of the Different Layers Covering the Cranium (Fig. 1)

A great deal of work has been carried out on the scalp [3, 8, 9, 12]. What
follows is a description of the various layers as they appear on dissection.

The scalp is made up of the skin, sub-cutaneous tissue and the galeal
aponeurosis.

The skin and subcutaneous tissue are extremely thick, representing one
of the special features of the scalp. The vessels and nerves are found at the
lower surface of the sub-cutaneous adipose tissue.

The innermost border of this sheet is delineated by the epicranium or
galeal aponeurosis. The terms epicranium or superficial musculoaponeur-
otic fascia are used to designate what is routinely referred to as the galea.
Strictly speaking, the galea corresponds to the aponeurotic compartment
of the fascia which incorporates a series of paired muscles (the frontal, auricular and occipital muscles) as well as an aponeurotic sheet. This sheet

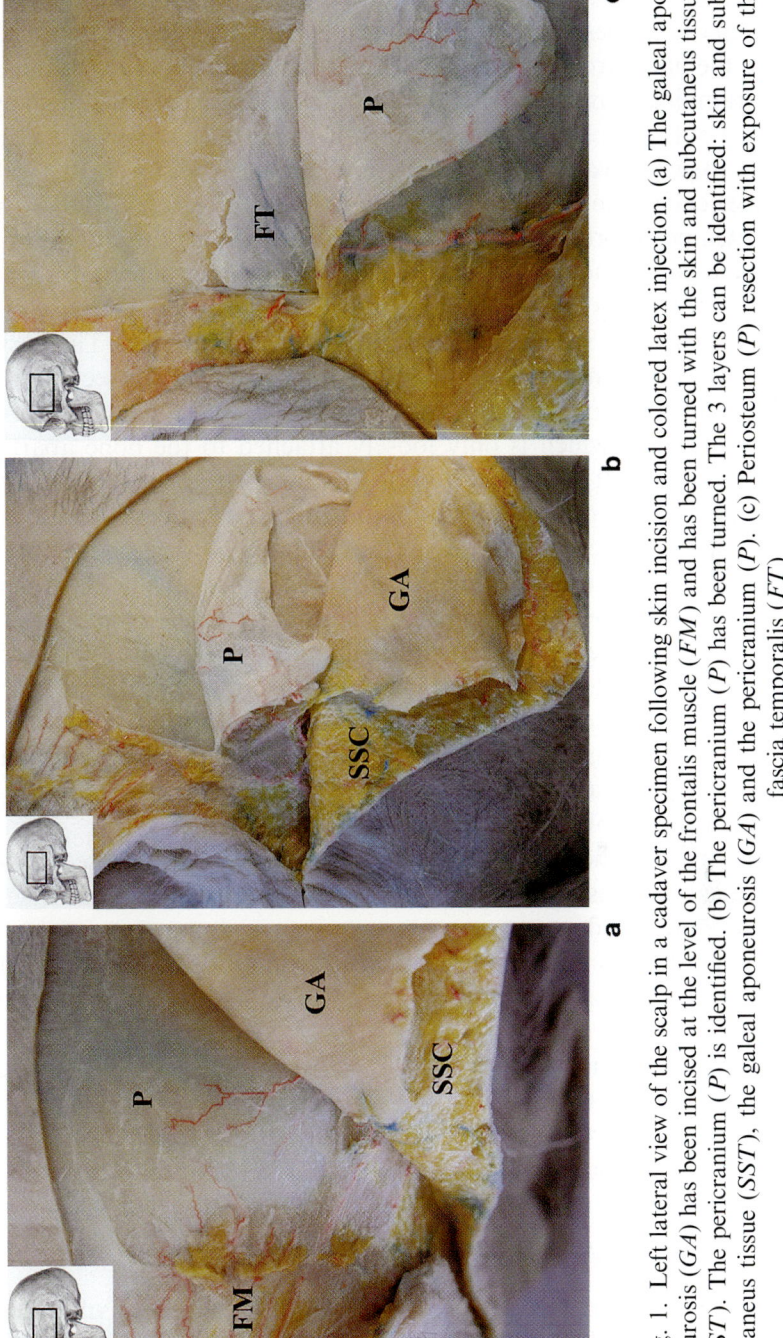

Fig. 1. Left lateral view of the scalp in a cadaver specimen following skin incision and colored latex injection. (a) The galeal aponeurosis (*GA*) has been incised at the level of the frontalis muscle (*FM*) and has been turned with the skin and subcutaneus tissue (*SST*). The pericranium (*P*) is identified. (b) The pericranium (*P*) has been turned. The 3 layers can be identified: skin and subcutaneus tissue (*SST*), the galeal aponeurosis (*GA*) and the pericranium (*P*). (c) Periosteum (*P*) resection with exposure of the fascia temporalis (*FT*)

extends from the frontal region back to the occiput, and from the vertex to the zygomatic arches. In front, it continues into the face. It is difficult to separate the galea from the sub-cutaneous fatty sheet because of the many arterial ramifications which originate in the vascular network of the galea and travel outwards. It is a dictum that when the galea moves, the skin and the fatty tissue moves with it [3, 12]. The lateral extension of the galea is sometimes called the temporoparietal fascia, and it is over this that the superficial temporal artery runs. This sheet is less dense. The frontal muscles arise in the galea and are inserted deep in the dermis. They are separated from one another by an extension of this fascia. The occipital muscles originate at the superior nuchal line and are inserted in the galea. The auricular muscles are very thin and are difficult to separate out by dissection.

The periosteum of the skull—also referred to as the pericranium—is a thin, fibrous sheet which is only loosely attached to the bone apart from along the sutures. It is easily lifted off the bone. It is conventionally said that the periosteum of the skull is continued at the temporal region by the aponeurosis of the temporal muscle (the temporalis fascia). This continues on down to attach at the zygomatic arch. The temporalis fascia is particularly strongly attached to the galeal aponeurosis. Thus, the temporalis muscle is situated between its fascia externally and bone internally. Blood and trophicity are supplied to this muscle by the deep temporal arteries which run across the muscle's internal surface.

Between the galea and the periosteum of the skull or the pericranium is found a tissue layer at which the scalp can be detached, namely the layer of subaponeurotic areolar connective tissue.

In the light of these anatomical considerations, it can be seen that the classic galeal flap described in the surgery of the anterior cranial base is, in reality, a periosteal flap.

In conclusion, in practical terms the surgeon needs to bear in mind that there are four successive layers between the skin and the periosteum of the skull:

– skin and subcutaneous tissue
– galeal aponeurosis (epicranium)
– subaponeurotic areolar connective tissue
– periosteum (pericranium)

Vasculature of the Scalp (Figs. 2–7)

The corrosion casting technique—the various possibilities of which are thoroughly described by Hill in 1981 [5]—reveals the richness of the vascular network at the cephalic extremity. It gives a picture of the intracranial and extracranial arterial and venous tree because it removes both soft tissues and the cranium.

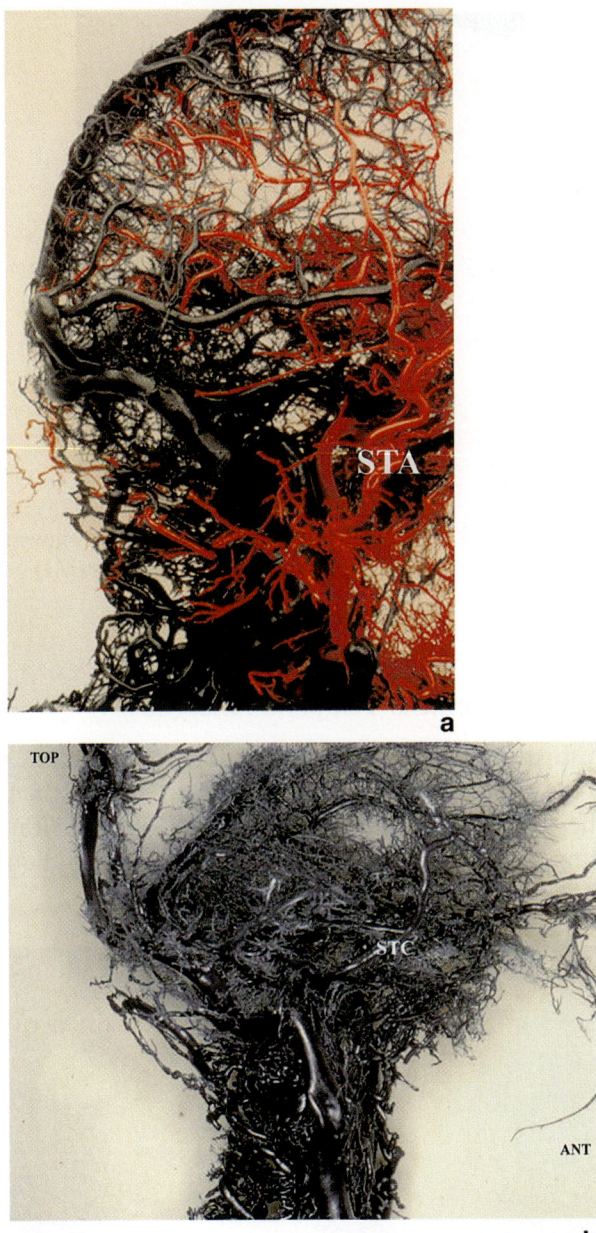

Fig. 2. (a and b) The arteries and veins to and from the head: Complete corrosion of the tissues leaves the cast of the intra and extracranial arterial and venous tree. Note the presence of a well identified superficial temporal artery (*STA*). The extracranial venous pattern is not systematized. See the superficial temporal collector (*STC*)

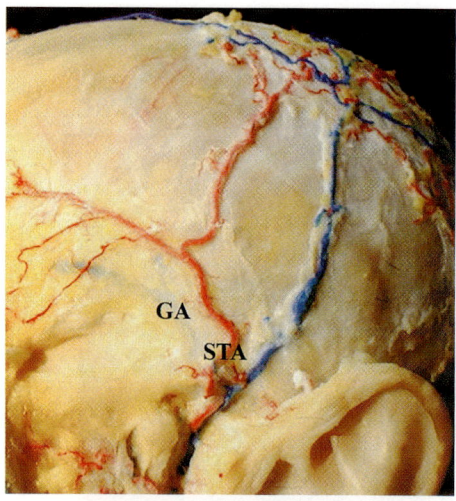

Fig. 3. Lateral view of a left superficial temporal artery (*STA*) (following colored latex injection) running over the galeal aponeurosis (*GA*)

Arteries

Superficial Temporal Artery (STA)

Extensive investigation has shown that the STA alone supplies not only the scalp but also the superior lateral half of the face, half of the parotid gland and part of the temporomandibular joint [6, 7, 10, 11]. Although in practice, interrupting its flow actually causes little damage to the scalp (mainly

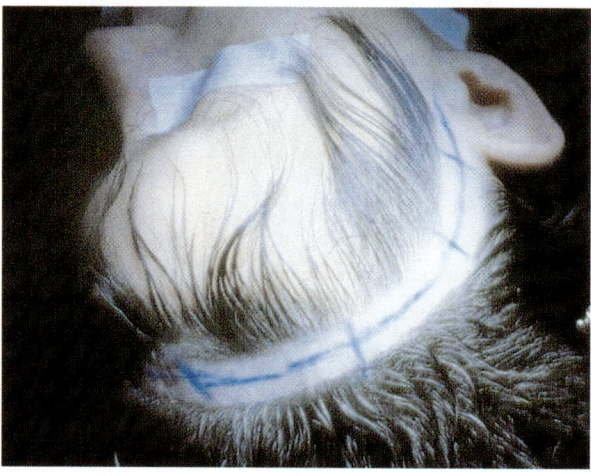

Fig. 4. The skin incision for the frontopterional approach

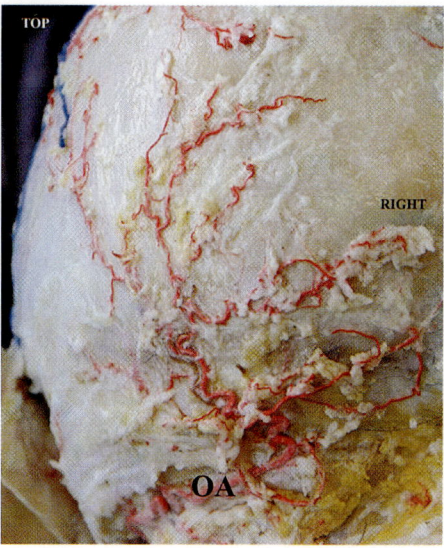

Fig. 5. Posterior view of the left occipital artery following colored latex injection

because of the existence of a vast network of anastomoses), it is nevertheless important when it comes to the creation of reconstructive flaps or
extracranial anastomoses with the petrous carotid artery in the course of
certain surgical procedures at the base of the skull.

Originating in the external carotid artery, the STA arises at the neck of
the condyle and then rises vertically in front of the tragus. At the zygoma,
it has an external diameter of the order of 2.2 mm [7]. It passes at an aver-

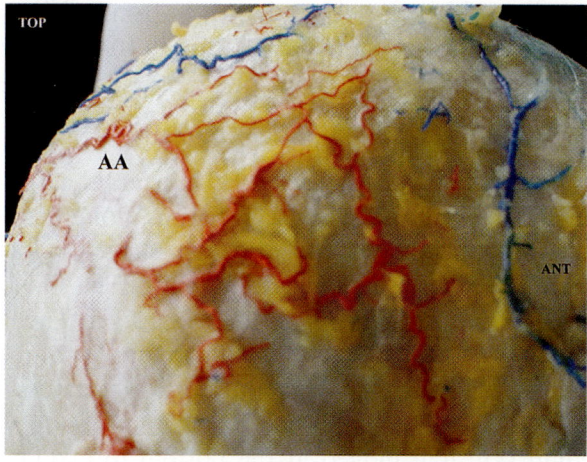

Fig. 6. Anastomosis (*AA*) between arteries of the scalp at the level of the vertex

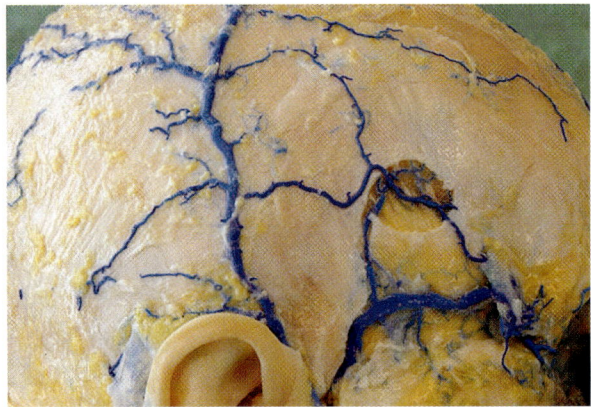

Fig. 7. Venous right sided injection of the scalp. Note the poor filling of the frontal area with a single pretragal collector

age of 0.94 ± 0.38 cm from the anterior edge of the tragus [11]. It runs along the internal surface of the sub-cutaneous tissue. Its average length between the zygoma and its branches is 31.7 mm [7]. In nearly 92% of the cases, the STA presents a bifurcation with branches to supply the frontal and parietal regions [7]. At this branch point, its diameter is no greater than 1.9 mm [7]. The frontal branch runs obliquely upwards and forwards. It forms numerous branches which form anastomoses. The parietal branch continues vertically up towards the vertex in a two-centimeter-wide band centered on the external auditory meatus and parallel to the frontal plane (landmark skin band) [6].

This architecture is important when it comes to frontopterional incisions at the hair line which can be made just in front of the tragus behind the STA [14]. Some surgeons make the incision in front of the STA [1].

Occipital Artery (OA)

The OA originates on the posterior side of the external carotid artery near the posterior belly of the digastric muscle. It runs along the digastric muscle as far as the mastoid region where it passes under the longissimus and splenius capitis muscles to form two terminal branches. The lateral branch carries on vertically upwards in the sub-cutaneous layer. The medial branch continues horizontally as far as the external occipital protuberance before turning through a right angle to carry on vertically across the trapezius muscle, ending in the sub-cutaneous tissue.

A paramedian skin flap hinged at the bottom to expose the posterior fossa can become necrotic if the stalk is too narrow.

Posterior Auricular Artery

This also originates at the posterior side of the external carotid artery from where it travels along the sternohyoid muscle to divide into two different branches at the mastoid aponeurosis. These branches supply the retroauricular regions and the pinna.

Arterial Anastomoses

Here, we will not discuss anastomoses between the territories of the internal carotid (via branches of the ophthalmic artery) and the external carotid: although familiarity with these is essential when it comes to embolization techniques, it is not relevant to installing skin flaps.

There are many anastomoses between the various territories of the external carotid artery, with the number apparently increasing as one gets closer to the vertex where circles of anastomotic density can be described. These anastomoses may be homo or contra-lateral.

For practical purposes, it is important to remember that:
– All the arteries of the scalp run into the sub-cutaneous tissue;
– All the arteries of the scalp flow upwards (so skin flaps must always be created with an inferior stalk);
– Anastomoses between the three main arterial territories of the scalp form from one side and with the opposite side, thereby justifying all the flaps envisaged by plastic surgeons.

Veins

The venous network has traditionally been seen as the poor cousin in the scalp's vascular system, and its extreme variability has probably put many off attempting to describe it. The idea that veins follow the path of arteries is particularly subject to doubt in this location, and this is probably the reason for the resounding failure of certain forms of flap. Although most drainage occurs via a superficial temporal collector, corrosion casts have revealed the great richness of the venous network. On the basis of the dissections that we have prepared, it appears that there are few venous collectors in the frontotemporal region which might explain why there is such extensive edema of the face sometimes following fronto-pteriono-temporal detachment of the scalp. This hypothesis was proposed by Lebeau in 1986 [6].

Innervation of the Scalp (Figs. 8 and 9)

Frontotemporal Branch of the Facial Nerve

The frontotemporal region is innervated by the superior or frontotemporal branch of the facial nerve. This branch leaves the parotid gland and

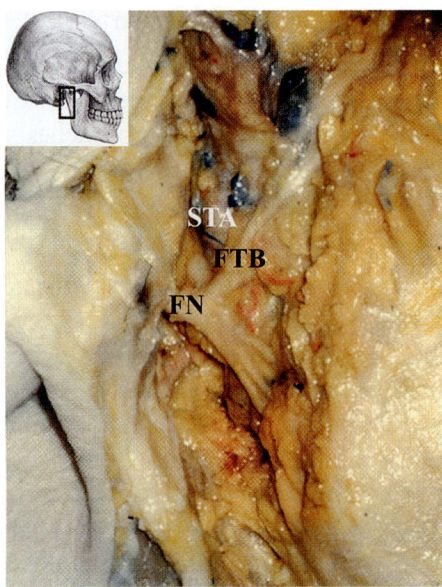

Fig. 8. Dissection of a right facial nerve (*FN*) in a cadaver specimen. Note the relationships between the frontotemporal branch of the facial nerve (*FTB*) and the trunk of the superficial temporal artery (*STA*)

climbs up across the lateral zygomatic arch at the internal surface of the sub-cutaneous tissue to the galeal aponeurosis or temporoparietal fascia which is particularly strongly attached to the temporalis fascia at this point. The zygomatic arch is crossed about one centimeter in front of the trunk of the STA [14].

This anatomical architecture has an important consequence: whatever method is used to lift the scalp following a pterional approach, the skin incision should not extend down as far as the zygoma and detachment of the galea should not be excessively anterior. It should stop no further than two fingerbreadths from the orbital arch and should carry the temporal fascia with it to avoid damaging the superior branch of the facial nerve. This has spurred the development of various techniques designed to overcome this problem [13, 14].

Yasargil developed the interfascial temporalis flap [14]. Dissection between the galea and the temporalis fascia is carried out from behind towards the front to stop two fingerbreadths (i.e. 3–4 cm) from the orbital arch. The temporalis fascia together with the muscle are incised vertically from the superior temporal line as far as the zygomatic arch. The temporalis fascia is then pulled forwards together with the muscle. Since there is

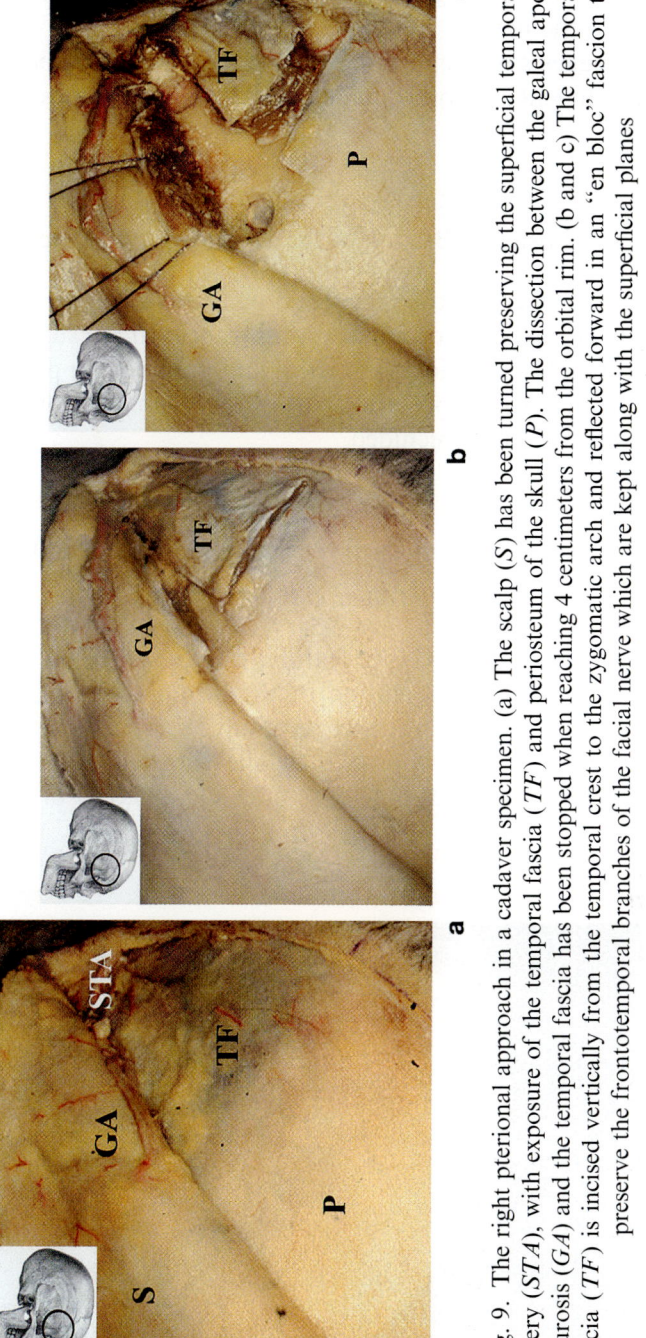

Fig. 9. The right pterional approach in a cadaver specimen. (a) The scalp (*S*) has been turned preserving the superficial temporal artery (*STA*), with exposure of the temporal fascia (*TF*) and periosteum of the skull (*P*). The dissection between the galeal aponeurosis (*GA*) and the temporal fascia has been stopped when reaching 4 centimeters from the orbital rim. (b and c) The temporal fascia (*TF*) is incised vertically from the temporal crest to the zygomatic arch and reflected forward in an "en bloc" fascion to preserve the frontotemporal branches of the facial nerve which are kept along with the superficial planes

no dissection between the galea and the temporalis fascia, there cannot be any damage to the nerve.

Spetzler suggests not separating the galea from the temporalis fascia at all with detachment of both cutaneous and muscle tissues achieved in a single step [13].

The Posterior Branch of the Second Cervical Nerve

This large nerve emerges from the trapezius muscle two centimeters below and lateral to the external occipital protuberance. In the sub-cutaneous tissue, it then branches out along a vertical and paramedian path. It provides the nervous supply to the posterior part of the scalp. Obviously, transverse incision in this region is precluded.

Identification of Surface Landmarks of the Skull

The Base of the Skull (Figs. 10 to 13)

Rather than describing the classic features of the base of the skull, we have chosen to focus on the orientation of the head in space and the consequences of its position during the operation. How the base of the skull is positioned for a pterional approach or one via the posterior fossa is extremely important and is going to determine the outcome of the procedure.

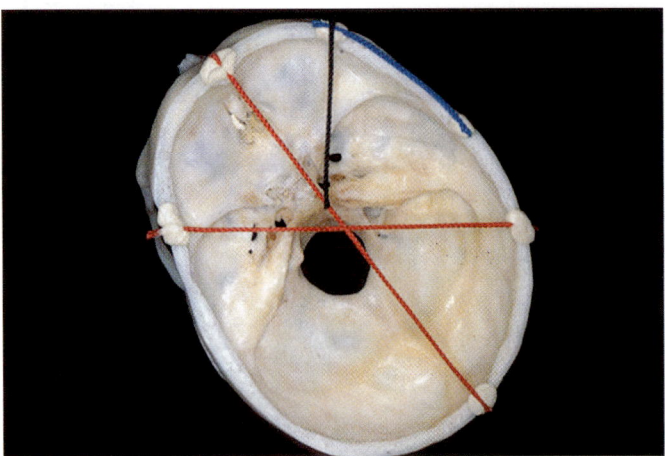

Fig. 10. Anatomical modifications after the patient's head has been rotated to the opposite side

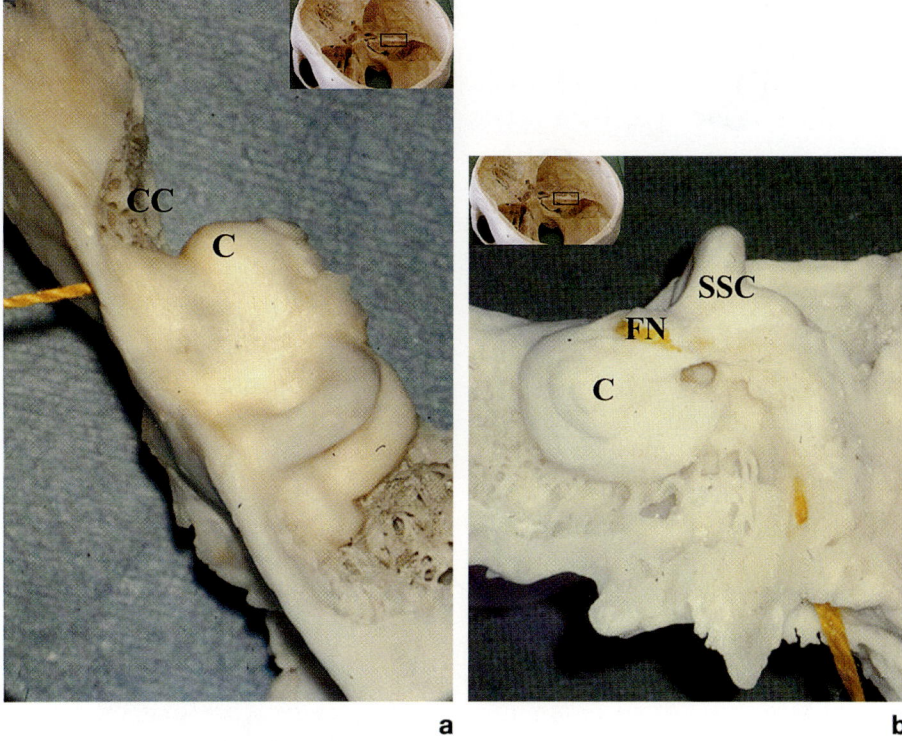

Fig. 11. Superior (a) and lateral (b) views of a drilled left dry bony labyrinth after resection of the middle ear. Note the carotid canal (*CC*) which is partially covered by the cochlea (*C*). Note the relationships between the facial nerve (*FN*) and the semi-circular canals (*SSC*)

The performance of the surgeon's approach begins with proper positionning of the patient's head and proper location of the bone flap.

Important Points to Remember

The greater the inclination of the cranial axis away from the vertical, the more vertical will be the direction of the sphenoidal crest and the more horizontal will be that of the petrosal crest. It applies for the sitting position also: if the head is rotated, the petrosal crest is in a more sagital position. The positions of these bony landmarks are fixed and they can constitute a useful aid to establish spatial relationships during a procedure, as long as the "dynamic" characteristics of the anatomical situation are understood. The advent of surgical neuronavigation tools has not in any way dispensed with the need for a profound understanding of anatomy, and laboratory-based teaching of the spatial relationships is crucial.

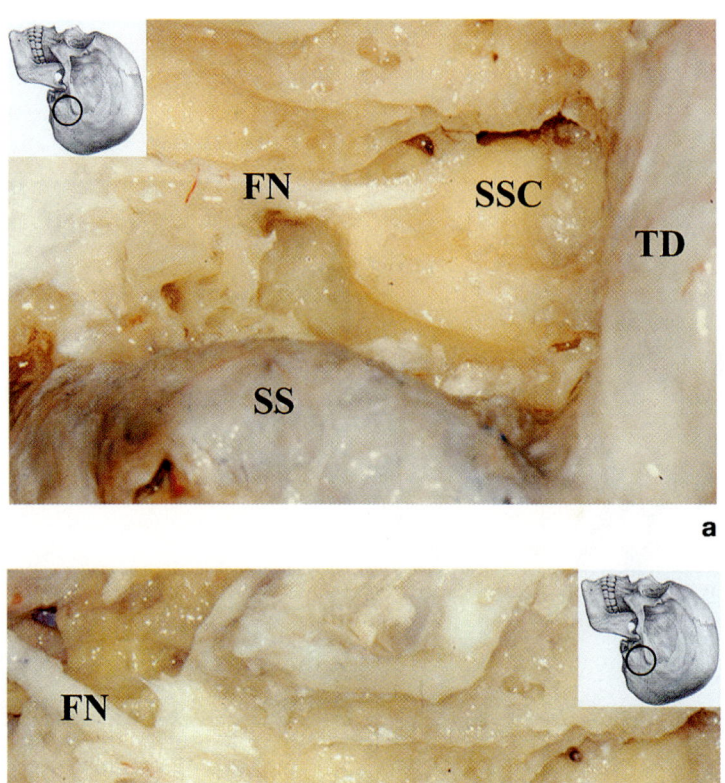

Fig. 12. (a and b) Drilling of a left bony labyrinth in a cadaver specimen to show the
relationships between the facial nerve (*FN*) and the bony structures. Note the location
of the sigmoid sinus (*SS*) and the temporal dura (*TD*)

Important Relationships at the Skull Base: The Bony Labyrinth,
the Facial Nerve and the Petrous Segment of the Carotid Artery

A comprehensive review of the anatomy of all the bones of the base of the
skull is beyond the scope of this chapter [4] but an awareness of certain
anatomical relationships is essential if the principle of the trans-petrous
approaches is to be grasped.

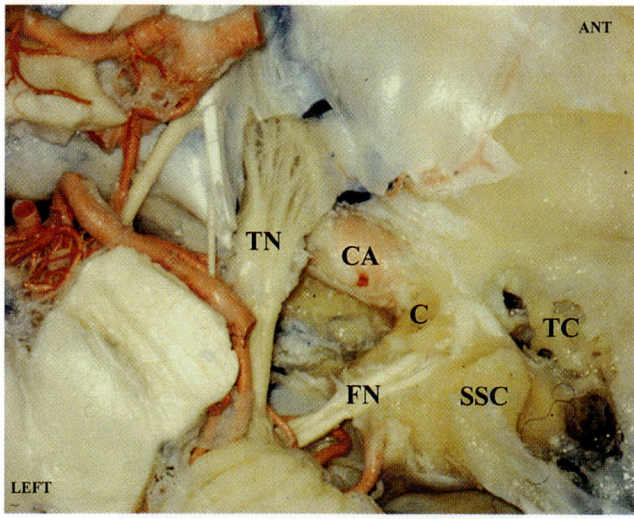

Fig. 13. Superior view of a drilled right middle cranial fossa in a cadaver specimen to show the relationships between the trigeminal nerve (*TN*), the carotid artery (*CA*), the facial nerve (*FN*), the cochlea (*C*), the semicircular canals (*SSC*) and the tympanic cavity (*TC*)

A dry specimen of a left sided drilled bony labyrinth can be used to show the ventral segment (cochlea) and the dorsal segment (vestibule and semicircular canals). Its length is about 2 cm located in the long axis of the petrous pyramid. The relationships between the lateral tract of the lateral semicircular canal and the second portion of the facial nerve, and between the ampullae of the superior and lateral semicircular canals and the genu of the facial nerve are to be emphasized. These landmarks are routinely used as a guideline when drilling the bone to avoid damage to the facial nerve during posterolateral approaches.

The cochlea lies near the cortical substance of the petrous apex. The carotid canal is usually partially covered by the cochlea so it is impossible to displace the carotid artery downwards from above without entering the cochlea. The petrous segment of the carotid artery may be or may not be covered by bone at the level of the middle cranial fossa. If not, it is separated from the trigeminal enlargement by a dural layer. If the floor of the middle cranial fossa is drilled, the facial nerve within the internal auditory meatus, the geniculate ganglion, and the middle ear will be exposed. Drilling the petrous apex medial to the carotid artery gives access to the posterior fossa along the inferior petrosal sinus.

The Key Surface Structures (Figs. 14 to 16)

The "keyhole" concept makes appropriate placement of the first burr hole of importance. Two of these are particularly important, the one in the pterional region and that around the sinuso-jugular axis.

The Pterional Approach

The hole should be made behind the lateral crest of the external orbital rim of the frontal bone. This will afford access to the inferior surface of the frontal lobe just above the roof of the orbit.

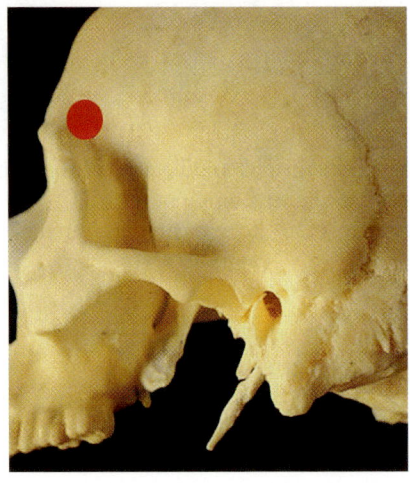

a

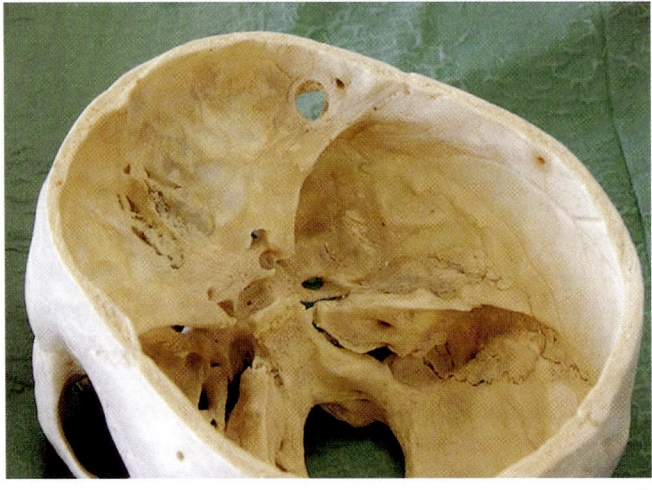

b

Fig. 14. (a and b) The pterional keyhole burr hole is show for the pterional approach

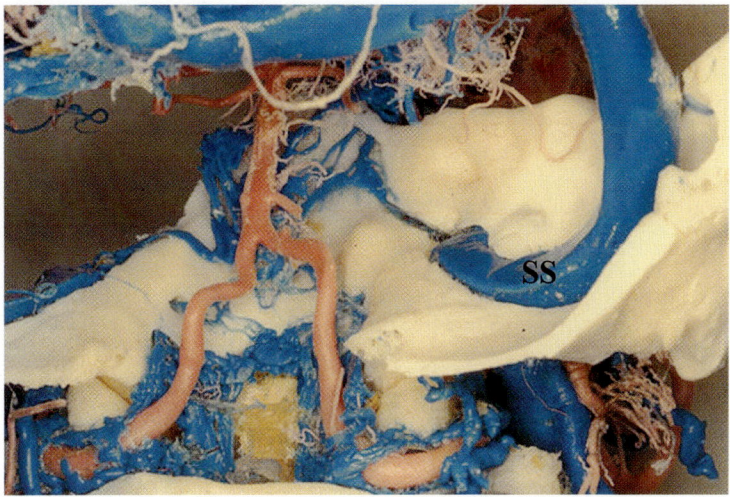

Fig. 15. Corrosion cast showing the relationships between the sigmoid sinus (*SS*) and the bony structures at the skull base

Venous Sinus Relationships to Surface Landmarks of the Skull

The sinuso-jugular axis is one of the keys to access to the base of the skull. It is dangerous (being surrounded by many other structures) but accessible (superficial, posterior, continuous). All corrosion casts show that the transverse sinus, the transverse-sigmoid junction and the sigmoid sinus are deeply located within a large bony groove which makes the first burr hole quite difficult and risky if placed just on the venous structure.

Placement of the burr holes for craniotomy must be compatible with the position of the transverse and sigmoid sinuses. Posterior surface landmarks have been proposed in the literature [2]. The asterion, the mastoid groove and the superior nuchal line were found to be the most valuable landmarks. Because the asterion could not be identified in almost 60% of the cadaver specimens, various methods have been proposed. We believe that the first burr hole can be placed in the angle between the asterion, the parietomastoid and the occipitomastoid sutures. The second burr hole can be placed in the angle between the asterion, the lambdoid and the occipitomastoid sutures. Doing so on both sides of the occipitomastoid suture, you stay below the venous structures. Making the holes on either side of the occipitomastoid suture ensures avoidance of the venous structures above. For combined exposure with via a supratentorial approach, third and fourth burr holes can be placed on both sides of the squamosal suture.

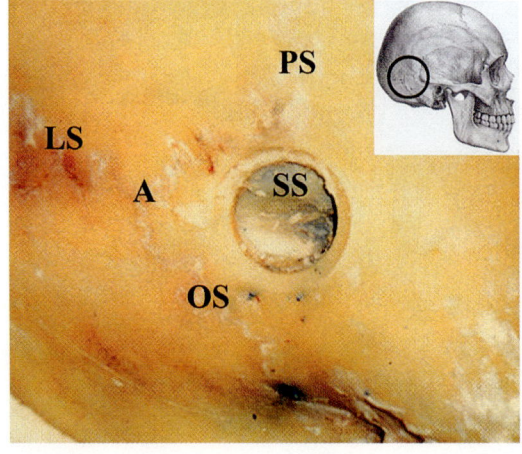

Fig. 16. (a and b) The keyhole burr holes are shown along the right sinuso-jugular axis; (*A* asterion; *LS* lambdoid suture; *OS* occipitomastoid suture; *PS* parietomastoid suture)

Conclusion

Although today's young neurosurgeons have at their disposal sophisticated surgical neuronavigation tools, they nevertheless still need the basic anatomical knowledge which is indispensable to understanding and executing a classic cranial approach. Teaching how to make the exposure to create a perfect, centered bone flap needs to be kept as a priority in the surgical training of even the youngest surgeons. The basic anatomy described in this chapter is by no means exhaustive but it is intended to serve as a working basis and to stimulate thought about the major role that the Anatomy Laboratory should play in any neurosurgery training program.

References

1. Al-Mefty O (1991) Clinoidal meningiomas. In: Al-Mefty O (ed) Meningiomas. Raven Press, New York
2. Avci E, Kocaogullar Y, Fossett D, Caputy A (2003) Lateral posterior fossa venous sinus relationships to surface landmarks. Surg Neurol 59: 392–397
3. Chayen D, Nathan H (1974) Anatomical observations on the subgaleotic fascia of the scalp. Acta anat 87: 427–432
4. Fournier HD, Mercier P, Velut S, Reigner B, Cronier P, Pillet J (1994) Surgical anatomy and dissection of the petrous and peripetrous area. Anatomic basis of the lateral approaches to the skull base. Surg Radiol Anat 16: 143–148
5. Hill EG, McKinney WM (1981) Vascular anatomy and pathology of the head and neck: method of corrosion casting. In: Carney AL, Anderson EM (eds) Advances in neurology, vol 30: diagnosis and treatment of brain ischemia. Raven Press, New York
6. Lebeau J, Antoine P, Raphael B (1986) Introduction angéiologique à la chirurgie du scalp. Ann Chir Plast Esthét 31: 321–324
7. Marano SR, Fischer DW, Gaines C, Sonntag VKH (1985) Anatomical study of the superficial temporal artery. Neurosurgery 16: 786–790
8. Marty F, Montandon D, Gumener R, Zbrodowski A (1986) Subcutaneus tissue in the scalp: anatomical, physiological, and clinical study. Ann Plast Surg 16: 368–376
9. Potparic Z, Fukuta K, Colen LB, Jackson IT, Carraway JH (1996) Galeopericranial flaps in the forehaed: a study of blood supply and volumes. Brit J Plast Surg 49: 519–528
10. Ricbourg B, Mitz V, Lassau JP (1975) Artère temporale superficielle. Etude anatomique et déduction pratiques. Ann Chir Plast 20: 197–213
11. Stock AL, Collins HP, Davidson TM (1980) Anatomy of the superficial temporal artery. Head Neck Surg 2: 466–469
12. Tolhurst DE, Carstens MH, Greco RJ, Hurwitz DJ (1991) The surgical anatomy of the scalp. Plast Reconstruct Surg 87: 603–612
13. Spetzler RF, Stuart Lee K (1990) Reconstruction of the temporalis muscle for the pterional craniotomy. Technical note. J Neurosurg 73: 636–637
14. Yasargil MG, Reichman MV, Kubik S (1987) Preservation of the frontotemporal branch of the facial nerve using the interfascial temporalis flap for pterional craniotomy. J Neurosurg 67: 463–466

Author Index Volume 1–31

Advances and Technical Standards in Neurosurgery

Subject Index Vol. 1–31

Advances and Technical Standards in Neurosurgery

SpringerNeurosurgery

Advances and Technical Standards in Neurosurgery

Volume 24

1998. XIII, 310 pages. 57 figures, partly in colour.
Hardcover **EUR 133,–**
ISBN 3-211-83064-2

Advances: • The Septal Region and Memory (D. Y. von Cramon, U. Müller) • The in vivo Metabolic Investigation of Brain Gliomas with Positron Emission Tomography (J. M. Derlon) • Use of Surgical Wands in Neurosurgery (L. Zamorano, F. C. Vinas, Z. Jiang, F. G. Diaz) **Technical Standards:** • The Endovascular Treatment of Brain Arteriovenous Malformations (A. Valavanis, M. G. Yasargil) • The Interventional Neuroradiological Treatment of Intracranial Aneurysms (G. Guglielmi) • Benign Intracranial Hypertension (J. D. Sussman, N. Sarkies, J. D. Pickard)

Volume 23

1997. XV, 278 pages. 89 figures, partly in colour.
Hardcover **EUR 133,–**
ISBN 3-211-82827-3

Advances: • A Critical Review of the Current Status and Possible Developments in Brain Transplantation (S. Rehncrona) • The Normal and Pathological Physiology of Brain Water (K. G. Go) **Technical Standards:** • Transfacial Approaches to the Skull Base (D. Uttley) • Presigmoid Approaches to Skull Base Lesions (M. T. Lawton, C. P. Daspit, R. F. Spetzler) • Anterior Approaches to Non-Traumatic Lesions of the Thoracic Spine (A. Monteiro Trindade, J. Lobo Antunes)• The Far Lateral Approach to Lumbar Disc Herniations (F. Porchet, H. Fankhauser, N. de Tribolet)

All prices are recommended retail prices
Net-prices subject to local VAT.

SpringerWien NewYork

P.O. Box 89, Sachsenplatz 4–6, 1201 Vienna, Austria, Fax +43.1.330 24 26, books@springer.at, **springer.at**
Haberstraße 7, 69126 Heidelberg, Germany, Fax +49.6221.345-4229, SDC-bookorder@springer.com, springer.com
P.O. Box 2485, Secaucus, NJ 07096-2485, USA, Fax +1.201.348-4505, service@springer-ny.com, springer.com
All errors and omissions excepted.

SpringerNeurosurgery

Advances and Technical Standards in Neurosurgery

Volume 26

2000. XVI, 346 pages. 83 figures, partly in colour.
Hardcover **EUR 179,95**
ISBN 3-211-83424-9

Advances: • Multiple Subpial Transection (C. E. Polkey) • Hemispheric Disconnection: Callosotomy and Hemispherotomy (J.-G. Villemure, O. Vernet, O. Delalande) • Central Nervous System Lymphomas (H. Loiseau, E. Cuny, A. Vital, F. Cohadon) • Invited Commentary: Treatment of Diseases of the Central Nervous System Using Encapsulated Cells, by A. F. Hottinger and P. Aebischer (Advances and Technical Standards in Neurosurgery Vol. 25) (A. E. Rosser, T. Ostenfeld, C. N. Svendsen) **Technical Standards:** • The Intracranial Venous System as a Neurosurgeon's Perspective (M. Sindou, J. Auque) • Reconstructive Surgery of the Extracranial Arteries (R. Schmid-Elsässer, R. J. Medele, H.-J. Steiger) • Surgical Treatment of Lumbar Spondylolisthesis (P. W. Detwiler, R. W. Porter, P. P. Han, D. G. Karahalios, R. Masferrer, V. K. H. Sonntag)

Volume 25

1999. XIV, 241 pages. 54 figures, partly in colour.
Hardcover **EUR 106,–**
ISBN 3-211-83217-3

Advances: • Treatment of Diseases of the Central Nervous System Using Encapsulated Cells (A. F. Hottinger, P. Aebischer) • Intracranial Endoscopy (G. Fries, A. Perneczky) • Chronic Deep Brain Stimulation for Movement Disorders (D. Caparros-Lefebvre, S. Blond, J. P. N'Guyen, P. Pollak, A. L. Benabid) **Technical Standards:** • Recent Advances in the Treatment of Central Nervous System Germ Cell Tumors (Y. Sawamura, H. Shirato, N. de Tribolet) • Hypothalamic Gliomas (V. V. Dolenc) • Surgical Approaches of the Anterior Fossa and Preservation of Olfaction (J. G. Passagia, J. P. Chirossel, J. J. Favre)

All prices are recommended retail prices
Net-prices subject to local VAT.

P.O. Box 89, Sachsenplatz 4–6, 1201 Vienna, Austria, Fax +43.1.330 24 26, books@springer.at, **springer.at**
Haberstraße 7, 69126 Heidelberg, Germany, Fax +49.6221.345-4229, SDC-bookorder@springer.com, springer.com
P.O. Box 2485, Secaucus, NJ 07096-2485, USA, Fax +1.201.348-4505, service@springer-ny.com, springer.com
All errors and omissions excepted.

SpringerNeurosurgery

Advances and Technical Standards in Neurosurgery

Volume 28

2003. XIV, 360 pages. 80 figures, partly in colour.
Hardcover **EUR 144,95**
ISBN 3-211-83803-1

Advances: • Recent Advances in Stem Cell Neurobiology (T. Ostenfeld, C.N. Svendsen) • Mapping of the Neuronal Networks of Human Cortical Brain Functions (S. Momjian, M. Seghier, M Seeck, C.M. Michel) **Technical Standards:** • The Management of Brain Abscesses • (S. Livraghi, J.P. Melancia, J. Lobo Antunes) • Respective Indications for Radiosurgery in Neuro-otology Surgery for Acoustic Schwannoma • (W. Pellet, J. Regis, P-H. Roche, C. Delsanti) • Commentary • (R. Macfarlane, D. Moffet) • Cerebral Revascularization • (H.J.N. Streefkerk, A. Van der Zwan, R.M. Verdaasdonk, H.J. Mansveld Beck, C.A.F. Tulleken) • Surgical Anatomy of the Temporal Lobe for Epilepsy Surgery • (M. Sindou, M.Guenot)

Volume 27

2002. XIV, 244 pages. 97 figures, partly in colour.
Hardcover **EUR 99,95**
ISBN 3-211-83605-5

Advances: • Multi-Modality Monitoring of Acute Brain Trauma (R. Kett-White, P. J. A. Hutchinson, M. Czosnyka, S. Boniface, J. D. Pickard, P. J. Kirkpatrick) • The Concept of Diffuse Axonal Injury (J. Sahuquillo, A. Poca) • Endoscopic Endonasal Transsphenoidal Surgery (E. de Divitiis, P. Cappabianca) **Technical Standards:** • Surgery of Temporal Lobe Epilepsy (M. Vapalahti) • Surgical Exposure of the Vertebral Artery - Application to Spinal and Skull Base Surgery (B. George) • Neurosurgical Management of Pineal Tumours (Y. Sawamura, N. de Tribolet)

P.O. Box 89, Sachsenplatz 4–6, 1201 Vienna, Austria, Fax +43.1.330 24 26, books@springer.at, **springer.at**
Haberstraße 7, 69126 Heidelberg, Germany, Fax +49.6221.345-4229, SDC-bookorder@springer.com, springer.com
P.O. Box 2485, Secaucus, NJ 07096-2485, USA, Fax +1.201.348-4505, service@springer-ny.com, springer.com
All errors and omissions excepted.

SpringerNeurosurgery

Advances and Technical Standards in Neurosurgery

Volume 30

2005. XVI, 289 pages. 40 figures, partly in colour.
Hardcover **EUR 125,–**
ISBN 3-211-21403-8

Advances: • Gene Technology Based Therapies in the Brain (T. Wirth, S. Ylä-Herttuala) **Technical Standards:** • Anatomy of the Orbit and its Surgical Approach (G. Hayek, Ph. Mercier, H. D. Fournier) • Neurosurgical Concepts and Approaches for Orbital Tumours (J. C. Marchal, T. Civit) • Endoscopic Third Ventriculostomy in the treatment of Hydrocephalus in Pediatric Patients (C. Di Rocco, G. Cinalli, L. Massimi, P. Spennato) • Minimally Invasive Procedures for the Treatment of Failed Back Surgery Syndrome (P. Mavrocordatos, A. Cahana) • Surgical Anatomy of Calvarial Skin and Bones – With Particular Reference to Neurosurgical Approaches (H. D. Fournier, V. Dellière, J. B. Gourraud, Ph. Mercier)

Volume 29

2004. XIV, 304 pages. 101 figures, partly in colour.
Hardcover **EUR 125,–**
ISBN 3-211-14027-1

Advances: • Disorders of Consciousness: Anatomical and Physiological Mechanisms (J. L. Valatx) • Advances in Craniosynostosis Research and Management (J. Guimarães-Ferreira, J. Miguéns, C. Lauritzen) **Technical Standards:** • Preoperative Clinical Evaluation, Outline of Surgical Technique and Outcome in Temporal Lobe Epilepsy (A. Immonen, L. Jutila, R. Kälviäinen, E. Mervaala, K. Partanen, J. Partanen, R. Vanninen, A. Ylinen, I. Alafuzoff, L. Paljärvi, H. Hurskainen, J. Rinne, M. Puranen, M. Vapalahti) • Motor Evoked Potential Monitoring for Spinal Cord and Brain Stem Surgery (F. Sala, P. Lanteri, A. Bricolo) • Motor Evoked Potential Monitoring for the Surgery of Brain Tumours and Vascular Malformations. (G. Neuloh, J. Schramm) • Functional Neuronavigation and Intraoperative MRI (C. Nimsky, O. Ganslandt, R. Fahlbusch) • Surgical Anatomy of the Insula (M. Guenot, J. Isnard, M. Sindou)

All prices are recommended retail prices
Net-prices subject to local VAT.

SpringerWienNewYork

P.O. Box 89, Sachsenplatz 4–6, 1201 Vienna, Austria, Fax +43.1.330 24 26, books@springer.at, **springer.at**
Haberstraße 7, 69126 Heidelberg, Germany, Fax +49.6221.345-4229, SDC-bookorder@springer.com, springer.com
P.O. Box 2485, Secaucus, NJ 07096-2485, USA, Fax +1.201.348-4505, service@springer-ny.com, springer.com
All errors and omissions excepted.

Springer-Verlag
and the Environment